AF304884

Models of Life

Dynamics and Regulation in Biological Systems

Reflecting the major advances that have been made in the field over the past decade, this book provides an overview of current models of biological systems. The focus is on simple quantitative models, highlighting their role in enhancing our understanding of the strategies of gene regulation and dynamics of information transfer along signaling pathways, as well as in unravelling the interplay between function and evolution.

The chapters are self-contained, each exploring key methods and principles for understanding quantitative aspects of life through the use of models. They focus, in particular, on connecting the dynamics of proteins and DNA with strategic decisions on the larger scale of a living cell, using *E. coli* and phage λ as key examples. Encompassing fields such as quantitative molecular biology, systems biology and biophysics, this book will be a valuable tool for students from both biological and physical science backgrounds.

End-of-chapter summary sections, along with questions placed throughout chapters, help to explain the key concepts to students. Solutions are available online at www.cambridge.org/sneppen.

KIM SNEPPEN is Professor of Physics at the Niels Bohr Institute and Director of the interdisciplinary Center for Models of Life (CMOL) at Copenhagen University, Denmark. Drawing on experience across several academic disciplines, his work explores the frontier between complex dynamic systems and living systems, and in his role at CMOL he promotes hands-on development of quantitative models of central biological processes. Sneppen is also co-author of *Physics in Molecular Biology* (Cambridge University Press, 2005).

Models of Life

Dynamics and Regulation in Biological Systems

KIM SNEPPEN
Niels Bohr Institute and Center for Models of Life,
Copenhagen University, Denmark

CAMBRIDGE
UNIVERSITY PRESS

CAMBRIDGE
UNIVERSITY PRESS

University Printing House, Cambridge CB2 8BS, United Kingdom

Cambridge University Press is part of the University of Cambridge.

It furthers the University's mission by disseminating knowledge in the
pursuit of education, learning and research at the highest international
levels of excellence.

www.cambridge.org
Information on this title: www.cambridge.org/9781107061903

First published 2014

Printed in Spain by Grafos SA, Arte sobre papel

A catalog record for this publication is available from the British Library

Library of Congress Cataloging in Publication data
Sneppen, Kim, author.
Models of life : dynamics and regulation in biological systems / Kim Sneppen,
Niels Bohr Institute and Center for Models of Life, Copenhagen University,
Denmark.
       pages   cm
ISBN 978-1-107-06190-3 (Hardback)
1. Life sciences–Mathematical models.   2. Biology–Mathematical models.
I. Title.
QH323.5.S62 2014
570.1′51–dc23   2014003456

ISBN 978-1-107-06190-3 Hardback

Additional resources for this publication at www.cambridge.org/sneppen

Contents

Preface *Page* **ix**

1   Life from a physics perspective   **1**

1.1   Life copies from itself and from other life   1
1.2   Biology is the result of a very long historical process   2
1.3   Self-assembly and the genetic code   3
1.4   Biological molecules have long backbones   6
1.5   Computation   6
1.6   Information gathering and communication   7
1.7   Feedback   7
1.8   Push–pull   8
1.9a   Life is modular   9
1.9b   Life is not modular   9
1.10   Stochastic processes play an important role in life   9
1.11   Life harvests energy and converts it to cyclic motion   10
1.12   Biological physics is "$k_B T$ physics"   13

2   *E. coli* as a model system   **17**

2.1   DNA, RNA and proteins   17
2.2   The *E. coli* cell   22
2.3   Transcription and translation numbers   24
2.4   *E. coli* content varies with its growth rate   32
2.5   Summary   35

3   Dynamics of regulatory links   **36**

3.1   Regulating a piece of DNA   36
3.2   Transcription regulation   41
3.3   Post-transcriptional regulation   44
3.4   Summary   48

4   Statistical mechanics of phage λ   **49**

4.1   Lifecycle of λ phage   49
4.2   Bacterial growth and counting   56

4.3	Chemical binding and counting	59
4.4	Chemistry and co-operativity as statistical mechanics	62
4.5	Distant DNA in gene regulation	68
4.6	Summary	78

5 Diffusion and randomness in transcription — **79**

5.1	Random walks and diffusion	79
5.2	Timescales for target location in a cell	83
5.3	Traffic on DNA	89
5.4	Bursty transcription initiation	94
5.5	Economy of operons: noise minimization	100
5.6	Summary	102

6 Stochastic genes and persistent decisions — **104**

6.1	Stochastic simulations	104
6.2	Maintenance of lysogeny in the λ system	109
6.3	The phage λ circuit beyond CI and Cro	117
6.4	Summary	122

7 *cis*-Acting gene regulation and epigenetics — **123**

7.1	Nucleosomes and their enzymes	123
7.2	Nucleosome-mediated epigenetics	127
7.3	A regulated two-state model	135
7.4	Coupled epigenetics in olfactoric differentiation	142
7.5	Other models for *cis*-mediated gene regulation	146
7.6	Summary	149

8 Feedback circuits — **150**

8.1	Small regulatory subnetworks	150
8.2	Negative feedback	151
8.3	Positive feedback	159
8.4	Combining feedback in small-molecule regulation	161
8.5	Combined feedback loops and spatial organization	167
8.6	Summary	176

9 Networks — **177**

9.1	Convergent computation	177
9.2	Connectedness	178
9.3	Large-scale molecular networks	190
9.4	Analysis of network topologies	193
9.5	Models for scale-free molecular networks	198
9.6	Summary	204

10   Signaling and metabolic networks   **205**

   10.1   Signaling across networks   205
   10.2   Enzyme kinetics and phosphorylation   212
   10.3   Adaptation   216
   10.4   Metabolic fluxes   219
   10.5   Summary   225

11   Agent-based models of signaling and selection   **226**

   11.1   Excitable agents as cells in a tissue   226
   11.2   Information spreading on social scales   227
   11.3   Tragedy of the commons   235
   11.4   Summary   241

12   Competition and diversity   **242**

   12.1   Phage worlds   242
   12.2   Phage–bacteria ecology   252
   12.3   Phage games: lysogeny or lysis   265
   12.4   Speciation and biodiversity of sessile species   272
   12.5   Summary   277

13   Evolution and extinction   **279**

   13.1   History   279
   13.2   Fitness landscapes   282
   13.3   Punctuated equilibrium and co-evolution   286
   13.4   Summary   295

Appendix   **296**

   Langevin versus Fokker–Planck equation   296
   Kramers' equation   297
   First passage for random walks   299

References   **301**
Index   **334**

Preface

This book is the result of outstanding collaborations with inspiring collegues and friends. A part of its content is based on an earlier book that I published with Giovanni Zocchi, with whom Mogens H. Jensen and I started the biological physics initiative at Copenhagen University. The major content of the book has been developed during the last decade, when I had the luck to be in charge of the Center for Models of Life at the Niels Bohr Institute. This center exists thanks to generous funding from the Danish National Research Council, who gave me the option to invite and employ inspiring collaborators.

Much of the content of this book is based on activities at the center. The book was also developed in parallel with our scientific endeavors and bachelor degree level courses on gene regulation, networks and complex systems for which I had the pleasure of lecturing. Accordingly, the big part of the book is devoted to classical models from molecular biology, spanning a wide spectrum of model systems and synthetic circuits as they unfold in classical model organisms like *Escherichia coli*.

I would hope that this book is useful for bachelor degree-level students in any quantitative discipline related to physics or biology. In fact, I believe the book also could be of inspiration for mathematically inclined high school and college students. If one just disregards the most boring and technical equations, the main messages should be conveyed by the frequent use of illustrations and provocative statements.

In general, the chapters of the book can be read without following any particular order, as they, to a large extent, provide alternative ways to work with quantitative aspects of life from the perspective of physics.

The focus of the book is quantitative, yet simple, models. A main criterion for selection is Occam's razor principle, where a model loses meaning as it gets too complicated. *Models of Life* is accordingly not aimed at giving a complete description of the studied systems. Rather it provides the reader with an understanding of interesting biology, facilitated by models developed within the criterion *that the quality of a model declines sharply with each added parameter*.

The proposed models will, occasionally, challenge this criterion, and in certain cases explore ways to hide the "parameter nightmare" with sampling over a large numbers of parameter sets. This will then leave the main messages to structural features of a model, like signs of feedbacks, or type of interaction that are essential to understand a given phenomenon.

Many friends and collegues have helped me both in scientic projects and inspiration. The most exceptional help was from Namiko Mitarai, who has been pivotal in many of the described models and finally took the effort to read and criticize the many mistakes I had in the nearly finished book. Major parts of the book are built on a beautiful and ongoing collaboration with Ian Dodd, who also taught me much of the biology I know today. Stanley Brown also deserves very special thanks as my mentor in biology, teaching me about the complexity of *E. coli*, and introducing me to the wonderfully quantitative scientists around the λ phage, including in particular Sankar Adhya, Donald Court, Barry Egan, Harvey Eisen, Max Gottesman, Keith Shearwin and Lynn Thomasen.

Important collaborators and tutors in biology also include Steen Pedersen and Genevive Thon from Copenhagen University, Kenn Gerdes from Newcastle University, Eric Masse from Sherbroke University and finally Szabolcs Semsey as our resident biologist in the center. Szabolcs also developed many models and helped with critical reading of the manuscript, pinpointing especially boring sections. I hope it has improved, and that the reader will find inspiration in most of the text.

All the PIs, post-docs and students that have resided at the center are also to be thanked, with very special thanks to Anne Alsing, Mikkel Avlund, Jacob Bock Axelsen, Sebastian Bernhardsson, Philip Gerlee, Jan Haerter, Silja Heilmann, Mogens Høgh Jensen, Sandeep Krishna, Ludvig Lizana, Joachim Matthiesen, Mille Micheelsen, Simone Pigolotti, Martin Rosvall and Ala Trusina, all of whom have been pivotal in the presented models. These models were also developed with physics friends from across the world, of whom, in particular, Per Bak, Stefan Bornholdt, Hiizu Nakanishi and Sergei Maslov contributed with central ideas and concepts.

I also thank the many students that took my courses through the years, of which my recent students Celie Feldager, Sirin Gangstad and Svend Steffensen took additional effort to give constructive comments to many of the currently used sections.

Finally I would like to express gratitude to my wife Simone and my children Albert, Eva, Thor and Ida, who had the patience to leave me in peace during the many evenings and weekends it took to write the book. In particular I am happy for discussions with my youngest son Albert, which, after careful reading of Darwin's original works, pinpointed many of the deep insights that this scientist developed on speciation and the enormous timescales of evolution.

1 Life from a physics perspective

In this first chapter we review some basic concepts of organization and dynamics in living systems, with emphasis on what makes them different from systems that one normally studies with the tools of physics. Subsequent chapters use model systems to illustrate how interactions between proteins, metabolites and DNA govern strategic decisions on the scales ranging from a living cell, to organisms to populations.

Life has provided scientists with a vast number of stories [1–9] reflecting facets of the dynamic interplay between material and memory, an interplay that is formed by dynamical processes from the noise on the molecular scale, to the exceptionally long memory of replicating DNA, to the exponential growth of populations [10, 11].

In short, life is self-reproducing, persistent and robust (we are nearly 4 billion years old). Living systems are open systems that have "memorized" how to channel energy into self-reproducing networks. Living cells are complex (after all more than 1000 different types of molecules are needed to make even the simplest cell work), "more" than the sum of its parts (arbitrarily dividing an organism kills it). Life harvests energy and it evolves. The essential processes for the workings of life take place from the scale of a single water molecule to balancing the atmosphere of the entire planet [7, 12].

Biology has provided us with some fundamental/universal mechanisms that one meets in various disguises and variations [13].

1.1 Life copies from itself and from other life

Life copies and repeats on all scales, DNA, cells, individuals, young animal copying their parents, to us humans who copy each other's behavior, organizations [14] and inventions (see Fig. 1.1). Copying is often good as it opens up the possibility of "conquering the world" by exponential (Malthusian [15]) growth (see Fig. 1.2):

$$1 \rightarrow 2 \rightarrow 4 \rightarrow 8 \rightarrow 16 \rightarrow \ ....$$

Copying is functional, especially when it replicates what has already been proven to work. Learning also includes copying, see Fig. 1.1. Learning is often thought of as perfect copying, but without mistakes and failed attempts at copying there would be little progress in life. Failed attempts are necessary when life is struggling to keep up

Figure 1.1 Copying is important in the learning process where children often copy behavior from their older peers.

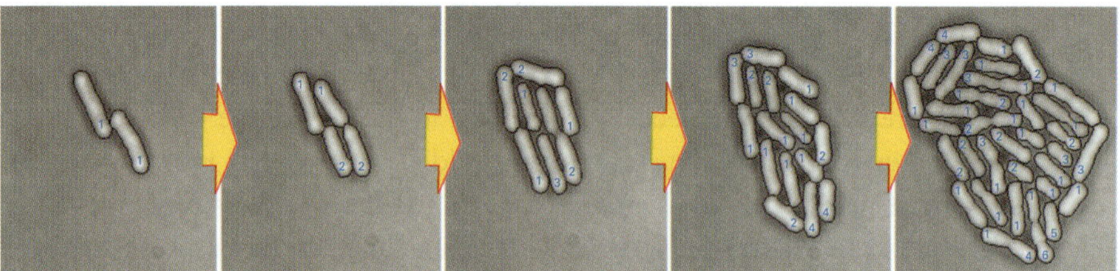

Figure 1.2 Copying and growth of E. coli cells that replicate every 30 minutes (micrographs courtesy of A. Trusina). The numbers refer to the age of cells in the number of cell generations.

with changes in its environment. A consequence of the power of exponential growth is the ability of any life form to explore scale, from the simplest single-celled organisms to algae bloom on the scale of an ocean; organisms replicate until they meet boundaries, set by other species or ultimately by the planet. The down side of exponential growth is that it is so powerful that it will often induce crashes, thereby making risk management essential for long-term sustainability.

1.2   Biology is the result of a very long historical process [1, 4, 16, 17, 18, 19]

A view highlighted first by Darwin, based on his observation of differences between species in similar but spatially separated habitats. Dependence on history implies that it is not possible to "explain" a biological system by applying a few fundamental laws in the same way it is done in physics. If one replays the tape of history, the hydrogen atom could not be different from what it is, but an *E. coli* cell could. In evolution, it is much easier to modify already existing mechanisms than to invent completely new ones [20]. On evolutionary timescales nearly everything changes by cutting, pasting and reshuffling of already working subsections. As a consequence, all living organisms on Earth are part of the same 3.6 billion-year-old "family" that share ribosomes and the genetic code.

Evolution is self-organization on a long timescale, a stochastic but biased process where survivors are primarily selected to best deal with the contemporary environment formed in part by other life forms. Such a relatively short-term optimization often conflicts with long-term survival when the environment experiences catastrophic changes. However, there may be selection for the ability to evolve fast enough to adapt to changes in their surroundings. Thus, evolution may select the ability to evolve, for example through modulated single nucleotide mutations or large-scale DNA reshuffling using recombinations and transposons [9].

Perhaps the most mind-boggling aspect of evolution is the enormous time spans involved. Understanding the evolutionary processes is often hampered by our lack of perspective of how vast time is. Enormous time spans can apparently be hugely creative, provided that selection and evolvability is combined with memories of earlier successes:

$$Time \approx God \tag{1.1}$$

an "equation" that primarily serves to highlight the power of time in self-organizing systems. Figure 1.3 illustrates evolution and the randomness of history, by following the births and deaths of ammonite species over a quarter of a billion years.

1.3　Self-assembly and the genetic code

Self-assembly is an inherent property of life, and is, in particular, essential to the production of new cells using information from their ancestors. On the smallest scale it includes the equilibrium relaxation of highly non-random chains of amino acids into their minimum-energy state, which is the native shape of a functional protein [23]. Another larger-scale self-assembly process is the development of a full-scale adult animal from a single fertilized cell. In general, self-assembly is the replay of a pre-designed plan according to some fixed rules. Rules that in life are set by the information stored in the genetic code, the macromolecules it encodes and various processes and feedbacks that these molecules execute together in biomolecular circuits and networks.

Self-assembly relies on information stored on a one-dimensional, double-stranded DNA molecule (Fig. 1.4). Its linear nature mirrors the one-dimensional "backbone" of other polymers that make life work. The complementarity of two DNA strands allows *copying* of each of them, by virtue of separating the double-stranded DNA into two single-stranded DNAs that each carry the full information. The actual copying is done

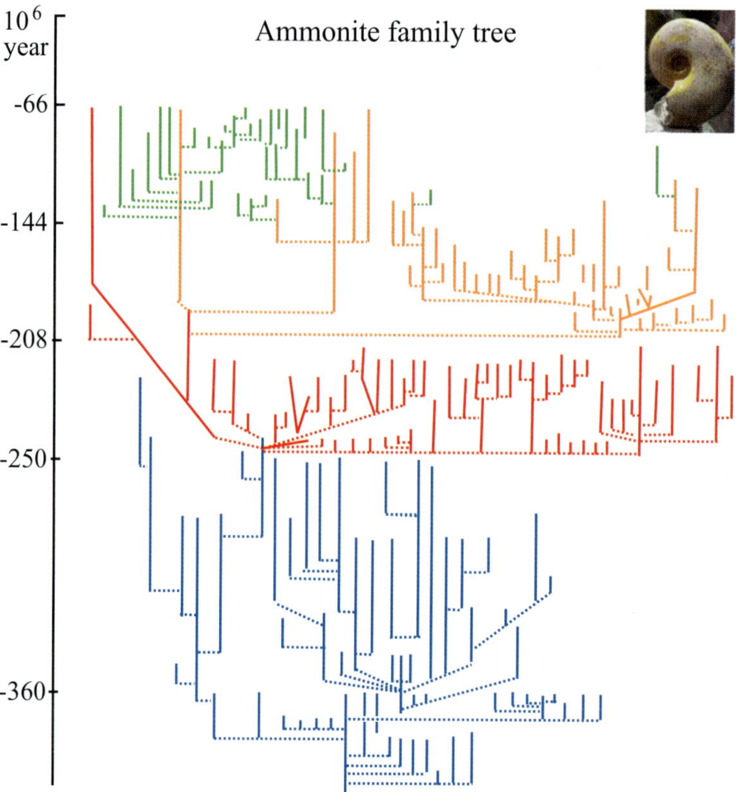

Figure 1.3 Ammonite family tree redrawn from [16]. Ammonites lived in water and fossilized well, leaving a fossil record of about 7000 ammonoid species. During their evolutionary history spanning 400 to 66 million years ago, one observes periods with fast speciation and simultaneous extinction of many different species. Today there are no living ammonites. Remarkably, in spite of a copying ability that would allow any species to take over the world in a few years, the individual species coexisted for million of years. A simple model showing cascades of co-evolutionary extinctions can be found in [17].

by DNA polymerase, a specialized protein machine that can copy 1000 base pairs in about one second.

DNA is transcribed into complementary RNA by RNA polymerase and subsequently "translated" into the amino-acid backbone of a protein by the ribosome. This last translation step again uses matching of complementary base pairs on mRNA to the corresponding tRNA, thereby translating codons consisting of three consecutive base pairs to the corresponding amino acid. Each codon can code for one of the 20 possible amino acids used by Life.[1] Inside the ribosome, these amino acids are subsequently added to an amino-acid chain that eventually will fold into the structure of a protein.

---

[1] In *E. coli* the translation takes of the order of 0.05 seconds per amino acid, and thus production of a typical 300 amino-acid protein takes of the order of 15 seconds.

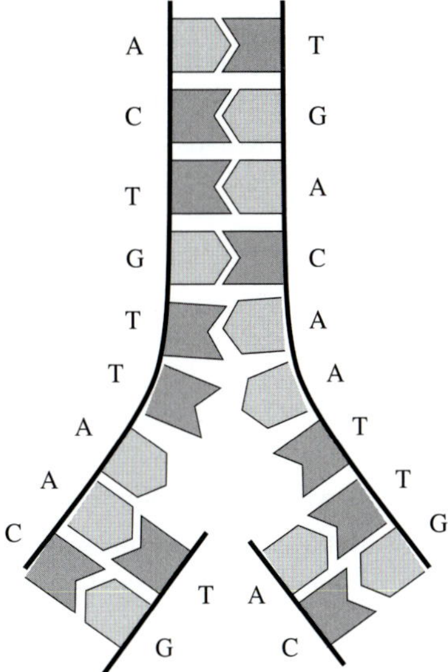

Figure 1.4 Information in life is maintained one-dimensionally by a double-stranded polymer called DNA. Each polymer strand in the DNA contains exactly the same information, coded in the form of a sequence of four different bases, creating pairs formed by bases that are complementary to each other. This complementarity is constructed using hydrogen bonds, with each base providing a particular pattern of hydrogen donors and hydrogen acceptors. Duplication is achieved by separating the strands and copying each one. This interplay between memory and replication allowed billions of years of complex history.

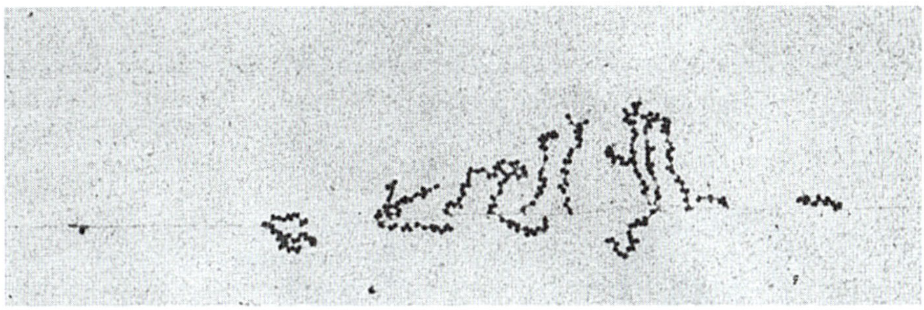

Figure 1.5 Transcription of messenger RNA (mRNA) and its subsequent translation into multiple identical proteins by ribosomes. The picture illustrates the flow of information from DNA, to mRNA, to proteins, and was taken using an electron microscope by Miller *et al.* [21] (reused with permission). Beautiful drawings of the interior of a living cell can also be found in [22].

The sequence of events leading to the assembly of a protein is usually summarized in terms of the central dogma for flow of information, from "memory," to "carbon copy," to "machine" (see Fig. 1.5.):

$$DNA \rightarrow RNA \rightarrow protein \tag{1.2}$$

This represents a simplified view of the actual situation. That is, if there were only DNA inside a cell then nothing would happen. To facilitate copying one needs several proteins to convert DNA to RNA, as well as an ample supply of energy in the form of ATP and GTP. A sustainable copying process in practice involves a complicated molecular network with feedback.

1.4   Biological molecules have long backbones

At the molecular scale life is made of polymers: DNA, RNA and proteins. Even membranes are built of nearly one-dimensional molecules with large aspect ratios. Perhaps mechanics at the nanoscale can only work with polymers, molecules that are kept together by strong (covalent) bonds along their backbones, while having the ability to form specific structures by utilizing much weaker bonds perpendicular to the backbone. These weak bonds are typically hydrogen bonds, where a hydrogen atom is shared in the same way as water is held together by sharing H atoms. The interplay between robustness and flexibility of polymers is illustrated in Fig. 1.6. Flexibility also allows folded polymers to work as complex molecular machines, in spite of their nano-scale size. Thus, functional molecules in life extensively use a seperation of energy scales: strong binding to maintain their identity, and weak binding to perform work.

1.5   Computation

A living cell is an incredible information-processing machine: a single *E. coli* cell can make about 3 000 000 proteins in 30 minutes, corresponding to a rate of about

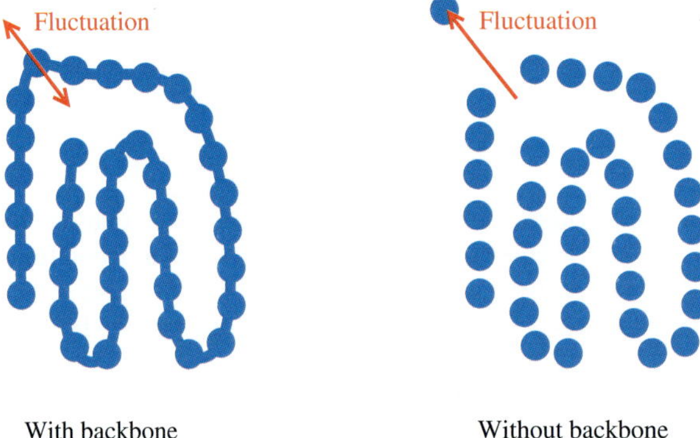

Figure 1.6 Illustration of the self-healing properties of a device with a polymer backbone. Thermal or other fluctuations may dislodge a single element, but if it is part of a polymer it typically moves back into the correct position.

3Mb/second information transfer, comparable to the speed of a typical internet connection today. And all this computation takes place within a very small volume of only $1\,\mu m^3 = (0.001\,mm)^3$, containing DNA with about $5\,000\,000$ base pairs. The information density far outnumbers any computer chip created by humans, and even one million *E. coli* cells occupy much less space than a modern CPU, thus beating it on computation speed as well. The information management of copying the genome is supplemented by "computation" associated with organisms responding to shifts in environment, conditioned by specialized regulatory proteins. Only the simplest organisms (such as the prokaryote *Mycoplasma pneumoniae* with 677 genes) can essentially manage without transcription control. More complex prokaryotes typically need a larger number of control units, a number that in fact grows with the square of the number of genes [24].

1.6   Information gathering and communication

Information gathering and communication is used when organisms have to prepare for changes in food or for sudden exposure to predators. *E.coli*, for example, senses nutrient gradients by use of receptors, and uses this information to navigate towards better conditions [546, 548, 550, 551]. Other bacteria sense the presence of each other through signaling molecules [390, 552] and sometimes use this to decide when to form protective biofilms [391] which involves the cooperative effort of billions of bacteria. In wider perspective, communication opens for a large scale coordination that may reduce dangers and improve the usage of available resources. Absence of information about eventual calamities favors timely bet-hedging, where part of the population prepares for the worst at the cost of growing slower when conditions are good [705, 706, 708, 709]. Interestingly, the long-term fitness gain by bet-hedging can be expressed in terms of the severity and probability of calamities (see Section 12.3).

1.7   Feedback

Feedback is a near-essential part of life, allowing biological organisms to adapt to a number of environmental changes by altering their behavior (see Fig. 1.8). Biological networks are complex systems with multiple feedbacks. Often these involve proteins called transcription factors that regulate production of other proteins, which in turn perform some functional task. When the task changes, a signal goes back to the transcription factor, which then alters the protein production rate, for example as illustrated in Fig. 1.7. Remarkably, feedback nearly always combines several types of network connections, for example regulation of protein production by binding of transcriptional regulators to small metabolic molecules. Negative feedback loops are sometimes combined with positive feedback loops. This is, for example, seen in temporal and

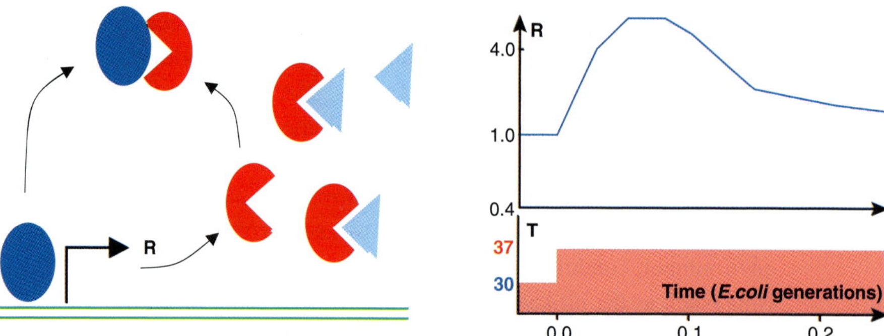

Figure 1.7 Basic negative feedback in biological networks where a regulatory protein (blue), directs production of many "worker proteins" (red). These "red" proteins bind to the "triangle molecule." If some "red" proteins are still available they bind to the "blue" protein and prevent it from initiating production of more "red" proteins. The right-hand panel shows an example of such a response, quantifying the activity of a heat-shock promoter after a 7 °C temperature shift in *E. coli* [25]. This response implies that the protein content of the cell changes and adapts to the new, higher temperature.

spatial decisions during embryonic development, as well as in spatio-temporal wave propagation [26].

1.8   Push–pull

Sustainability and stability of biological systems, from a folded protein, to cellular homeostasis, to the population of a species, often relies on pairs of counteracting "forces." In protein folding, the total binding energy of hundreds of $\mathrm{kcal\,mol^{-1}}$ is counteracted by almost equally strong entropic forces, resulting in the net stability of a folded protein of only about $10\,\mathrm{kcal\,mol^{-1}}$. To maintain the cell in its optimally functioning state, proteins are constantly activated and de-activated to keep the cell primed for responding to external changes on a fast timescale. To achieve sensitivity, enzymes and signaling proteins are constantly phosphorylated and de-phosphorylated [27]. Maintaining epigenetic memory of differentiated cells involves hundreds of molecules that are reshuffled or changed on a fast timescale, while maintaining the overall state of the system.[2] Even pathogen–host interactions often result in an exceedingly long lasting battle, in spite of a fast turnover of individual viruses in the host, as seen, for example, with HIV [28] or trypanosomes [29]. Perhaps many of our other diseases also reflect a disturbed balance between the large and nearly equally strong opposing drives of repair and decay; for example, is Parkinson's a disease associated with aggregation that occurs within days outside the cell, but takes decades to unfold when cellular machinery can

---

[2] Other examples include toxin–anti-toxin systems in bacteria, as well as restriction-modification systems in bacteria, both examples that allow their "host" to prime themselves against external challenges.

counteract it [30]? An example of a push-pull in regulation is presented by phage λ, where the decision to integrate into the *E. coli* genome involves a sequence of promoters that alternately favors each of the two pathways: PRM $-$ PR $+$ PRE $-$ PR$'$ $+$ PAQ (Fig. 4.5).

1.9a   Life is modular

Life is built of parts that in their turn are built of even smaller parts, over a wide range of scales. This facilitates robustness: if a process doesn't work, there are many ways in which parts can be modified or replaced in order to fix it [32]. Molecular-scale examples include secondary, tertiary and quaternary protein structures (the latter being complexes of multiple proteins). Network modules that each facilitate an appropriate response to external stimuli include subcellular compartments such as the nucleus, cytoplasm, etc. Most importantly, the minimal independent (or nearly independent) module of life is a single-celled organism.

The living cell is in itself a compact, complex system, with a large number of macromolecules squeezed into a minimum volume. These components are connected by networks orchestrating cellular behavior and response. As with computers, the smaller the cell, the faster the response. Therefore small cells are often competitive, with some macromolecules with copy numbers as low as one; a small number for sure, but one that allows the cell to respond sensibly to changes in external conditions.

1.9b   Life is not modular

Life is more than the sum of its parts. Removing a single gene often leads to the death of an organism. Another observation is that the number of regulatory genes for prokaryotes increases as the square of the number of genes that should be regulated [488, 489]. In other words, doubling the size of a system requires four times more regulators. Thus the regulatory network of a living cell is a well-integrated system and not modular in any simple sense of this word (see Fig. 1.8).

1.10   Stochastic processes play an important role in life

Randomness influences processes from molecules [33, 34] to cells [35, 36, 37] to life and death of individuals or entire ecosystems [3]. At the smallest scale, stochasticity in life encompasses Brownian motion of molecules, mutations, trial-and-error strategies and individuality of genetically identical cells in bacterial populations. An example of the trial-and-error mechanism is microtubule growth, attachment and collapse [34]. This particular molecular process is important in the construction of connections between neurons in our brain.

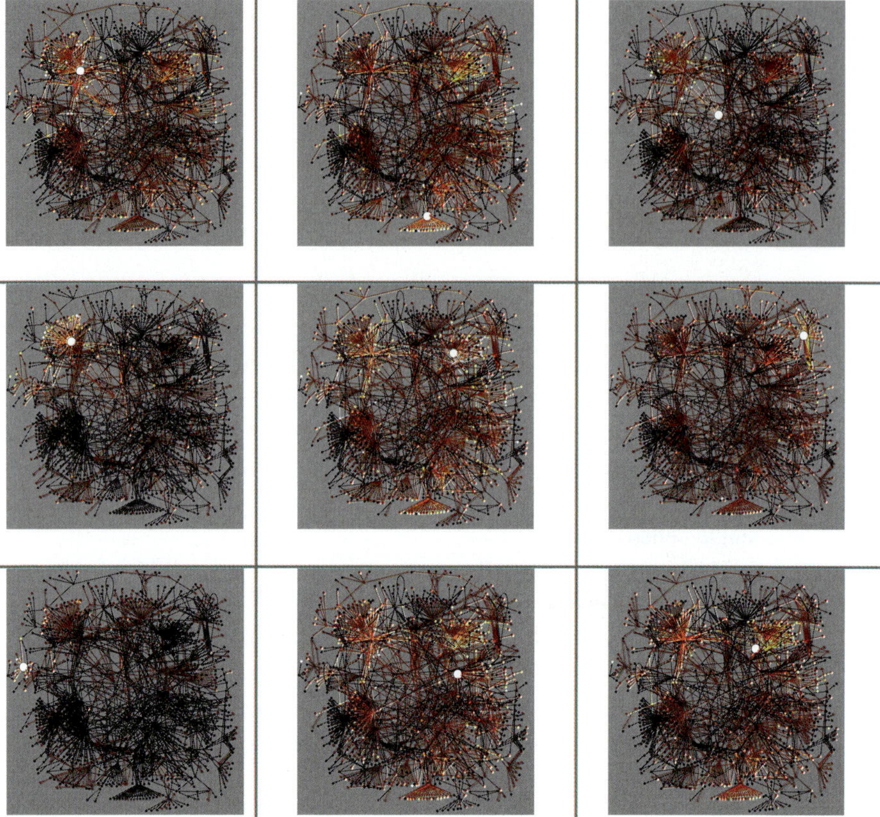

Figure 1.8 Responses in protein expression of the gene regulatory network of *S. cerevisiae* to various types of external environments/shocks [31]. One can see that the response is mostly localized (light color means large change in gene expression). The network only includes transcription regulation, and thus systematically misses the feedback links.

The individuality of cells has been probed by single-cell measurements of gene expression. Variability of cell fate has been attributed to these fluctuations (see Fig. 1.9). In fact, large variability in gene expression from individual to individual is often a good strategy to hedge against sudden environmental changes.[3] Without phenotypic and genetopic variations evolution could not have taken place.

1.11   Life harvests energy and converts it to cyclic motion

Energy allows life to maintain itself in an excited state, thereby allowing control of directional processes, including mechanical work and sorting of useful molecules from

---

[3] Examples of bet-hedging include persistent state of bacteria [40], sporulation [37] and lysogenic state of temperate phages [38, 39].

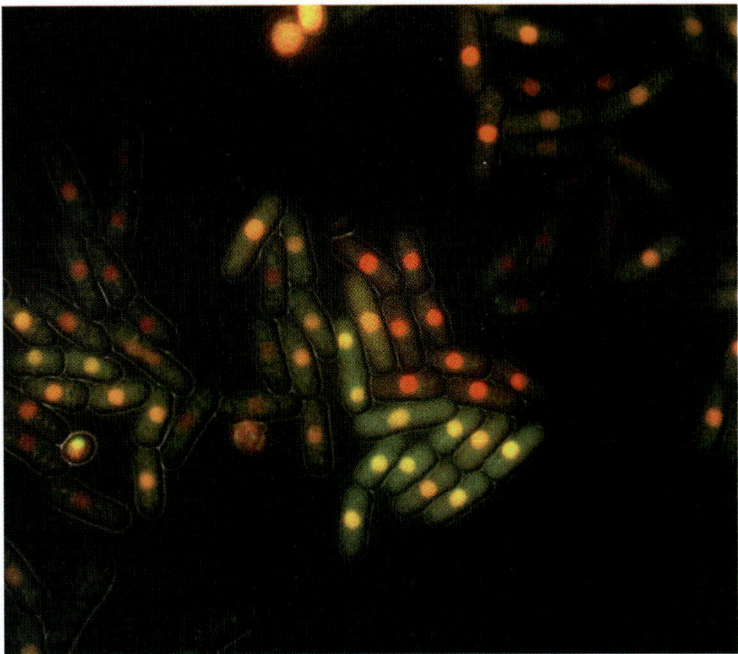

Figure 1.9 Genetically identical yeast cells displaying different expressions of certain green and red fluorescent proteins. The expression is sufficiently discrete that cells obtain individuality due to stochastic single-molecule events (micrograph courtesy of G. Thon and A. Trusina).

useless molecules. Stored energy is used to transport, sort and arrange many different molecules of life into correct positions and configurations.

The energy is harvested from food or from light, converted to ATP or GTP and subsequently released when these molecules bind and lose a phosphate group (ATP→ADP+P). It is then used to direct particular proteins to perform mechanical work, in the classical physical meaning of a force that drags an object a certain distance. This work includes directed movement of motor proteins illustrated in Fig. 1.10, where about $10\,\mathrm{kcal\,mol^{-1}}$ of energy is converted to ~ 10 nm steps and a mechanical force of a few piconewtons $(10^{-12}\,\mathrm{N})$.[4] Energy is also used to activate proteins in signaling or regulatory networks, using these to channel information towards new processes in the organism.

The conversion of energy to work typically involves directional cyclic change of states, in its simplest form

$$1 \to 2 \to 3 \to 1 \to 2 \to 3 \to 1 \to \dots \tag{1.3}$$

where at least one of the transitions demands energy input.

Cycling is an inherent part of error correction, where amino acids are added to an elongating peptide chain after being checked repeatedly for whether they match a given

---

[4] The thermal energy $1\,\mathrm{k_B}T \sim 4\,\mathrm{pN \cdot 1nm}$ means that an energy of $1\,\mathrm{k_B}T$ can be used to move an object 1 nm opposing a force of 4 piconewtons.

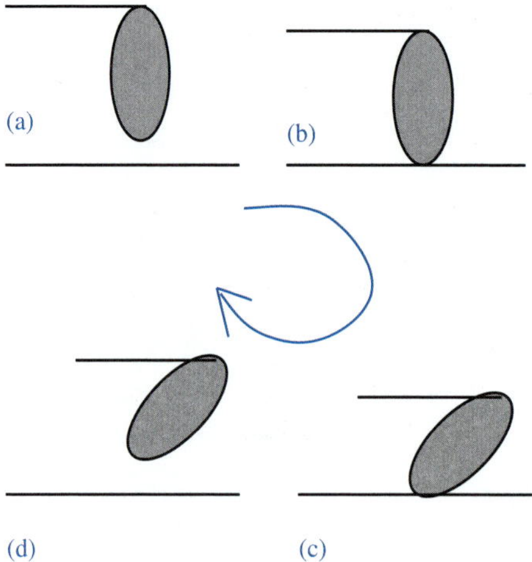

Figure 1.10 States of a motor protein as it walks, for example along a microtubule. Each of the four panels show the "head" of a motor protein above the substrate it binds and moves along. The protein may be kinesin walking on a microtubule or myosin walking on actin. The "stroke state" ($b \to c$) is associated with release of the products P and possibly ADP from $ATP \to ADP + P$. The "rigor state" (c) is the state where the protein stalls upon death, when no ATP is available. The release from the substrate ($c \to d$) is associated to binding of ATP. For myosin, approximate rates/energetics of the transitions can be found in [42]. A simple model can be found in [43].

codon on the translated mRNA (see Fig. 1.11). A larger-scale example is the excitable medium [26, 41] that facilitates propagation of signaling waves, providing spatially coupled transitions between a relaxed, an excitable and an excited state. Another example is found in the polymerase chain reaction (PCR), where DNA is amplified through cycling between double-stranded DNA at low temperature and single-stranded DNA at high temperature, which subsequently is moved to single-stranded DNA at low temperature again. The cycle is completed when the single DNA strands are each copied to form double-stranded DNA at low temperature, where raw material and energy is supplied for the work needed to make new copies.

In addition to the "governing" directed and cyclic processes, equilibrium has a role, for example in the binding and unbinding of proteins to DNA (bound $\rightleftharpoons$ free). Another example of near equilibrium behavior is found in the first step in translation, where the best match between a codon and a tRNA is found on the basis of its equilibrium binding energy (see Fig. 1.11). Equilibria contrast with cyclic processes by their bi-directionality. That is, in contrast to sequences of events that allow directionality, the equilibrium processes involves states 1 and 2 connected by transitions

$$1 \to 2 \to 1 \to 2.... \text{ equivalent to } (1 \rightleftharpoons 2)$$

that in themselves do not give directionality and work. In short, alternating between two states in equilibrium allows for transiently excited states that subsequently decay, whereas counting to three allows life.

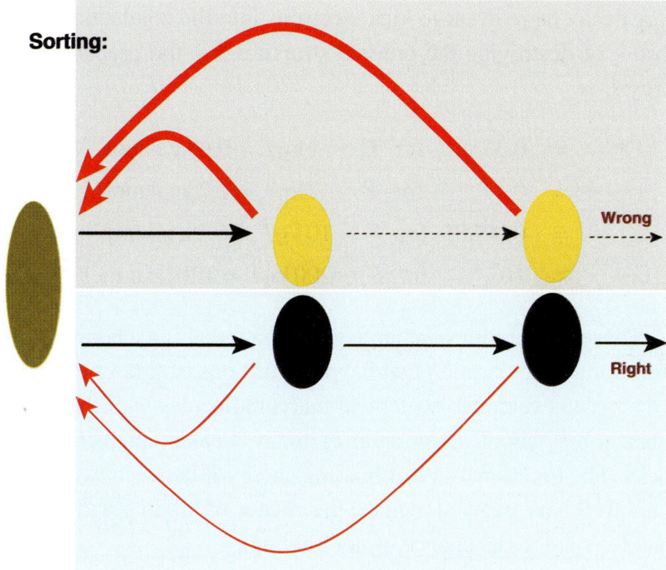

Figure 1.11 Error correction using the sorting associated with a cyclic process. In the first step a tRNA+amino acid is binding a given codon with a probability given by a on-off rates from pure equilibrium considerations. Subsequently, the bound amino acid is converted into a highly excited state, using energy. A state where a wrong amino acid decays fast from, thereby providing a second sorting step. Error correction was first proposed by Hopfield in terms of the proofreading that occurs within the DNA polymeraze [44]. Error correction at various levels minimizes errors in DNA copying and mRNA translation processes. It has been found in various other disguises, including the DNA-mismatch repair system [45], and the immune system [46, 47].

1.12　Biological physics is "$k_B T$ physics"

The relevant energy scale for molecular interactions that control biological mechanisms is $k_B T$, where T is the room temperature and k_B is the Boltzmann constant.[5] This is not true for most of the systems described in a typical physics curriculum. For example, the energy scale for a covalent bond such as C–H, C–C or C–N is of the order of $\sim 5\,\text{eV}$.[6] In comparison, the typical energy of an atom at room temperature is $\sim k_B T_{\text{room}} \approx 0.025\,\text{eV}$.

The approach to physics of biological systems with their "$k_B T$-strength" binding is therefore different from the way one would study, for example, solid-state physics. In solid-state systems one starts with a given structure and calculates energy levels.

---

[5] Notice that $k_B N_A = R$, with $N_A = 6.02 \cdot 10^{23}$, Avogadro's number, and R the gas constant. Thus the energy per molecule is about 1 $k_B T = 0.6$ kcal mol^{-1} at the absolute temperature $T = 300$ K, a comfortable temperature for most life forms. In standard physics units 1 kcal mol$^{-1} = 4184\,\text{J}/6 \cdot 10^{23} = 7 \cdot 10^{-21}\,\text{J}$.

[6] 1 eV = 1 electron volt, which is the energy that one electron gains by being moved across one volt in an electric potential.

Thermal energy may be relevant to kick electrons into the conduction band, but $k_B T$ is not on the brink of destroying the ordered structure, i.e. the probability of breaking a bond of energy E is

$$
\begin{aligned}
e^{-E/k_B T} \quad &= \quad 0.37 \qquad &\text{for } E \sim 1\,k_B T = 0.6\,\text{kcal mol}^{-1} \\
&= \quad 0.14 \qquad &\text{for } E \sim 2\,k_B T = 1.2\,\text{kcal mol}^{-1} \\
&= \quad 0.00005 \quad &\text{for } E \sim 10\,k_B T = 6.2\,\text{kcal mol}^{-1} \\
&\approx \quad 10^{-87} \qquad &\text{for } E \sim 200\,k_B T = 120\,\text{kcal mol}^{-1} = 5\,\text{eV}
\end{aligned}
$$

per attempt.[7] These probabilities are large given that breaking a bond is attempted every time the atom vibrates, a vibration that happens millions of times per second. Typically the binding energy between two biological macromolecules is of the order 10–$20\,k_B T$, a quantity that comes about as the sum of many weaker interactions between their building blocks. The functional size of binding energy E between two biological macromolecules may well have evolved to make the chance of binding within a cell of volume V approximately equal to the chance of not binding:

$$
e^{-E/k_B T} \sim \frac{V}{v}
$$

Here v is the volume associated with the bound complex of two molecules.[8]

Soft-matter systems often self-assemble in a variety of structures (e.g. amphiphilic molecules in water form micelles, bilayers, vesicles, etc.; amino-acid chains fold to form globular proteins). These ordered structures exist in a fight against the disruptive effect of thermal motion, an order that is maintained by the combination of a polymer-like backbone and multiple weak interactions perpendicular to the backbone, as illustrated in Fig. 1.6.

The quantity that encompasses the disruptive effects of thermal motion is the entropy S, which is a probability measure of microscopic disorder. That is, if one considers a number of (micro)states with energies E_1, E_2,... the combined probability of these states will be proportional to

$$
e^{-E_1/k_B T} + e^{-E_2/k_B T} + ... = e^{S/k_B} \cdot e^{-E/k_B T} \equiv e^{-F/k_B T}
$$

where E is the average energy, e^{S/k_B} counts their effective number and F is the total free energy of these states. For binding between biomolecules, both E and S play a large role; the entropy S varies so much with temperature that life becomes hugely sensitive to a

---

[7] The exponential comes because the probability of collecting $1\,k_B T$ in one chemical bond is $1/e = 0.37$. The probability of collecting $2\,k_B T$ is $1/e^2$ because each of the $1\,k_B T$ units has to be present for it to happen simultaneously. The probability of collecting many $k_B T$ units in one specific bond is correspondingly a product of many factors of $1/e$.

[8] For *E. coli* $V = 1\,\mu\text{m}^3$, whereas the convention for binding energies uses densities in molar (M), corresponding to binding volumes $v \sim 1/\text{M} = 0.63\,\text{nm}^3$, giving $e^{-E/k_B T} \sim 1000^3 \sim 10^9$ and a balanced binding energy $E \sim 21 k_B T = -13\,\text{kcal mol}^{-1}$.

change of even a few percent in the absolute temperature, a variation that is associated with the huge changes in the properties of water with temperature.[9]

To set the frame for the coming chapters you may visualize the human society as a gigascale implementation of a living cell, a system that has ample capacity to copy, with factories as devices of manufacturing essential ingredients, a system that evolved to its present state through a long historical struggle, occasionally interrupted by erratic catastrophes, but mostly with long periods of stasis or even progress. The human social system also maintains memory, by aligning letters on long strings of text. Strings of text that are copied by other, less elegant mechanisms than in the cell, but copied and propagated nonetheless. The parallels goes on, with maybe the largest difference being the size of the noise, where the fluctuation set by Brownian motion in the cell is replaced by some underlying randomness in human activity.

1.12.1 Questions

1.12.1 One *E. coli* bacterium divides in 30 minutes, if the temperature is good and food is plentiful. One bacterium fills a volume of 1 μm^3. How long would it take for one *E. coli* to reach the volume of the Earth (which has the radius of $\sim 6.000\,$km)? Repeat the calculation for human replication (each having a volume of about 0.1 m^3), assuming a doubling time of 30 years.

1.12.2 There are 10^{14} cells in a human body and every day 10^{11} cells die. How much information needs to be re-generated inside a human to keep it functional over a lifetime of 80 years (human DNA contains 3×10^9 base pairs)? There is damage to DNA $\sim 10^5$ times per cell per day in humans, every one of which could lead to cancer. How many damages need to be repaired during human lifetime?

1.12.3 Life is highly specific. Assume that you need to say hello to everybody in a city with 10^6 people. How long would this take if saying hello takes 1 second? What is the maximum time unit allocated to non-specific binding between pairs of random proteins in a *E. coli* cell, if any particular pair meets at least once within the 30 minutes generation time of an *E. coli* cell. There are of the order of 3 000 000 proteins in *E. coli*.

1.12.4 Consider the four-step cycle of a molecular motor modeled in terms of two variables $\varphi_1 = 0, 1$ and $\varphi_2 = 0, 1$ and an energy function:

$$H = A \cdot (\varphi_1 + \varphi_1 \varphi_2) \tag{1.4}$$

where the degeneracy $g = 10$ of the $\varphi_i = 0$ states is larger than that of the unique $\varphi_i = 1$ states, implying that $\varphi_i = 0$ is 10 times more likely than $\varphi_i = 1$

---

[9] For example, the viscosity of H_2O changes by a factor of two when T changes 30 °C, implying that the friction of moving in water close to 0 °C is much larger than when moving at large temperatures. An increase in viscosity reflects increased rigidity of the hydrogen bonds between the oxygen atoms.

when $A = 0$. Let A alternate between 0 and 1. For $A = 0$, the state $(0,0)$ is by far the most likely, and the $(1,1)$ state will decay to the $(0,0)$ situation. For $A = -1$, on the other hand, the $(1,1)$ state is favored. Define motion $x \to x + 1$ respective to $x \to x - 1$ in terms of the sequence of shifts in these variables, and show that directed motion can be possible [43].

1.12.5 Continue parallels between the human society and physical aspects of life outlined in this chapter.

2 *E. coli* as a model system

2.1 DNA, RNA and proteins

The macromolecular building blocks of living cells are DNA, RNA, proteins, polysaccharides and lipids. DNA stores the information needed to make proteins, RNA participates in the assembly of the proteins and the proteins themselves are the final devices that perform the tasks. Lipids form membranes and thereby allow living systems to form separate compartments and cells.

Let us first list the sizes of these building blocks. We do this in units of equivalent information content. DNA is used as the main information storage device. The information resides in sequences of four bases: A,T,C and G. Three subsequent bases form a codon, which codes for one of the 20 amino acids (aa). The genetic code is shown in Table 2.1. The weight of a codon is larger for DNA (because it is double stranded) than for RNA (single stranded), which in turn is heavier than the amino acids it codes for:

DNA: 2000 daltons per codon,
RNA: 1000 daltons per codon,
Proteins: 110 daltons per amino acid.

Here 1 dalton = $1 \, \mathrm{g} \, \mathrm{mol}^{-1}$ = $1.7 \cdot 10^{-27} \, \mathrm{kg}$ ($\sim$ the mass of one hydrogen atom). This hierarchy of masses does not reflect the amount that is present in a cell, because of the amplification in copy number at any step in the conversion of information from DNA to mRNA to proteins.

The peculiar structure of biological macromolecules represents the existence of two distinct energy scales. The polymer backbone is held together by covalent (strong) bonds, i.e. energies of the order of several electron volts ($\sim 40\text{–}200 \, \mathrm{kcal} \, \mathrm{mol}^{-1}$). The secondary and tertiary structures, which for the DNA is a double helix, and for a protein its folded form, as well as recognition processes between these big molecules, depend on the much weaker hydrogen bonds, which each are about $1\text{–}2 \, \mathrm{kcal} \, \mathrm{mol}^{-1}$. As room temperature corresponds to an energy of about $k_B T = 0.617 \, \mathrm{kcal} \, \mathrm{mol}^{-1} = 4.2 \cdot 10^{-20} \mathrm{J}$ per molecule,[1] the structure and mutual interactions of large biomolecules depend on many weak bonds.

---

[1] Throughout this book we use the notation of $k_B T$ and other energies in units of $\mathrm{kcal} \, \mathrm{mol}^{-1}$, where $1 \, \mathrm{kcal} \, \mathrm{mol}^{-1}$ means 1 kcal per $6 \cdot 10^{23}$ units or $0.7 \cdot 10^{-20} \mathrm{J}$ per unit.

5′ end	T	C	A	G	3′ end
T	Phe (F)	Ser (S)	Tyr (Y)	Cys (C)	T
T	Phe (F)	Ser (S)	Tyr (Y)	Cys (C)	C
T	Leu (L)	Ser (S)	STOP	STOP	A
T	Leu (L)	Ser (S)	STOP	Trp (W)	G
C	Leu (L)	Pro (P)	His (H)	Arg (R)	T
C	Leu (L)	Pro (P)	His (H)	Arg (R)	C
C	Leu (L)	Pro (P)	Gln (Q)	Arg (R)	A
C	Leu (L)	Pro (P)	Gln (Q)	Arg (R)	G
A	Ile (I)	Thr (T)	Asn (N)	Ser (S)	T
A	Ile (I)	Thr (T)	Asn (N)	Ser (S)	C
A	Ile (I)	Thr (T)	Lys (K)	Arg (R)	A
A	Met (M)	Thr (T)	Lys (K)	Arg (R)	G
G	Val (V)	Ala (A)	Asp (D)	Gly (G)	T
G	Val (V)	Ala (A)	Asp (D)	Gly (G)	C
G	Val (V)	Ala (A)	Glu (E)	Gly (G)	A
G	Val (V)	Ala (A)	Glu (E)	Gly (G)	G

Table 2.1 The genetic code with amino acids corresponding to each of the 64 different triplet [51] codons. These are "read" from the so-called 5′ end to the 3′ end of one strand of the DNA. The codon ATG=AUG also codes for the start of the translation of an mRNA into a protein. The colors refer to translation speeds for codons in *E. coli* grown in minimal medium [52, 53]: red refers to slow $\sim 4\,\mathrm{codons\,s^{-1}}$, green to intermediate $\sim 9\,\mathrm{codons\,s^{-1}}$, and blue to fast $\sim 35\,\mathrm{codons\,s^{-1}}$. Under these conditions the average translation speed is $12.5\,\mathrm{codons\,s^{-1}}$.

2.1.1   DNA

A single strand of DNA consists of a negatively charged deoxyribose-phosphate[2] backbone and one of four different bases (A, T, G, C) attached to each unit. The backbone is connected by strong covalent bonds. Each of the two strands of DNA has a direction: there is a 5′ end and a 3′ end of each strand. Two complementary DNA strands come together (with anti-parallel orientation) to form the double helix: A pairs with T, and G pairs with C.

In the double helix, the sugar phosphate backbone is on the outside and the bases on the inside, stacked like the steps of a ladder, see Fig. 2.1. The bases pair up by using two hydrogen bonds for A–T, and three hydrogen bonds for G–C. The binding energy for a base pair is 1–$2.5\,\mathrm{kcal\,mol^{-1}}$, whereas the energy associated with initiating a bubble inside a DNA double helix is about $5\,\mathrm{kcal\,mol^{-1}}$.

The persistence length[3] of double-stranded DNA (dsDNA) is about $l_p \sim 60\,\mathrm{nm}$, while single-stranded DNA has a persistence length of $\sim 1\,\mathrm{nm}$. For the dsDNA (B-DNA) the helix diameter is $18\,\mathrm{Å}$ and there is one helix turn every $34\,\mathrm{Å}$, corresponding to 10.4 base

---

[2] In contrast to RNA, which has a ribose-phosphate backbone. Notice that ribose is a sugar molecule, and thus the information to make life is maintained by sugar.

[3] Persistence length is the measure of how stiff a polymer is and measures how long the polymer needs to be in order for sizeable bending due to thermal fluctuations to be observed.

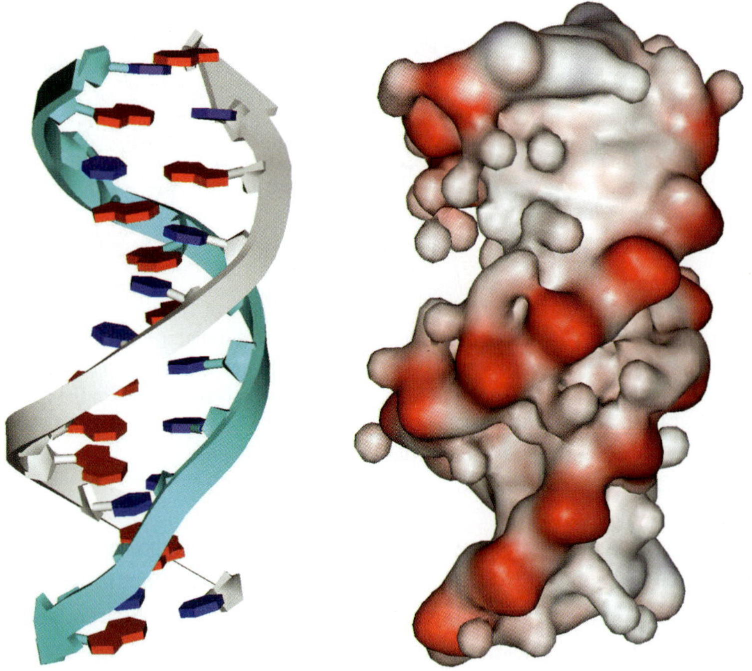

Figure 2.1 A little more than one turn of the DNA double helix. The left panel shows the two strands with bases that match the complementary bases. The right panel shows the DNA external surface that is accessible for a water molecule. Figure from PDB ID: 1BNA [48]

pairs (bp). The DNA has a major groove and a minor groove. The basis for recognition of specific sequences by DNA-binding proteins is through exposing the bases within the grooves: each of the four pairs (AT, TA, GC and CG) expose a unique spatial pattern of hydrogen-bond acceptors and donors, as well as a single positive charge.

A DNA-binding protein mainly recognizes the unique pattern of H-bond donors and acceptors associated with a short DNA sequence (of about 6-9 bp). More subtle effects involving local sequence-dependent conformations and rigidity may also play a role. It is remarkable that the protein α-helix typically fits well in the major groove of DNA, thus achieving the ideal exposure of amino-acid side chains to the DNA base pairs. In fact, the major groove exposes all four different bases, whereas proteins can only identify AT=TA and CG=GC in the minor groove. Therefore the major groove has higher information content [54, 55] and is associated with specifically binding proteins, whereas the minor groove is often used for more non-specific DNA-binding proteins, like HU or FIS [56].

2.1.2 RNA

RNA carries information in a similar way to DNA, but, like proteins, some RNA can perform particular catalytic functions. RNA consists of a covalently connected

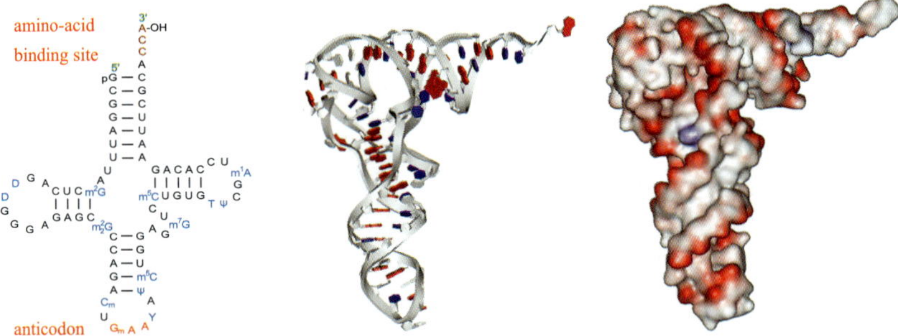

Figure 2.2 Secondary structure of RNA molecules, here a transfer RNA (tRNA) that is the carrier of the genetic code. Note that the shown "cover leaf structure" folds up to the final L shaped tertiary structure characteristic of tRNAs, with recognition site placed in the lower end and the amino acid bound to the upper right end. The rightmost panel shows the surface of the folded tRNA that is exposed to water. Fig. from PDB ID: 1EHZ [49]

ribose-phosphate backbone with a sequence of the four bases A,G,C and U. Thus T from DNA is replaced by U in RNA. This has the effect that in RNA G can pair with both C and U, thereby allowing more sliding between RNA strands than DNA would allow. This helps single-stranded RNA from being captured in metastable states. The RNA backbone also has an additional OH group, which makes it more flexible.

Importantly, RNA is used as a transient information carrier, and the lifetime of messenger RNA (mRNA) is relatively short and exposed to active biological regulation/degradation [57].

Some RNA molecules have important enzymatic functions in the cell, including transfer RNA (tRNA), which carries aminoacids to the ribosome, and ribosomal RNA (rRNA) that forms a large part of the ribosome. RNA folding is more complicated than with DNA because it easily forms hairpins and loops. Therefore, RNA may fold into elaborate three-dimensional structures, see, for example, Fig. 2.2. In contrast to proteins, these structures can be estimated fairly well by calculations because most of the energy in RNA is associated with its secondary structure [58, 59], defined in terms of one binding partner, maximum, for each base.

2.1.3   Proteins

A protein is a heteropolymer build of a sequence of amino acids connected along a covalently connected backbone. A typical protein consists of ~ 300 aminoacids which amounts to a globule of about 5 nm in diameter. The chemical structure of one amino acid is shown in Fig. 2.3. Centrally there is an aminogroup and a carboxyl group, separated by a carbon (C) atom. The 20 different amino acids used in proteins differ by the side chain R. The simplest sidechain is just a hydrogen (H) atom, corresponding to the smallest amino acid, glycine. Amino acids come in sizes between $0.06 \, \text{nm}^3$ and $0.23 \, \text{nm}^3$ [60], with affinities for water between $-3.4 \, \text{kcal mol}^{-1}$ and $+3.0 \, \text{kcal mol}^{-1}$ [61].

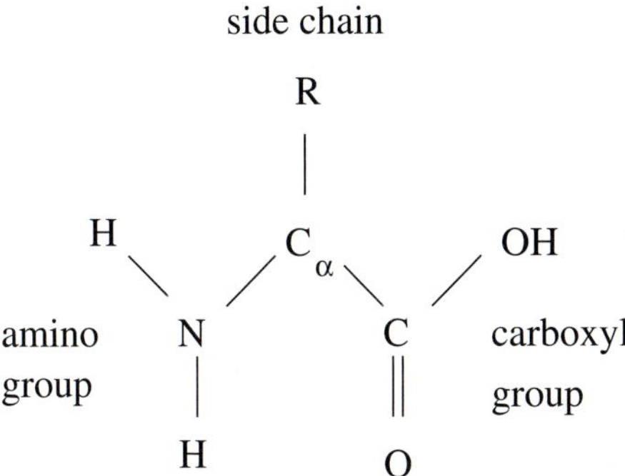

Figure 2.3 A single amino acid, consisting of a CCN backbone and a side group R that can take one of 20 different forms. The simplest are $R = H$ (glycine) and $R = CH_3$ (alanine).

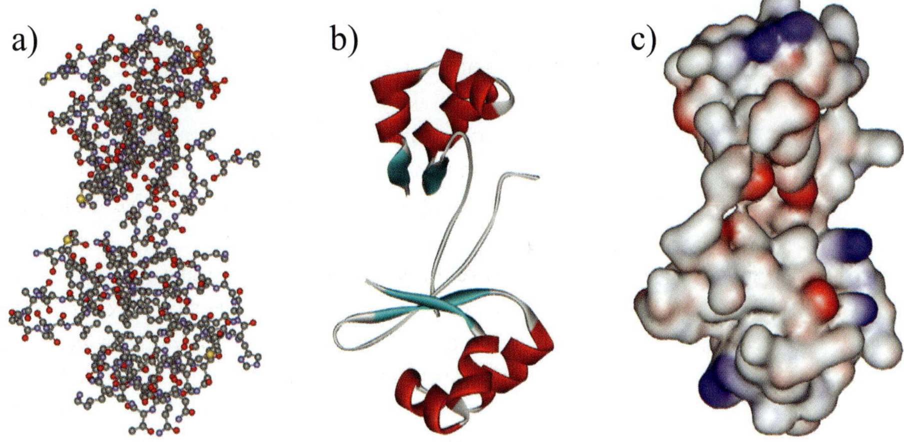

Figure 2.4 Dimer of Cro proteins from phage lambda: (a) shows all atoms, (b) emphasizes secondary structures, consisting of alpha helix elements in red, (c) the surface of the dimer that is accessible by a water molecule. Figure from PDB ID: 5CRO [50]

The sequence of amino acids defines the primary structure of the protein (see Fig. 2.4). Often, a protein will spontaneously fold into a well-defined rigid three-dimensional structure that is also determined by its sequence of amino acids [23]. This folding takes place on timescales from fractions of seconds to several seconds [62]. Proteins are the molecular machines that perform all the tasks necessary for the living cell, often by making directed movements or by catalytic activity while using energy. Proteins often regulate each other's production or activity, while yet other proteins build or manipulate the structure of the cell. Most proteins are not actively degraded, but some are degraded by proteases and have lifetimes that can be as short as 1 minute.

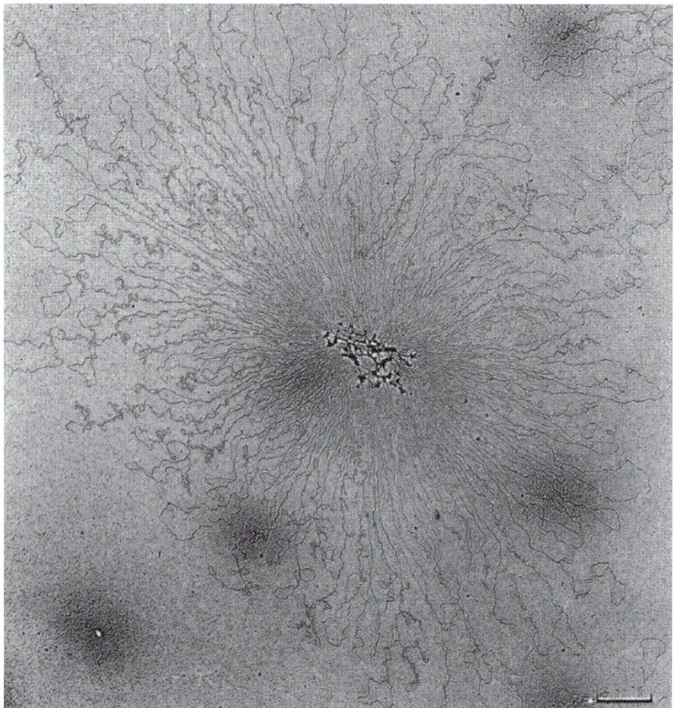

Figure 2.5 Image shows a bacteria that lyse (break up) with its DNA leaving the cell. Primarily one sees that the DNA can expand far beyond the borders of the *E. coli* cell (from Kavenoff *et al.* [90]). (Each chromosome is about 1000 times longer than the length of the *E. coli*). Image is reproduced with kind permission from Springer Science and Business Media.

2.2   The *E. coli* cell

Biological molecules connect to form functional systems inside any living cell. As a cell prototype and model organism, we primarily consider the prokaryote *Escherichia coli*. From a physics perspective, the *E. coli* cell may be viewed as a small "bag" of DNA, RNA and proteins, surrounded by a membrane (see Figs. 2.5 and 2.6). The "bag" has a volume of about $1\,\mu m^3$ that varies as the cell grow and divides. In addition, fast-growing cells are large, whereas poor growth media causes *E. coli* to reduce cell size and at the same time replicate more slowly [67].

The interior of the cell is crowded with about 30–40% of its weight in proteins and other macromolecules, and only about 60% as water. Further, the water contains a number of salt ions[4], in particular K^+, Cl^- and Mg^{2+}, each of which influences the stability of different molecular complexes.

---

[4] Total ion concentration is about $300\,mM \sim 300\,000\,000$ per *E. coli* cell, with K^+ concentration of about 200–250 mM.

Volume: total, cytoplasm	$1\,\mu m^3$, $0.7\,\mu m^3$
Surface area	$6\,\mu m^2$
Length of DNA	$1550\,\mu m$
Average protein size (in amino acids)	360
Average diameter of protein	5 nm
Concentration of proteins	5–8 mM
Diameter of ribosome	20 nm
Volume fraction of proteins	17%
Volume fraction of all RNA	6%
Volume fraction of mRNA	0.2%
Volume fraction of DNA	1%
Volume fraction of ribosomes	8%
Number of mRNA molecules	4000
Number of rRNA molecules	18 000
Number of tRNA molecules	200 000
Number of lipids	25 000 000
Number of all proteins	3 500 000
Number of water molecules	23 000 000 000
Number of ions	120 000 000
Number of ATP molecules	2 000 000
Number of free amino acids	6 000 000
Number of ATP molecules to make 1000 nt RNA	2000
Number of ATP molecules to make a 360 aa protein	1500
Number of ATP molecules to make one *E. coli* cell	$\sim 15\,000\,000\,000$
Number of glucose molecules to make one *E. coli*	$\sim 500\,000\,000$

Table 2.2 *E. coli*, from the cyper cell database (CCDB) [63, 64] accumulating data from a number of sources, reflecting growth conditions between 30-60 minutes cell generation time. The volume of a protein in Å^3 is approximately its mass in daltons multiplied by two. For cell to cell variation, see [65]. The ATP numbers are from [66].

The chemical composition of an *E. coli* cell under good growth conditions with a cell generation time of about $\tau_{\text{gen}} \sim 30\text{–}60$ min is listed in Table 2.2. For some of the small molecules, like amino acids and ATP, the table contains both the total number of molecules consumed in cell generation and the number of freely available molecules at any particular time. If m is the number of free small molecules, the free concentration develops as

$$\frac{dm}{dt} = P - \frac{m}{\tau} \tag{2.1}$$

with τ being the residence time in the free state. In the steady state $m = \tau \cdot P$, whereas the total production in cell generation $m_{\text{total}} \sim P \cdot t_{\text{gen}}$. As a result

$$\tau = t_{gen} \cdot \frac{m}{m_{total}} \tag{2.2}$$

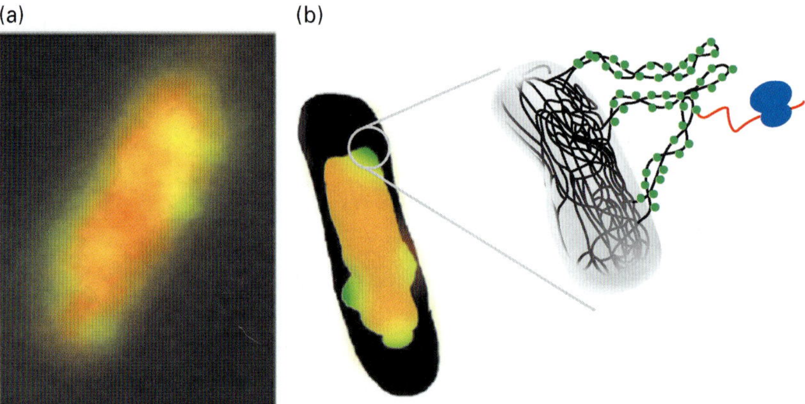

Figure 2.6 Microscopy of *E. coli*, with RNA polymerase (RNAP) shown in green and DNA shown in red [71], reveals that transcriptional activity is clustered in subsections of the *E. coli* around the ribosomal genes. (A) Real picture using fluorescent proteins [71]. (B) Interpretation of [71], where the green dots indicate RNA polymerases and the blue area a ribosome.

which for ATP gives a recycling time of $\tau \sim 1.800 \cdot 15 \cdot 10^9/(2 \cdot 10^6)$ s ~ 0.1 s. A similar estimate for the $1\,000\,000\,000$ amino acids (aa) in an *E. coli* cell suggests that its free pool of $6\,000\,000$ aa is replaced within a time $\tau = 1800 \cdot (6 \cdot 10^6)/10^9$ s ~ 9 s.

The genome of *E. coli* is a single DNA molecule with $4.6 \cdot 10^6$ base pairs (total length of about 1.5 mm). It codes for the 3500–4200 different proteins (dependent on the *E. coli strain*) and a number of RNA molecules. Pictures of *E. coli* are shown in Figs. 2.5 and 2.6 emphasizing, respectively, the huge amount of DNA that there is inside the cell, and the spatial organization that is found even within this very small living organism [68, 69, 70]. In most of this book we entirely disregard this spatial organization.

2.3   Transcription and translation numbers

Transcription and translation are performed by huge molecular motors, which each work their way along specific one-dimensional strings, using them as templates for generating another one-dimensional string. This, of course, demands energy, which in the cell takes the form of ATP, recycled at an enormous rate under good growth conditions.

Production demands the presence of building blocks: the bases to make RNA units, and the amino acids to build the protein polymers. Each of the 20 different amino acids can only be assembled into peptide strings when they are first attached to the end of a corresponding transfer RNA (tRNA). The other end of the tRNA is complementary to an mRNA triplet, matched by the ribosome. Subsequently, the amino acid is covalently linked to the previous amino acid in the growing peptide chain. We now put some numbers on this sequential production machinery, involving first RNAP and then the ribosome.

2.3.1 RNAP

RNAP is a protein complex, consisting of four protein types ($\alpha_2\beta\beta'\sigma$), and in *E. coli* it has a total mass of about 400 kDa = 400 000 Da (where single amino acids have masses between 90 and 200 Da). When RNAP binds to a promoter sequence on the DNA [72], it occupies 65–75 bp. After transcription initiation, the RNAP transcribes the DNA with a speed that varies between $\sim$ 30bp s^{-1} and 90bp s^{-1}, where the ribosomal genes are transcribed fastest [73, 74]. The transcribing RNAP is stopped by terminator signals, which in *E. coli* involve a DNA sequence that codes for an mRNA sequence forming a short hairpin.

Dependent on growth conditions, *E. coli* cells have 1500 (τ_{gen} = 100 min) to 11 000 RNAPs (τ_{gen} = 24 min) [75], of which between 200 and 3300 are actively transcribing.[5] The free concentration of RNAP is estimated to be 17% [76]. The remaining RNAPs are either bound to promoters or non-specifically bound to the DNA.

We remind the reader that 1 nM (=1 nanomole per liter, with 1 mole = $6 \cdot 10^{23}$) corresponds to about one molecule per *E. coli* cell, more precisely to 0.6 molecules per μm^3:

$$1/1\,\mu\text{m}^3 = 1/(10^{-15}\,\text{l}) = 10^{15}\text{l}^{-1} = 1\,\text{mol} \cdot \text{l}^{-1}/(6 \cdot 10^8)$$
$$= 1/(6 \times 10^8)\,\text{M} = 1.6\,\text{nM} \sim 1\,\text{nM}$$

2.3.2 Transcription

RNAP binds to DNA at certain binding sites (called promoters) and initiates transcription. After binding, the RNAP may subsequently transcribe the DNA in a direction determined by the orientation of the promoter, see Fig. 3.1. Each gene has one or more corresponding promoters, and some promoters initiate transcription of several genes (in prokaryotes, not in humans). When DNA is transcribed then RNA is produced.

The activity of a promoter is a result of three consecutive steps [77, 78, 79] illustrated in Fig. 2.7:

(1) RNAP binds to a promoter with rate k_b to form a closed complex.[6] Especially important are short sequences around the -10 (TATAAT) and the -35 bp (TTGACA) positions where position $+1$ is the start site for the RNA production and + numbers refer to downstream nucleotides to be transcribed. Binding may be transient because the RNAP can fall off the promoter with a rate k_u.

(2) Open complex formation is also called the "isomerization step" and takes place with a forward rate k_o. This involves unwinding of about 12–18bp of

---

[5] Of which 24% (at τ_{gen} = 100 min) to 79% (at τ_{gen} = 24 min) are transcribing stable RNA, i.e. rRNA or tRNA. The corresponding rate of mRNA sythesis varies from $0.4 \cdot 10^6$ nucleotides min^{-1} cell^{-1} to $2.3 \cdot 10^6$ nucleotides min^{-1} cell^{-1}.

[6] RNAP target location may be facilitated by RNAP binding to nearby DNA and subsequent sliding to the promoter site.

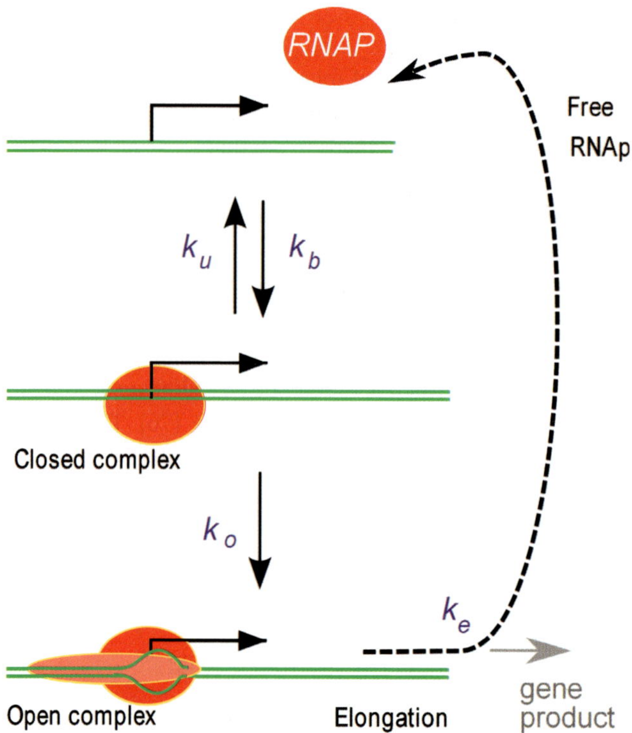

Figure 2.7 Hawley–McClure's three-step model of transcription initiation: first an RNAP binds to form a closed complex at the promoter with the on-rate $k_b \cdot [RNAp]$ and dissociates with off rate k_u. While in closed complex it can form an open complex with rate k_o (also called the forward rate, as this is assumed to be a directed non-equilibrium process). For each binding, the likelihood of the forward reaction is the ratio of k_o to $k_o + k_u$. Finally, the open complex starts elongating to initiate transcription with rate k_e. When the RNAP has left the promoter, a new RNAP may bind.

    double-stranded DNA in order to form an "open bubble" that exposes the DNA bases for transcription.[7]

(3)   Transcription is initiated with rate k_e, after which the RNAP elongates with a speed of about 30–90bp s^{-1}. The DNA is transcribed from the 5′ end to the 3′ end. When transcribing, either the RNAP rotates around the DNA, or more likely, the DNA rotates while the RNAP remains fixed.

Given the arrows between the states, the time between subsequent firings will be a sum of tree times

$$\tau = \frac{1}{k_e} + \frac{1}{k_o} + \frac{1}{k_b \cdot [RNAP]} \cdot \left(1 + \frac{k_u}{k_o}\right) \tag{2.3}$$

---

[7] The transcription bubble forms between positions –9 to +3bp. The bound RNAP occupies $\sim$ 65 bp in the DNA region from –55 to the +20 position [80].

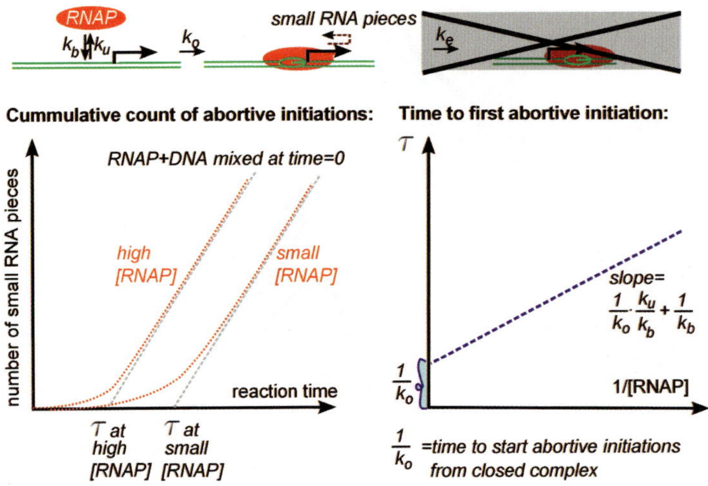

Figure 2.8 *In vitro* dissection of a promoter using a transcription assay where one nucleotide is missing in the medium, which in turns prevent RNAP from leaving the promoter. When RNAP is in open complex it starts a cycle of abortive initiations, each leading to a small RNA segment the accumulation of which can be measured. Notice that this persistent recycling is faster than normal elongation initiation, implying that the production rate = slope of the left-hand figure is not useful in itself. However, by analyzing the intersection time τ shown in the left-hand panel as a function of RNAP concentration, one can determine the isomerization rate k_o and the dissociation constant k_u/k_b [81].

representing respectively the times associated with the two irreversible steps (open complex and elongation initiation), plus the time associated with equilibrium binding.[8] $1/k_o$ and $1/k_{on} = \frac{1}{k_b \cdot [RNAP]} \cdot \left(1 + \frac{k_u}{k_o}\right)$ can be measured *in vitro*, by studying the binding of RNAP to a promoter and the subsequent production of small nucleotide sequences, see Fig. 2.8. The key observable is the ongoing formation of very small RNA sequences, which are produced when a missing nucleotide forces RNAP in an open complex to repeatedly stop and reinitiate its transcriptional activity [78, 79, 81].

The experiment is started at time $t = 0$ when RNAP is mixed with DNA pieces containing the promoter. Dependent on RNAP concentration, there will be a delay after which all RNAP have entered open complexes, leading to the constant accumulation of small nucleotides. The "intersection" $(1/k_o)$ and the "slope" in Fig. 2.8 are determined by varying the concentrations of RNAP. The overall activity of *E. coli* promoters varies by nearly four orders of magnitude [79], consistent with the finding of k_o rates between 0.001 and 0.1 s^{-1}, whereas the [RNAP] dissociation constant $k_u/k_b \in [1\,\text{nM}; 1\,\mu\text{M}]$, see [79].

The initiation rates of mRNA synthesis for genes that we consider in this book include about one per second for ribosomal genes, one every 18 seconds for the promoter PR

---

[8] Formally Eq. 2.3 can be derived by assigning the occupation probability of the open complex θ_o, of closed complex θ_c and then solve for steady state $\frac{d\theta_o}{dt} = k_o\theta_c - k_e\theta_o = 0$ and $\frac{d\theta_c}{dt} = k_b \cdot [RNAP] \cdot (1 - \theta_o - \theta_c) - k_o\theta_c - k_u\theta_c = 0$. The rate of promoter initation is then $k_e \cdot \theta_o$.

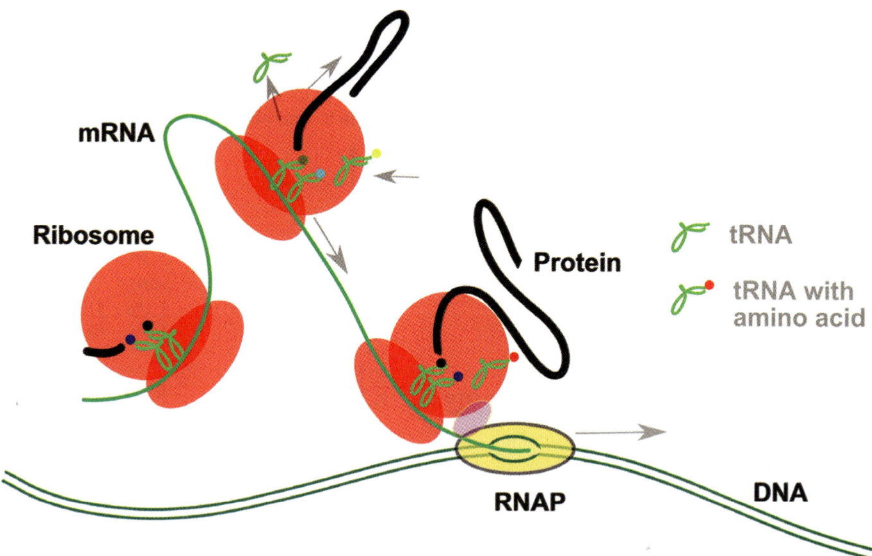

Figure 2.9 RNAP moves along the DNA while transcribing it into mRNA, which subsequently is translated into a protein by the ribosome. The same mRNA may easily be translated into several proteins, as shown here, where three ribosomes act simultaneously on the mRNA. The first ribosome is directly coupled to the transcribing RNAP through the Nus protein, supporting the motion of RNAP by dragging the mRNA out of it and thereby lowering friction/clogging [83, 84].

in phage λ, one every 250 seconds for the unstimulated promoter PRM in phage λ and 1 per cell generation for the LacI promoter. The initiation frequency is a primary determinant for the concentrations of the proteins in the cell. However, transcription factors may influence any of the promoter parameters, for example by causing activation by lowering k_u in a type of regulation that is called recruitment [82].[9]

2.3.3 Translation

It is inside the ribosome that the "information world" of RNA is converted to the "machine world" of proteins. Ribosomes translate the nucleotide sequence to amino acids, using the genetic code. Each amino acid is attached to a specific tRNA molecule that contains the triplet codon [51] corresponding to an amino acid as illustrated in Fig. 2.9. In the ribosome this codon is matched to the complementary mRNA codon, thus translating a sequence of codons to a sequence of amino acids that subsequently "folds" to make a protein.

The ribosome is a large molecular complex, that in *E. coli* consists of 56 different protein subunits, plus some fairly large RNA subunits [86]. In bacteria it has a mass

---

[9] By lowering the off-rate of RNAP binding to a transcription factor, the RNAP is effectively present at the promoter longer, and therefore there is a greater probability of initiating transcription.

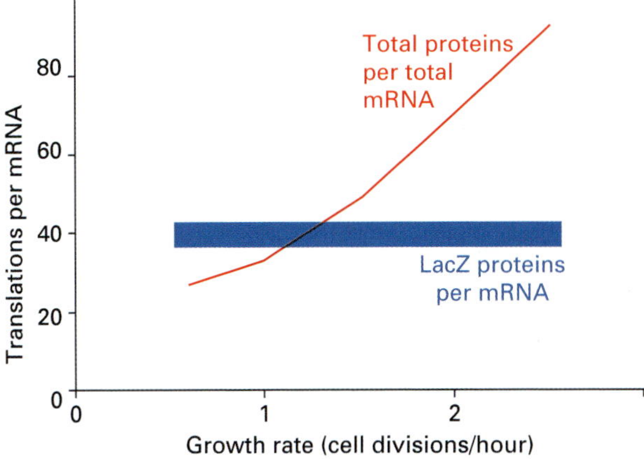

Figure 2.10 Number of proteins per message typically increases with growth rate of *E. coli* because of increased ribosome concentration, whereas proteins per message for LacZ stays around 40, independent of the growth rate. Data on overall translations per mRNA are from [75], which also reports that distance between translating ribosomes decreases from 40 codons to 20 codons at fast growth. Data on LacZ are from [85], which also reports that the major determinant for increasing overall protein per message is increased translation initiation (from 16 min^{-1} at 0.6 doublings per hour to 55 min^{-1} at 3 doublings per hour).

of $2.5 \cdot 10^6$ daltons, of which one-third is proteins. In a rich medium, the production of ribosomal RNA (rRNA) amounts to about half of all RNA production in the *E. coli* cell. At any time the rRNA occupies 80–90% of the total RNA content of *E. coli* and tRNA about one-sixth of that number, whereas mRNA only constitutes about 3% of the RNA.

Dependent on growth conditions *E. coli* have 7000 ribosomes (at generation time of 100 min) to 70 000 (at 24 min) [75], of which only 5% are idle [87]. Ribosome abundance is regulated [88, 89, 90, 91], and the number of ribosomes depends on the growth rate. In particular, when exposed to amino-acid starvation, the protease Lon degrades ribosomes and thereby recycles their amino acids [92].

Once initiated, the translation proceeds at a codon-specific speed of 4–35 codons s^{-1} giving an average translation speed of 12.5 codons s^{-1} in minimal media at 37 °C [53, 94]. This average speed grows to a maximum of 20 codons s^{-1} for rich media [75].

A transcribed mRNA is in fact translated by ribosomes many times [75], as illustrated in Fig. 2.9, and explored in a traffic model in Fig. 2.11, a model that could be adjusted to overall growth conditions using the ribosome initiation rate and average speed from [75]. The average number of translations per mRNA in *E.coli* is estimated to be ∼30 at slow growth and up to ∼90 at very rapid growth [75], see Fig. 2.10.

The number of proteins per transcribed mRNA depends on the strength of the ribosome binding site and on the lifetime of the mRNA. The best ribosome binding site on the mRNA is the Shine–Delgarno sequence (AGGAGGU), located about 10 base pairs upstream from the translation start signal [95]. The degradation of mRNA can be regulated through RNase enzymes, which actively degrade the mRNA [96]. The overall

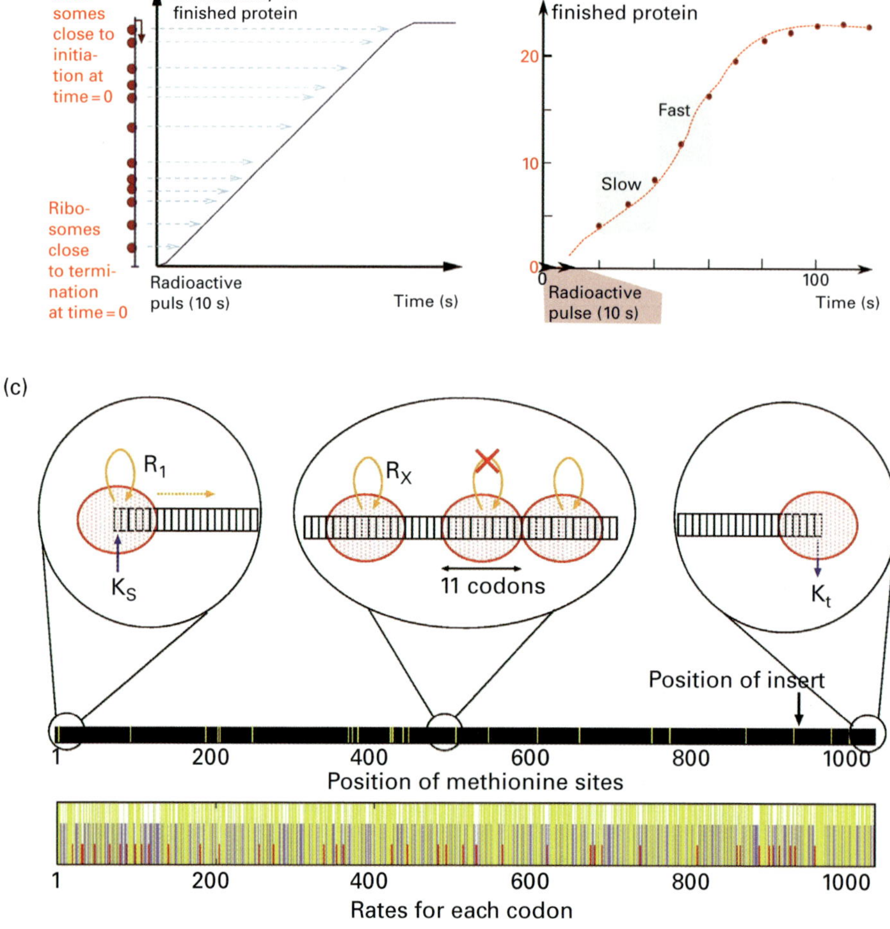

Figure 2.11 (A) The in vivo measurement of the speed of ribosomes as they translate the LacZ mRNA. The progress of ribosomes is measured by the amount of incorporated radioactive methionine in finished proteins. Radioactive methionine is only during a short pulse at Time = 0. A linear increase in measured methionine reflects a constant speed. (B) Data is taken at a doubling time of one hour. The data allow us to identify regions where the mRNA is translated at low and respectively high rates. (C) A traffic model of ribosomes that translate the LacZ mRNA [53]. Each ribosome covers 11 codons, and a ribosome cannot move if the one ahead is blocking the motion. Codons with speed of 35 codons s^{-1} are yellow, 9 codons s^{-1} are blue and 4 codons s^{-1} are red [93].

distribution of mRNA lifetimes is shown in Fig. 2.12, exhibiting a pronounced difference between the time an mRNA can be translated (average lifetime 2.5 min) and the time that remnants of the mRNA exist in the cell (average 7 min). Noticeably, the distribution of the number of mRNAs across all genes is very broad, fitted by the distribution $P(number) \propto 1/number^{1.7}$ [97].

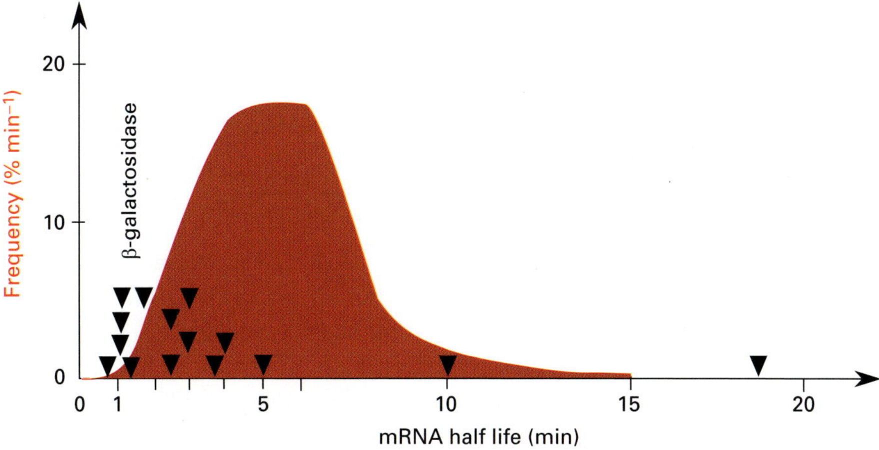

Figure 2.12 Genome-wide study of chemical mRNA lifetimes in *E. coli* [99] for growth in rich medium (LB) at 30 °C where cell generation time was 30 minutes. The lifetime distribution at three times slower growth was found to be similar [99]. The measurement was based on DNA microarrays, which count mRNA, irrespective of whether this can be translated. Functional half life measured by [57] is shown with black triangles. The average lifetime of 22 measured mRNAs is only 2.5 min [57]. Remarkably, for these mRNAs there was no correlation between abundance of a protein and the lifetime of its mRNA [57].

In *E. coli*, translation and transcription in fact occur at the same speed, a coordination that may depend on physical contact between the ribosome and the transcribing RNAP [83, 98].

2.3.4　Questions

2.3.1　Assume an *E.coli* cell with a volume of 1 μm^3, containing 25% proteins by weight. An amino acid weighs on average 100 Daltons (1 Da = $1.7 \cdot 10^{-27}$ kg), and amino acids have the same density as water. How many amino acids are incorporated in the proteins in the cell?

2.3.2　Assume that an average protein consists of 360 amino acids. How many proteins are there in the above *E. coli* cell?

2.3.3　Assume that one bacterial generation takes 30 minutes, and that it contains 40 000 active ribosomes. What is the average rate of translation in the cell (codons per second per ribosome).

2.3.4　Assume that at any time there are 1000 RNA polymerases (RNAP) in an *E. coli* cell that transcribe mRNA. Assume further that transcription and translation occur at the same rate. Calculate the average number of proteins produced per mRNA.

2.3.5　What time does it takes to clear a promoter site before a new RNAP can bind (RNAP occupies 75 bp on the promoter)? What is the expected theoretical maximum activity for a promoter?

2.4   *E. coli* content varies with its growth rate

Between 1940 and 1950, Monod [100] observed that the growth rate (λ) of *E. coli* is remarkably reproducible, and obeys so-called Michaelis–Menten kinetics

$$\Lambda = \Lambda_0 \cdot \frac{\text{Lactose}}{\text{Lactose} + K} \tag{2.4}$$

with a maximum growth rate of $\Lambda_0 \approx 2.5\,\text{h}^{-1}$. The variation in growth speed comes with an overall change in the content of *E. coli* listed in Table 2.2. For example, an *E. coli* cell contains between $0.6 \cdot 10^9$ aa at cell generation $\tau_{gen} = 100$ min to $2.5 \cdot 10^9$ aa at $\tau_{gen} = 24$ min [75]. The corresponding RNA content varies from 37 to 390 million bases, and the genome number increases from 1.6 to 3.8 [75]. The variation in the main physiological quantities are shown in Fig. 2.13.

References [102, 103] derived Eq. (2.4) by partitioning the total protein mass T into ribosomes R, metabolic proteins M^{10} and overheads Q:

$$T = Q + R + M \tag{2.5}$$

Here the ribosomes R work at speed γ' to produce the total proteins T with rate Λ:

$$\Lambda \cdot T = \gamma' \cdot R \tag{2.6}$$

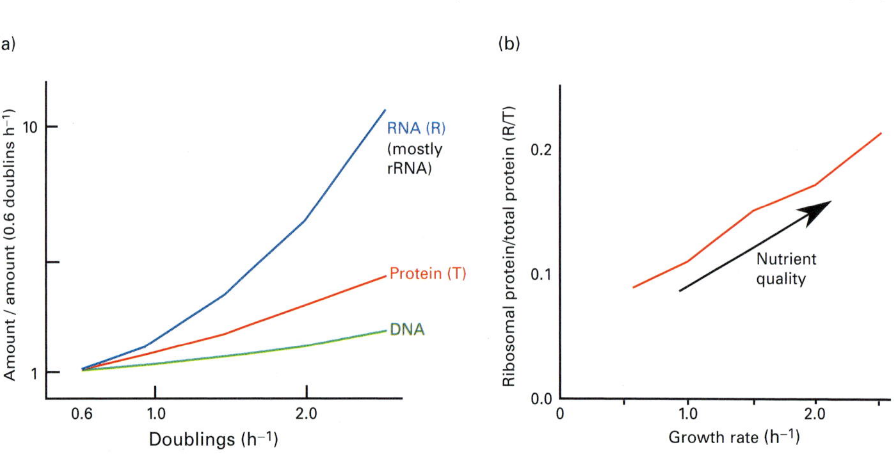

Figure 2.13 Relative composition of the *E. coli* cell as a function of its growth rate [75]. The total mass and protein content $\propto 2^{no.\ of\ doublings\ per\ hour}$ [101], whereas the ratio of biomass to water is constant. The increase in ribosome content makes the typical mRNA translation activity change from $16\ \text{min}^{-1}$ to $55\ \text{min}^{-1}$ at the highest growth rates [85]. Notice that the doubling per hour μ is given by growth rate Λ through $exp(\Lambda/\mu) = 2 \Rightarrow \Lambda = \mu\ln(2) = \ln(2)/\tau$, where τ is the cell generation time. The regulation of cell size with growth rate is discussed in [67].

---

[10] M denotes proteins that are used to convert food molecules to useful amino acids, sugars and lipids.

where γ' is proportional to the translation rate per ribosome,[11] taking into account that a constant fraction of ribosomes does not translate. Metabolic proteins M fuel the growth in a way that is proportional to the rate v at which they can convert nutrients to amino acids:

$$\Lambda \cdot T = v \cdot M \tag{2.7}$$

Combining these equations:

$$\frac{Q}{T} + \frac{\Lambda}{\gamma'} + \frac{\Lambda}{v} = 1 \;\Rightarrow\; \frac{1}{\Lambda} \propto \frac{1}{\gamma'} + \frac{1}{v} \tag{2.8}$$

$$\Rightarrow \Lambda \propto \frac{\gamma' \cdot v}{\gamma' + v} \tag{2.9}$$

which is equivalent to the Monod equation for the growth rate (using $v \propto$ [Lactose]). Eq. (2.8) implies that the cell generation $\ln(2)/\Lambda$ is proportional to the time it takes ribosomes to replicate themselves, $\ln(2)/\gamma'$, plus the time it would take the metabolic genes to generate their own biomass, $\ln(2)/v$.[12] The resulting growth is therefore limited by both translational and metabolic capacity [103]. In particular, notice that the limits on growth for large [Lactose] implied from Eq. (2.4) are primarily caused by a finite translation rate.

One may express Eq. (2.6) in terms of the translation rate γ of the active ribosomes:

$$\Lambda \cdot T = \gamma' \cdot R = \gamma \cdot (R - R^*) \tag{2.10}$$

where R here is again total ribosome count and R^* is the inactive ribosomes. Thus the ribosomal protein mass fraction [107]

$$\frac{R}{T} = \frac{1}{\gamma} \cdot \Lambda + (R^*/T) \tag{2.11}$$

implying that the ribosome fraction increases with Λ. This growth rate, on the other hand, is in practice changed by increasing v, by either increasing food quantity or food quality. Equation 2.11 was first deduced by [107] from the experiments of [101] (see Fig. 2.13):

$$\frac{\mathrm{RNA}}{\mathrm{protein}} = a \cdot \Lambda + b \tag{2.12}$$

To connect the two last equations, reference [107] assumed that the RNA to protein ratio is proportional to the fraction of ribosomes in the cell. This is reasonable because nearly all RNA in a cell is ribosomal RNA.

---

[11] Assuming that no proteins are degraded, and cell generation $\tau = \ln(2)/\Lambda$:
(number of amino acids in cell) = (number of active ribosomes)$\cdot$ (translation rate) $\cdot \tau \Rightarrow$
$(\lambda/\ln(2))\cdot$ (total protein mass)/(mass of one amino acid) =
(translation rate) $\cdot$ (mass of active ribosomal proteins)/(protein mass of one ribosome) $\Rightarrow$
ratio of protein mass in active ribosomes to total protein: $(R - R^*)/T = \Lambda/\gamma$ with
$1/\gamma = \dfrac{1}{\text{(translation rate)}\cdot\ln(2)} \cdot \dfrac{\text{(protein mass of one ribosome)}}{\text{mass of one amino acid}}$

[12] For explanation of the $\ln(2)$ factor see Figure caption 2.13.

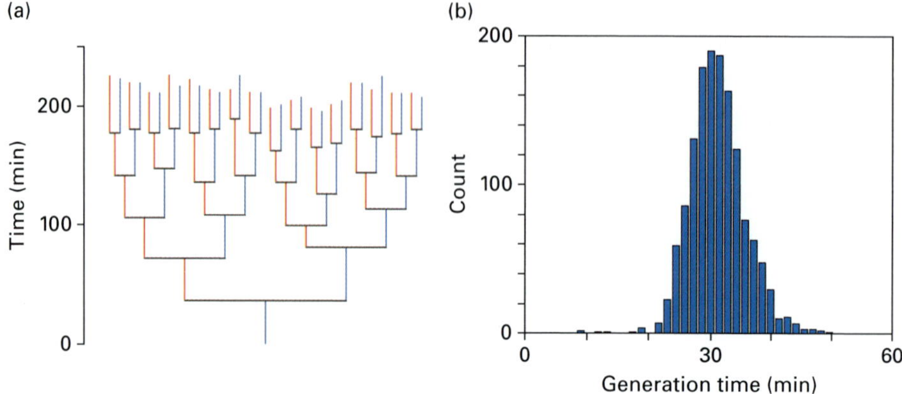

Figure 2.14 Variation in *E. coli* growth at 37 °C, as recorded by S. Vedel, H. Nuns and A. Trusina. The cell-to-cell variation in *E. coli* growth is quite broad. Reference [104], for example, reports a standard variation of 12 min when *E. coli* was grown with an average generation of 54 min. In later chapters we will discuss noise in gene expression, where one component of cell-to-cell variation will be these extrinsic variations in cell-to-cell growth rates. Part of the cell-to-cell variability could be due to bacterial ageing [105, 106].

The above considerations reflect the fact that the *E. coli* cell is a fairly efficient machine, replicating optimally given available resources. However, it is noteworthy, that even replication on the scale of a full *E. coli* cell is prone to large fluctuations, as illustrated in Fig. 2.14.

2.4.1 Questions

2.4.1 Show that the form of Eq. (2.9) is maintained when one considers that both γ and μ can be represented by Michaelis–Menten growth curves:

$$v = v_{max} \cdot \frac{[\text{nutrient}]}{[\text{nutrient}] + K} \quad \text{and} \quad \gamma = \gamma_{max} \cdot \frac{\Lambda}{\Lambda + \Lambda_g} \tag{2.13}$$

Here γ_{max} corresponds to a max translation speed of 20 aa s^{-1} and $\Lambda > \Lambda_g = 0.5$ doublings h^{-1} (from [75]).

2.4.2 From Eq. (2.9), and inserting Eq. (2.7)

$$\frac{R}{T} = 1 - \frac{Q}{T} - \frac{M}{T} = 1 - \frac{Q}{T} - \frac{\Lambda}{v} \tag{2.14}$$

one may investigate RNA per protein by changing γ through reduction of translation rates by certain antibiotics [102]. Sketch the corresponding RNA per protein versus Λ dependencies in Fig. 2.13 and identify v and overheads Q from the curves. Interpret the ribosome expression response to reduced translation ability of individual ribosomes.

2.4.3   Growth may be limited by more than nutrients, for phytoplankton in the ocean by light and photosynthesis, suggesting an extension:

$$T = Q + R + M + L \qquad (2.15)$$

where L is the part of the protein mass allocated to photosynthesis. Assume in addition to Eqs. (2.5), (2.6), (2.7) that $\Lambda T = \alpha L$ and express $1/\Lambda$ as function of $1/\alpha$.[13] With subdivided metabolism, the final growth can be limited with respect to any of these nutrients.

2.5   Summary

- The prokaryotes need of the order of 10^3 different types of proteins to function, and a total amount of about 10^6 proteins.
- In prokaryotes, translation and transcription occurs in parallel with a production time for an average protein of 15 to 30 s.
- Monod's growth law, $\Lambda = \Lambda_0 \cdot \frac{[\text{Food}]}{[\text{Food}] + K}$, expresses that the growth rate of bacteria is limited by the production cost of the translation machinery.
- As growth rate Λ is increased, the relative fraction of ribosomes in the cell increases as $R/T \propto \Lambda + \text{offset}$.

---

[13] For plankton in the ocean, one may subdivide their metabolism into carbon, nitrogen and phosphorus, which in the ocean are found in the so-called Redfield ratio C:N:P = 106:16:1[108]. In addition, iron, Fe, is often also limiting.

3 Dynamics of regulatory links

3.1 Regulating a piece of DNA

The information content of the genome is primarily managed by regulatory proteins through their interactions with the DNA, with each other, with RNA or with various small metabolic molecules. Some proteins, called transcription factors, regulate the production of other proteins by binding upstream of the genes for these proteins, thereby continually adjusting the metabolism and response of the cell to optimize future growth and survival.

Figure 3.1 shows two ways that a transcription factor (TF) can regulate the transcription of a gene. The right-hand panel shows a specific example of a regulatory protein that is bound to the DNA. The CAP (or CRP) protein is in fact the TF that regulates most genes in *E. coli*, and as such is the largest hub in its genetic regulatory network.

This chapter introduces a few quantitative models for genetic regulation. The basic process here is the binding–unbinding events of proteins, both to each other and to the operator DNA that controls transcription initiation. Thus, for readers that are not familiar with chemistry, we will now go through basic chemistry and co-operativity. In this section we follow the standard approach quite closely with on- and off-rates, and the consequences of co-operativity. In the next chapter we will venture into a more detailed discussion of how to treat binding as a combination of operators, and through this we will demonstrate how statistical mechanics provide us with a generic framework.

Consider a titration experiment that examines the bound fraction of a complex, as a function of the total concentration of one of its constituents. Let the reaction between the regulator R and an operator site O on a small DNA piece be:

$$R + O \rightleftharpoons RO \tag{3.1}$$

This reaction is in equilibrium when the rates of the reactions

$$R + O \rightarrow RO \quad \text{with} \quad \text{rate}_\rightarrow = k_\rightarrow [R][O] \tag{3.2}$$

$$R + O \leftarrow RO \quad \text{with} \quad \text{rate}_\leftarrow = k_\leftarrow [RO] \tag{3.3}$$

balance ($\text{rate}_\rightarrow = \text{rate}_\leftarrow$):

$$K = \frac{k_\leftarrow}{k_\rightarrow} = \frac{[R][O]}{[RO]} \tag{3.4}$$

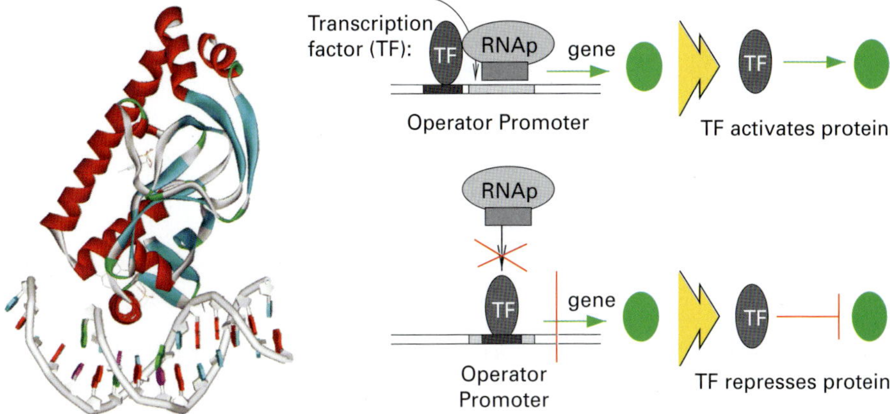

Figure 3.1 The left-hand panel shows the transcription factor CAP (or CRP) binding to a piece of DNA [109]. The binding is facilitated by an α-helix that fits into the major groove of the DNA, which is a structural feature that is found for many DNA-binding proteins. The right-hand panel illustrates the respective positive and negative regulation by a transcription factor (TF). The TF is a protein that binds to a region on the DNA called an operator (black area). Negative regulation occurs when the operator is placed such that the bound TF prevents the RNAP from binding to the promoter [72, 110]. Positive regulation [111] can be facilitated by recruitment [112], a mechanism where the TF binds an RNAP and thereby increases the frequency that the RNAP starts transcription. The far-right-hand panels show how elementary regulation in regulatory networks is represented.

K is called the dissociation constant; all concentrations refer to free concentrations in the solution, and thus not to total concentration. For example, the total concentration of operator sites is $[O]_T = [O] + [RO]$ and the fraction of occupied operator is

$$\frac{[RO]}{[O_T]} = \frac{[R]}{K + [R]} \tag{3.5}$$

provided that the concentration $[R] \gg [O_T]$. Here the constant K is in units of concentration and in fact specifies the concentration of R where the operator O is half occupied.

In molecular biology K is typically found to be $\sim 10^{-8\pm2}$ M. Such an experiment is illustrated in Fig. 3.2. Notice that the free concentrations $[R]$ and $[O]$ are less than the total concentrations $[R_T]$, $[O_T]$ because $[R_T] = [R] + [RO]$. The experiment is most easily analyzed when the operator concentration is kept much lower than all other concentrations ($[O] \ll [R]$), such that $[R_T] \approx [R]$. This situation is also typical inside the *E. coli* cell, where there is of the order of one operator of any particular type, whereas regulator proteins typically come in copy numbers of order 100.

3.1.1 Co-operative chemistry

Often in gene regulation it is desirable to switch between two extreme expression values, while only using a relatively small change in the concentration of the controlling protein.

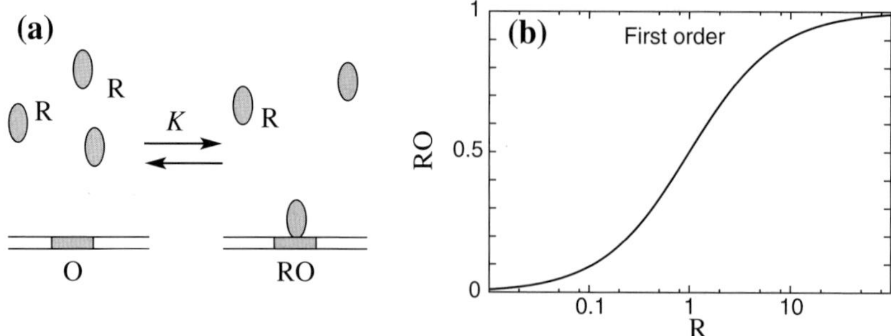

Figure 3.2 (a) Elementary chemistry illustrating the first-order reaction between a dimer R and one isolated operator site O. (b) Occupancy as a function of R concentration (for the first-order reaction shown: $[RO]/([O_T]) = [RO]/([RO] + [O])$). The experiment to determine K requires measurements of the amount of RO complex for different concentrations of R. This demands a separation of bound and free DNA, a separation that can be achieved by their different displacements on a gel filter when exposing them to electrophoresis. Alternatively one may use "foot-printing," where the parts of the DNA that are bound to the protein are protected against a DNAase that degrades all unbound DNA. The fraction of surviving DNA fractions is then measured by gel electrophoresis.

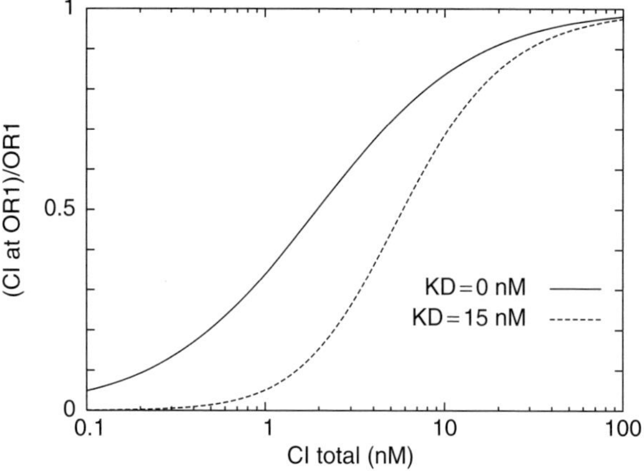

Figure 3.3 The occupancy of the operator OR1 in phage λ as function of total R = CI concentration (an example we explore further in the next chapter). The two curves illustrate the effect of the finite dimerization constant of R: dimerization makes the transition between the on- and off-states occur over a narrower interval of R values.

In molecular biology this is obtained by sequence of co-operative bindings, the simplest being dimerization.

Often a DNA-binding protein R only binds significantly to the DNA in the form of a dimer (see Fig. 3.3). Binding to DNA then involves two steps, a dimerization

$$(R)_M + (R)_M \rightleftharpoons R_2 \quad \text{with} \quad K_D = \frac{[(R)_M]^2}{[R_2]} \tag{3.6}$$

followed by binding to the operator

$$R_2 \ + \ O \rightleftharpoons R_2O \ \text{with a bound fraction} \ \frac{[R_2O]}{[O_t]} \ = \ \frac{[R_2]}{K + [R_2]} \tag{3.7}$$

where $[R_2]$ is the dimer concentration, K_D the dimerization constant, and $K = [R_2][O]/[R_2O]$ the dissociation constant for binding of the dimer to the operator. Expressed in terms of the monomer concentration, the bound fraction is:

$$\frac{[R_2O]}{[O_T]} \ = \ \frac{[(R)_M]^2}{K \cdot K_D + [(R)_M]^2} \tag{3.8}$$

which exhibits its largest change at a particular free monomer concentration:

$$[(R)_M] \ \approx \ \sqrt{K \cdot K_D} \tag{3.9}$$

Before plotting behavior versus concentration one should be aware that $[(R)_M]$ is the free concentration of monomer form of R. To express everything as a function of total concentration $[R_T]$ one must solve Eq. (3.6):

$$\frac{2[(R)_M]^2}{[R_T] - [(R)_M]} \ = \ K_D \tag{3.10}$$

where the total concentration of R molecules is given by the contribution from the free dimers $[R_2]$ plus the free monomers $[(R)_M]$, $[R_T] = 2 \cdot [R_2] + [(R)_M]$, ignoring any R dimer that may be bound to the operator. This gives:

$$(R)_M \ = \ -\frac{K_D}{4} + \frac{K_D}{4}\sqrt{1 + \frac{8}{K_D}[R_T]} \tag{3.11}$$

The dimer concentration:

$$\begin{aligned}
[R_2] &= \frac{1}{2}([R_T] - [R_M]) \\
&= \frac{1}{2}\left([R_T] + \frac{K_D}{4} - \frac{K_D}{4}\sqrt{1 + \frac{8}{K_D}[R_T]}\right) \\
&\sim \frac{1}{2}\left([R_T] + \frac{K_D}{4} - \frac{K_D}{4}\left(1 + \frac{4}{K_D}[R_T] - \frac{64}{8K_D^2}[R_T]^2\right)\right) \\
&= \frac{[R_T]^2}{K_D}
\end{aligned} \tag{3.12}$$

where the approximation ($\sim$) requires that $[R_T] \ll K_D$. When $[R_T] \gg K_D$, the dimer concentration simply becomes

$$[R_2] \ \sim \ \frac{1}{2} \ [R_T] \tag{3.13}$$

Dependent on the value of total R concentration relative to K_D, the fractional occupancy of an operator is (from Eq. 3.8):

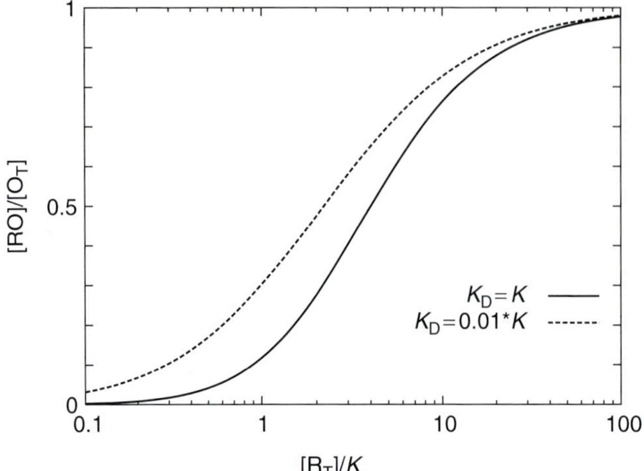

Figure 3.4 Fractional occupancy $Y = [R_2O]/[O_T]$ of one single operator, where the R molecules first form dimers in solution, and then bind to the operator. When the dimerization is constant, $K_D << K$, the reaction is effectively first order. On the other hand, when $K_D > K$, the occupancy curve switches in a narrower concentration interval.

$$\frac{[R_2O]}{[O]_T} \approx \frac{[R_T]^2}{2K \cdot K_D + [R_T]^2} \quad \text{for} \quad [R_T] << K_D \tag{3.14}$$

$$\frac{[R_2O]}{[O]_T} \approx \frac{[R_T]}{2K + [R_T]} \quad \text{for} \quad [R_T] >> K_D \tag{3.15}$$

representing regulation with Hill coefficients $h = 2$ and $h = 1$ respectively[1] (see Fig. 3.4). In general, co-operative effects come into play when concentrations are smaller than K_D, whereas saturated dimer concentration implies that all R are in the dimer form. The biggest co-operativity is therefore obtained when K_D is much larger than K.

3.1.2 Questions

3.1.1 Consider a titration experiment that varies R concentration with a fixed operator concentration in order to probe the first-order chemical reaction $R + O \leftrightarrow RO$. Compare the fractions of bound [RO] to [O] as a function of $[R_T]$, for $[O_T] = 10\,K$ and $K/10$.

---

[1] One may consider a process where n protein P molecules only bind when they simultaneously associate with a DNA segment O that is in much smaller concentration than P: $nP + O \leftrightarrow P_nO$ with $K^n = [P]^n[O]/[P_nO]$ and $[O_T] = [O] + [P_nO]$. Then:

$$\frac{[P]^n([O_t] - [P_nO])}{[P_nO]} = K^n \quad or \quad \frac{[P_nO]}{[O_T]} = \frac{([P]/K)^n}{(1 + ([P]/K)^n)}$$

where K has the molar units and $n = h$ is the Hill coefficient.

3.2   Transcription regulation

Regulation of transcription involves repeated association and dissociation of regulators (R) with operator sites on DNA. As the regulator concentration typically is much larger than the operator concentration, we can use Eq. (3.5) to express the probability that an operator is occupied by R

$$\text{bound fraction} = \frac{R/K}{1 + R/K} \tag{3.16}$$

where K is the dissociation constant. This equation can be extended to include co-operative effects by

$$\text{bound fraction} = \frac{(R/K)^h}{1 + (R/K)^h} \tag{3.17}$$

where simultaneous binding of multiple Rs is represented by Hill coefficients $h > 1$.

A common type of directed link in a regulatory network (RN) is a regulator R that represses [110] or activates [111] a promoter. If R activates the promoter, its activity is proportional to the bound fraction Eq. (3.17):

$$\text{Activity} \sim \frac{(R/K)^h}{1 + (R/K)^h} \tag{3.18}$$

If R represses the promoter, then the activity is given by the unbound fraction

$$\text{Activity} \sim \frac{1}{1 + (R/K)^h} \tag{3.19}$$

A large Hill coefficient makes activation more sensitive to R, and the activity approaches a step function around K, when h is large. Eqs. (3.18) and (3.19) do not model the binding of RNA polymerase and the subsequent steps that lead to production of mRNA. For a single promoter in steady-state conditions, this is not a serious limitation since non-equilibrium effects can be absorbed into an effective K [113].

The dynamics of the concentration, C, of a protein produced from a gene repressed by a regulator can be modeled using the following differential equation:

$$\frac{dC}{dt} = \text{leak} + \frac{\text{capacity}}{1 + (R/K)^h} - \frac{C}{\tau} \tag{3.20}$$

The first two terms count production: *leak* accounts for cases where repression cannot reach 100%, whereas *capacity* sets the maximum production rate of the protein over the base level *leak*. The lifetime of the protein is parameterized by τ, which takes into account both active degradation and dilution due to cell growth. Using a degradation rate that is proportional to C implicitly assumes that the degradation is not enzyme limited.[2]

---

[2] Where degradation is enzyme limited, the degradation term could be replaced by its Michaelis–Menten form $C/\tau \rightarrow k_{\text{cat}} \cdot C \cdot E/(C + K)$ with $1/\tau = k_{cat} \cdot E/K$.

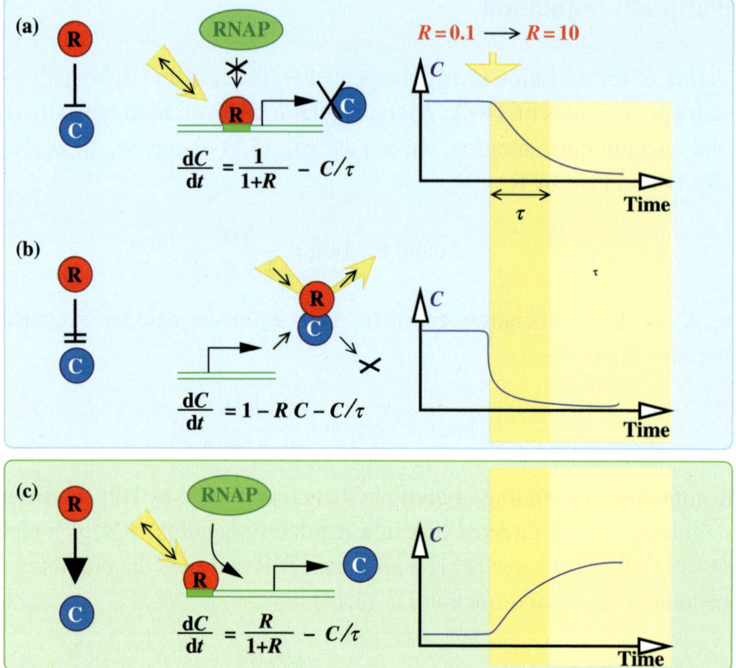

Figure 3.5 Regulating the concentration of a protein, C, through negative and positive control by regulator R. In all cases the protein concentration is measured in units of its binding constant. (a) R represses transcription. (b) R catalyzes degradation of C [114]. (c) R is a transcription activator. In each case, the adjacent plot shows how C responds to an instant 100-fold change in the concentration of R. Notice that active degradation, case (b), can be much faster than transcriptional repression, case (a), assuming of course that such a sudden activation of the protease is possible. For example, 50% of LexA can be degraded within 1 min after activation of RecA in *E. coli* [115]. Such a quick response would be impossible if RecA worked by repressing the transcription of LexA instead.

The parameter τ is particularly important in dynamics: a small value of τ implies a fast adjustment to the new steady state, but also increases the protein turnover and thus the metabolic cost. The cell has to balance its need to respond quickly to changes in the environment with the long-term cost during long periods when everything is unchanged. In practice, most proteins are not actively degraded, and therefore $\tau = 1/\lambda$ simply becomes equal to the overall dilution associated with the cell growth rate λ.

Figure 3.5a shows how C responds to a sudden 100-fold rise in R, from $K/10$ to $10K$ (that is, from a regime where the operator-bound fraction is close to zero, to one where it is close to unity). Note that τ similarly sets the timescale for response when R activates production (Fig. 3.5c). In Fig. 3.5b, the regulator is not bound to the DNA, but instead acts by binding to C and catalyzing its degradation. We will return to this important type of regulation in the next section.

The strategies shown may be combined, for example protein C may be both positively regulated by another protein, and negatively autoregulated. This is illustrated in Fig. 3.6.

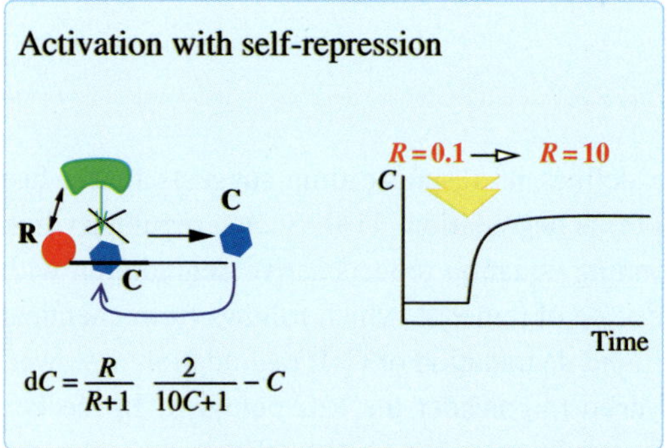

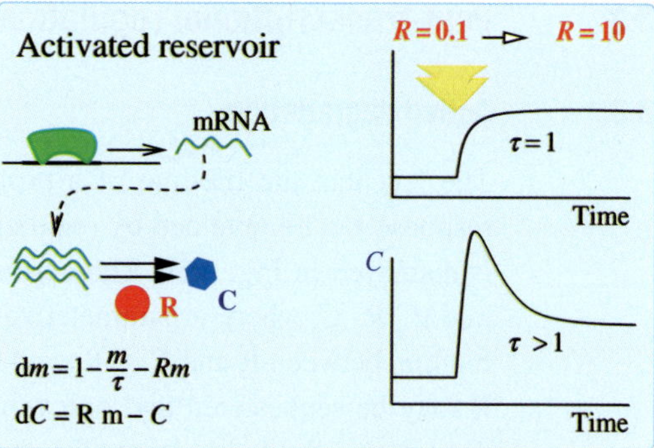

Figure 3.6 The left-hand panel shows that activation is faster when the activated gene represses itself. Self-repression was studied in synthetic designs by [116], and is observed in R self-repression in phage λ. The right-hand panel examines another way to obtain a fast response, by using a pool of inactive, but long-lived, $m = mRNA$, which is activated by the regulator R. This type of response is found in the unfolded protein response [117] as reviewed in [118]. Remarkably, for $\tau \to \infty$ this type of response motif is identical to the model suggested for perfect adaptation by [119].

As we will learn in Chapter 4, the equations used in the left-hand panel of Fig. 3.6 reflect a promoter regulated by two operators, one that can bind R and another one that can bind C; a regulatory setup that allows for a fast response, because the initial increase in C appears as if it could increase to much higher values than its self-repression will allow in the end [116].

The right-hand panel in Fig. 3.6 explores another interesting way to obtain a fast and adaptive regulatory response. Here the regulator R activates a pool of passive mRNA, which is subsequently translated and degraded. This leads to a transient spike in production of the encoded protein C, with a response that increases with the size of the passive pool of m, which is determined by τ relative to the speed at which R converts m to its active form [118].

3.2.1 Questions

3.2.1 Simulate the three different regulatory systems in Fig. 3.5 by integrating the temporal evolution. Hint: for simple repression this amounts to $C(t + dt) = C(t) + (\frac{1}{1+R} - C) \cdot dt$, with $dt = 0.01$ from $t = 0$ to $t = 10$. Start by using $R = 0.1$ and initialize C at its steady-state value $C = \tau/(R+1)$, $C = 1/(R + 1/\tau)$ and $C = \tau R/(R+1)$. At time $t = 10$ increase R by a factor of 100 to inspect the relaxation dynamics to the new steady state (integrating from $t = 10$ to $t = 20$). What parameter is important for determining the characteristic time of the relaxation, and what parameters define the steady state of C?

3.2.2 Simulate the two regulatory systems in Fig. 3.6. For the right-hand panel use $\tau = 1$, and investigate also $\tau = 10$.

3.3   Post-transcriptional regulation

3.3.1   Active degradation

The fact that the lifetime of a protein defines its response time suggests that a fast response can be obtained by regulation of its degradation [114]. Such a regulatory link is displayed in Fig. 3.5b. The corresponding equation reflects active degradation with rate $\gamma \cdot R \cdot C$, where γ parameterizes the rate of removal, which follows from chemical binding between R and C, followed by rapid degradation of C. If degradation is slower, R may be sequestered and one would need to consider the R:C complex. In the fast degradation limit, $1/\gamma R$ sets the lifetime of the protein C, a time that is then also the response time for adapting to a new value of R (see Fig. 3.5b).

Direct inactivation allows for a fast response to changes in R, while maintaining a low degradation rate otherwise (when R is low). This is true whether C is a protein or an mRNA. The same setup is in fact commonly used to inactivate a target mRNA (taking the role C from before) by small or microRNA (the regulator R from before). Such inactivation normally involves anti-sense pairing and subsequent degradation [120, 121, 122, 123, 124]. Active degradation or sequestration of proteins or mRNA is a central part of many metabolic and stress response systems.

3.3.2   Small RNA regulation

Small RNA regulation is mediated by small RNA pieces that in *E. coli* often act by facilitating active degradation of their target mRNA. Furthermore, this degradation is often associated with degradation of *both* the sRNA (s)and its target mRNA (m). In this case, the equation in Fig. 3.5b can be augmented as follows [120, 121, 122, 123, 124, 125]:

$$\frac{dm}{dt} = 1 - \gamma \cdot m \cdot s - m/\tau \tag{3.21}$$

$$\frac{ds}{dt} = \alpha - \gamma \cdot m \cdot s - s \tag{3.22}$$

where α quantifies the regulatory input and γ the mutual interaction strength [121, 122, 123, 124]. High γ or α allows for a high (regulated) degradation rate of m. The sharpness and speed of response after an increase in α will increase with the value of γ.

Figure 3.7b shows ultra-sensitive behavior of the steady-state values of m, as a function of α, provided that γ is high [122]. The all-or-nothing behavior of the system reflects an m-dominated state for $\alpha < 1$, shifting to an s-dominated state for $\alpha > 1$. At steady state, Eqs. 3.21 and 3.22 give (for $\alpha > 1$):

$$m = -\frac{1}{2}\left((\alpha - 1)\tau + \frac{1}{\gamma}\right) + \sqrt{\frac{1}{4}\left((\alpha - 1)\tau + \frac{1}{\gamma}\right)^2 + \frac{\tau}{\gamma}}$$

$$\sim \frac{\tau}{((\alpha - 1)\tau\gamma + 1)} \tag{3.23}$$

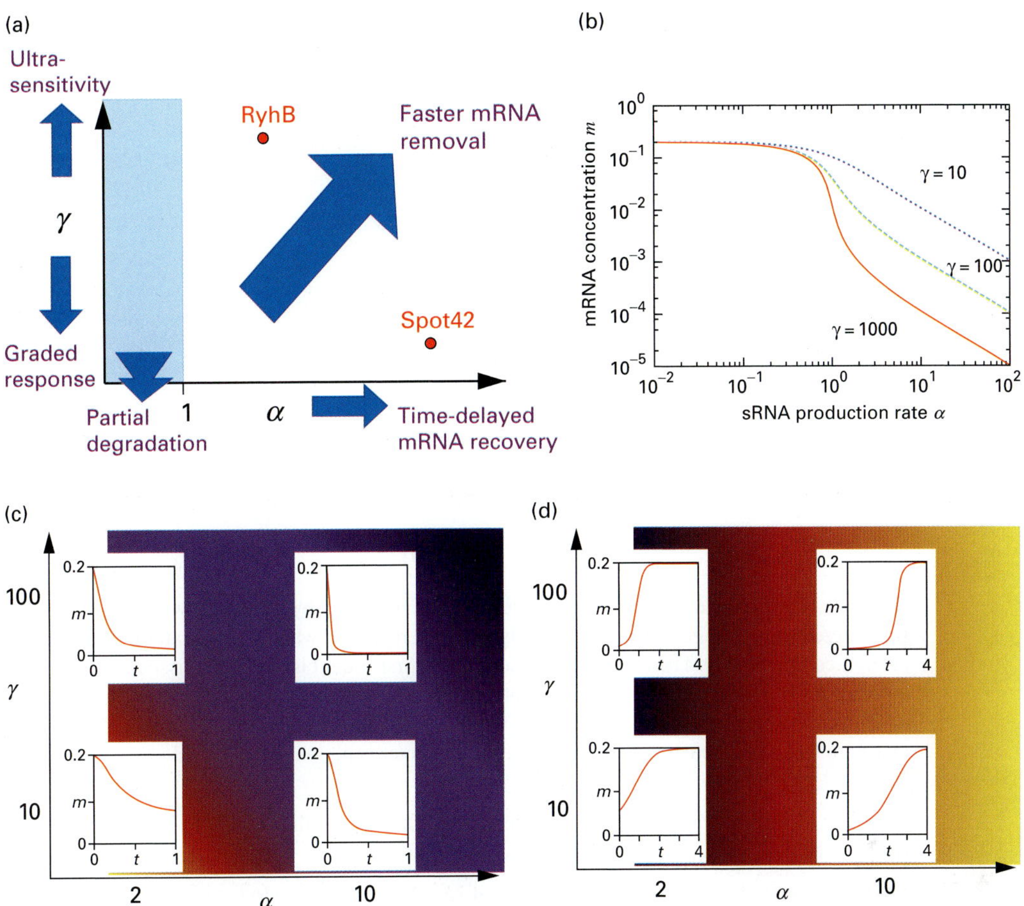

Figure 3.7 Behavior of sRNA regulation on a downstream target m is determined by two parameters, the excess production of sRNA and the strength of the mutual coupling. Notice that the mutual degradation allows the ultra-sensitive response of m with changing sRNA production (panel b). Also notice that the response times are very different when considering sRNA up-regulation, or sRNA down-regulation (when α is large in panels c,d). The figure is from [124].

where the approximation requires that $\gamma \tau >> 1$. This equation shows that for large γ, m drops drastically as α is increased above 1.

Examples of small RNA regulation are found in both the galactose uptake system and the Fe uptake system in *E. coli*. For the Fe uptake system, the RyhB–SodB interaction is characterized by a maximum of $\alpha = 4$ and $\gamma = 400$, reflecting a large mutual interaction [124]. In contrast, the Spot42–galK interaction in galactose metabolism has a maximum $\alpha = 18$ and $\gamma \sim 2$, giving a large potential for over-production of sRNA [124]. In turn, α can be regulated, as it simply reflects the production rate of an sRNA that is produced from a promoter. For the Fe uptake system, the sRNA is regulated by the transcriptional regulator Fur.

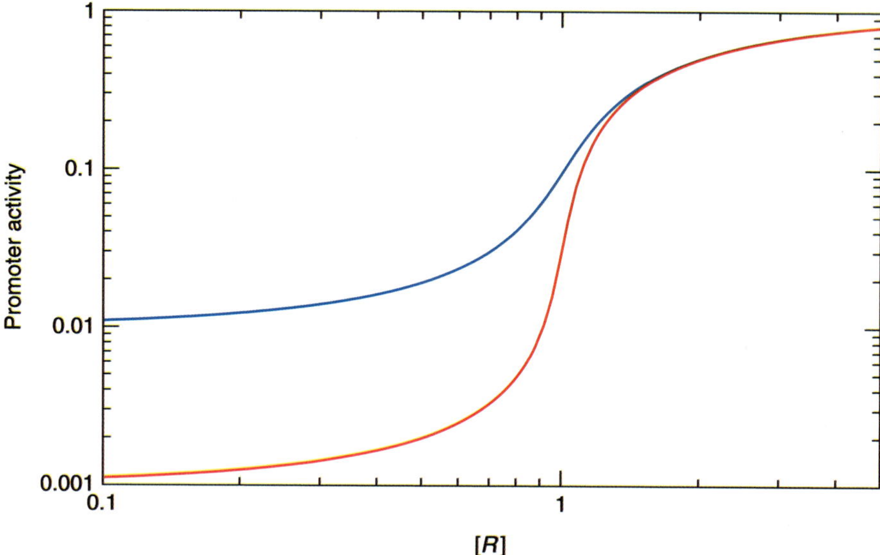

Figure 3.8 The predicted activity of a promoter that is repressed by protein C, which in turn is sequestered by a protein R. The protein only acts as a repressor when it is free and is assumed to act non-co-operatively. The repression is modeled as first-order binding with dissociation constant $K_o = 0.01$. C and R bind to each other with dissociation constants $K = 0.01$ (blue) and 0.001 (red) respectively. In the plot shown, we fix the total concentration of C to be *1*, and therefore when $[R]$ increases beyond $C = 1$, the promoter is no longer repressed.

In Fig. 3.7c, the value of α is changed from an initial value $\alpha = 0$ to its final value at time $t = 0$. In contrast, Fig. 3.7d examines the opposite shift, where $\alpha \to 0$ at time $t = 0$. One can see that a large maximal value of $\alpha \gg 1$ is associated with a large transient reservoir of sRNA after its production has stopped. This provides a time delay before the removal of sRNA production is sensed by the downstream mRNA, as illustrated in the far right-hand subpanels of Fig. 3.7d.

3.3.3   Protein sequestration as a regulatory link

Gene regulation or regulation of mRNA activity are not the only ways to regulate the free concentration of a protein. A commonly used regulatory strategy is to target the free concentration by degradation, as shown in Fig. 3.5 or by sequestration of the protein by binding to another target. Regulation by binding can be fast, as it relies primarily on the timescale at which one may change the concentration or the binding affinity of the sequestering "regulator." If C is the total concentration of the regulated protein, and R the total concentration of the regulating protein, the free concentration $C_f = C - [CR]$ is determined by:

$$K \cdot [CR] = (C - [CR]) \cdot (R - [CR])$$

$$\Rightarrow [CR] = \frac{C + R + K}{2} - \sqrt{\frac{(C + R + K)^2}{4} - C \cdot R} \qquad (3.24)$$

making $[CR] \sim \min(C, R)$ for strong binding $K \to 0$. For strong binding the free concentration of C therefore switches from $C_f \sim C$ for $R < C$ to $C_f \sim 0$ for $R > C$. Strong binding accordingly allows a very high effective Hill coefficient as a function of input.

Often protein–protein binding is combined with genetic regulation, thereby allowing an ultra-sensitive response to a moderate change in concentration of some input. Consider the indirect regulator R that binds to a transcriptional repressor C. Assume that C only acts as a repressor when it is not bound to R. With C denoting the concentration of the transcriptional repressor, the promoter activity:

$$\text{Activity} = \frac{1}{1 + C_f/K_o}$$

$$C_f = \frac{1}{2} \cdot (C - R - K) + \sqrt{\frac{1}{4} \cdot (C + R + K)^2 - C \cdot R} \qquad (3.25)$$

where K_o is the binding strength of free C to the operator regulating the promoter. Notice that even for the assumed non-co-operative regulation of the promoter, the resulting response to a change in total R is ultra-sensitive. This is illustrated in Fig. 3.8. As we will discuss later, protein sequestration can work as an efficient regulatory link, a link which may even support bi-stable expression states and epigenetics inside a living cell [126].

3.3.4 Questions

3.3.1 Consider the sRNA regulations from Eqs. (3.21) and (3.22) with parameter $\gamma = 10$, $\gamma = 100$, respectively $\gamma = 1000$. Calculate the steady-state concentration mRNA at $\alpha = 0.1$, then shift to $\alpha = 4$ and simulate the response. Use $\tau = 1$.

3.3.2 Simulate activation and de-activation of a small RNA on the expression of its target mRNA with $\alpha = 0.1 \to \alpha = 4 \to \alpha = 0.1$, and $\gamma = 100$, $\tau = 1$. Repeat the simulation for $\alpha = 0.1 \to \alpha = 40 \to \alpha = 0.1$ and $\gamma = 100$, $\tau = 1$.

3.3.3 Examine the indirect regulation caused by a protein R that sequesters a transcriptional repressor C as a function of the concentration of R with fixed $C = 1$. Assume that C only represses the promoter in its free form. This can be done by plotting the activity of the repressed promoter as one varies R around the critical value $C = 1$, for different values of K_o and K, with $K \cdot K_o = 10^{-4}$ being fixed.

3.3.4 Consider a simplified toxin–anti-toxin system of two proteins T and A, where free A (A_f) is a repressor of T:

$$\frac{dT}{dt} = \frac{1}{1 + A_f/K_o} - T/10 \qquad (3.26)$$

where A_f is given by the total toxin amount T with an A:T binding constant $K = 0.01$. Set $K_o = 0.01$ and investigate fixed points for the above equation when total $A = 0.1$, $A = 1$ and $A = 10$. For literature on TA systems see [127, 128].

3.4   Summary

- The statistical weight of any particular state of operator and promoter occupancy is proportional to the concentrations of any molecule that is bound in that configuration: a simple equation for the gene activity of a regulator R that activates a promoter is accordingly:

$$P \propto \frac{R}{1+R} \tag{3.27}$$

  where R is concentration in units of the dissociation constant of R with the operator, $R = [R]/K$.
- Co-operative binding means that the probability of occupying a state increases more than linearly with the concentrations of the binding molecules.
- Protein–protein binding allows ultra-sensitive change of the free protein concentration, when varying the total protein concentration of one of the binding partners.

4 Statistical mechanics of phage λ

4.1 Lifecycle of λ phage

4.1.1 Microbial perspectives

With about $5 \cdot 10^{30}$ prokaryotes on Earth [129], these small single-celled organisms dominate the global biomass.[1] However, the world is not safe even for bacteria. Bacteria are exposed to their own parasites, the bacteriophages [131, 132, 133, 134], which is the name for viruses that attack prokaryotes. Current estimates set the phage population to be about an order of magnitude larger than that of bacteria [133] and their ability for genetic engineering therefore makes these profound information carriers an important force in shaping the global ecosystem.

Molecular biology has, based on an entirely random contingency, chosen one of the many phages for detailed study: bacteriophage λ. This particular bacteriophage, or phage, has provided us with insight into both genetic regulation and the genetic code itself. The choice of λ as a model organism was not because it is particularly abundant in nature (in fact it is rare), but rather because it was the first bacteriophage found to have a genetic switch [135]. The scientific choice of model system was thus a true accident of history, but an accident which provided numerous useful insights because at least some gene regulatory mechanisms are universal across all taxonomic kingdoms of life [112].

4.1.2 Model system for epigenetics

The λ phage is a model system for decision-making in living systems; modulated by various inputs it stochastically decides between one of two developmental paths [112, 136, 137, 138, 139, 140]. In fact, it can be sustained in two distinct phenotypic states, characterized by widely different gene expression [142]. This phenomenom is called epigenetics. In more general terms, the phenomena of epigenetics is defined as:

---

[1] Prokaryotes contain $\sim$500 gigatons of carbon, which is slightly less than the total C in all other life put together. Prokaryotes further contain $\sim$100 gigatons of N and $\sim$10 gigatons of P, which is about a factor of 10 larger than in all other life put together. The upper 200 m of oceans contain about $5 \cdot 10^5$ cells ml^{-1}, or about 1% of the total prokaryotes on Earth [129], but with a generation time of $\sim$6$\rightarrow$25 days [129, 130] it constitutes the main fraction of the estimated $\sim 2 \times 10^{30}$ cell divisions per year. The majority of prokaryotes on Earth are found below the bottom of oceans, whereas soil constitutes most of the remaining reservoir [129].

Epigenetics = "a change in the state of expression of a gene that does not involve a mutation, but is nevertheless inherited in the absence of the signal (or event) that initiated the event. [141]." = bistability where each state typically "survives" cell divisions.

Epigenetics is inherent to cell differentiation, and thus essential for multicellular life. Epigenetics is sustained on a molecular level by positive feedback that drives differentiation into well-separated states [143, 144, 145, 146]. In a eucaryote each such state may correspond to a cell type [144]. In a bacterium, different gene expression allows proliferation in widely different environments [143]. In the λ phage, epigenetics allow the phage to adopt two different survival strategies: short-term proliferation, or longer-term "bet-hedging" by hiding in the bacterial genome.

4.1.3   Phage infecting the bacteria

The λ phage is a bacterial virus that uses a specific receptor protein on the bacterial surface to enter the cell. This membrane protein is the receptor for maltose, and is thus important for uptake of this nutrient. If there are no maltose receptors on the surface, the corresponding *E. coli* strain is resistant to the λ phage [149]. With the receptor present, the phage can bind and inject its DNA into the cell (see Fig. 4.1).

After infecting an *E. coli* cell, the λ phage enters either into an explosive lytic state, where finally the bacterium bursts and many copies of λ are released, or into a lysogenic state where λ integrates its DNA into the host cell DNA (see Fig. 4.2). In the lysogenic state, the resident phage (prophage) can be passively replicated for many generations of the *E. coli*. A phage that has this ability to enter into lysogeny is called temperate.

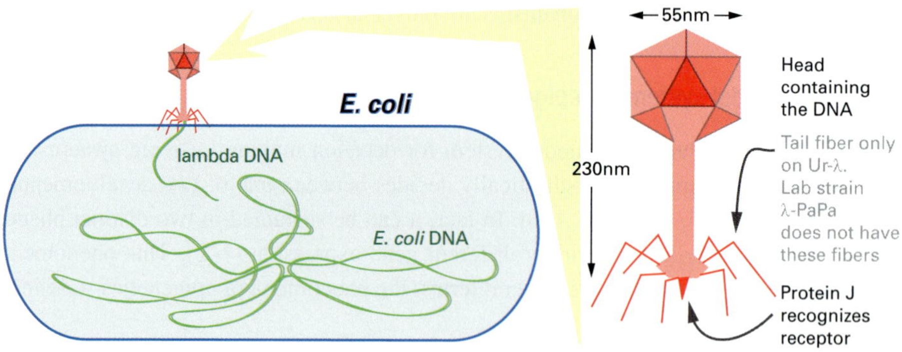

Figure 4.1 *E. coli* cell with λ phage injecting its DNA through the membrane via the maltodextrin porin (maltose receptor) of the *E. coli* cell. The *E. coli* cell has a length of about 1–2 μm, the full length of the λ head plus tail is 0.2 μm. The right-hand panel illustrates the λ phage in its original form, called ur-λ. The contemporary λ-PaPa phage that is used in all λ research lacks the tail fibers, which results in larger plaques and easier experiments [147]. It is the protein J that recognizes the maltose receptor on *E. coli* [148].

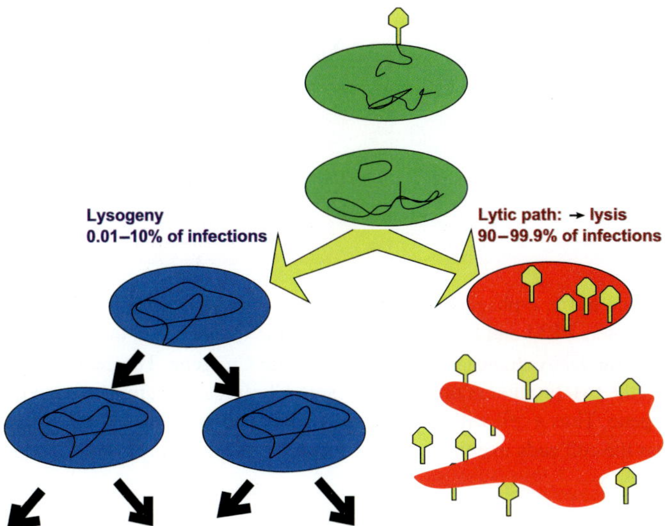

Figure 4.2 *E. coli* cell after infection by a phage λ either enters the lysogenic pathway (left), or the lytic pathway (right). Lysogeny lets the phage integrate into the *E. coli* chromosome, and be passively replicated with the host for hundreds of thousands of generations. Lysis leads to death and of the order of hundred released phages. Noticeably, for an exponentially growing bacteria, lysogeny is only selected in a small percentage of cases [150].

4.1.4 Selecting the lysogenic state

The initial decision whether to integrate or to lyse is taken through the interplay between two proteins that both have their production initiated from promoter PR: the protein Cro [136],[2] which degrades slowly (with a half-life $\sim$ 30 minutes [151]), and the protein CII whose degradation is faster (half-life down to $\sim$ 1.5 minutes at low concentrations [152]). CII degradation is reduced when the cell starves, because of repression of the *E. coli* protease HflB, which degrades CII[3]. Cro favors lysis while CII favors lysogeny. The lysogenic state is primarily selected when the cell has a slow metabolism[4] or when more than one λ infects a given cell [153].

Once a phage has established the lysogenic state, PR is repressed by CI [154, 155] (see Fig. 4.3). CI thereby both maintains lysogeny and makes the cell immune to infection by other λ phages. This immunity is secured through the CI, which represses Cro and essentially all other genes of the infecting phage. Thereby a newly entered phage

---

[2] The existence of Cro was inferred by a series of beautiful experiments by H. Eisen [136], using a variant of a λ lysogen with a temperature-sensitive CI. As this CI^{857} did not dimerize above $\sim$ 37 °C, shifting a lysogen from a low temperature to 42 °C and back again for various time intervals allowed Eisen and co-workers to pinpoint the role of Cro in the switch.

[3] HflB is a central player in bacterial heat shock. However, it is named after its influence on λ. Thus *E. coli* without HflB has a High Frequency of Lysogeny.

[4] Starvation increases lysogeny frequency by a factor of 100 [139, 150].

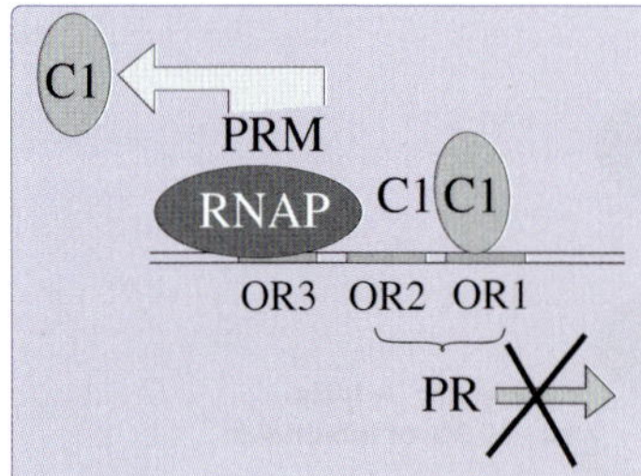

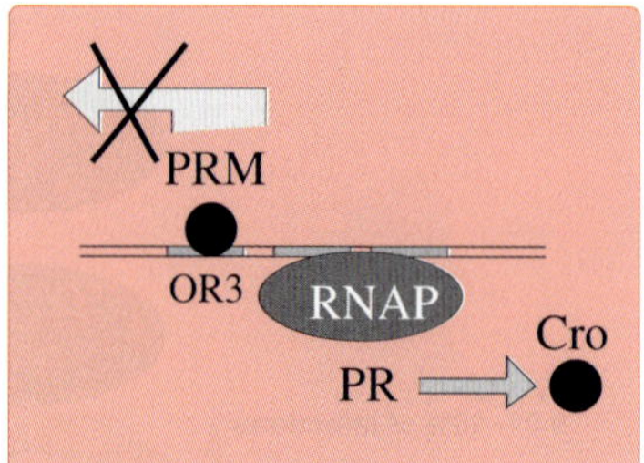

Figure 4.3 The core of the λ-phage switch, with arrows to indicate transcription directions for the two promoters PRM and PR. Transcription of the *CI* gene is in the opposite direction (and on the opposite DNA strand) to transcription of the *Cro* gene. The geometry of operator sites relative to the two promoters makes the protein CI repress the cro gene, and the protein Cro repress the cI gene. Thus a cell with both CI and Cro will be unstable and the positive feedback will ensure that the cell ends in either a CI-dominated state (lysogeny), or a Cro-dominated state (lysis).

cannot replicate, nor integrate into the host chromosome (it cannot express the genes needed for integration) except by host-determined recombination events.[5]

4.1.5   Leaving the lysogen

Upon radiation with UV light, the *E. coli* DNA is damaged, and the SOS system activated. The SOS system is responsible for repair of single-stranded breaks of DNA through a series of steps that is initiated by activating the DNA-binding protein RecA. Activated RecA in turn directs cleavage of both a specific gene regulator in *E. coli* and of the the central gene regulator CI in phage λ [156]. CI is a two-domain protein, one part responsible for DNA binding, the other for dimerization. If CI is cleaved into its two domains, it cannot form dimers and therefore cannot bind sufficiently strongly to DNA to suppress the lytic promoter PR. Thus Cro is produced and the lytic pathway is initiated. When lysis is induced, the *E. coli* is killed, and of the order of 100 phages are released.

4.1.6   Operator details & regulatory network

Using the genetic switch visualized in Figs. 4.3 and 4.4 we now want to repeat the above story in more quantitative terms. Each of the proteins comes from mRNA that was transcribed by an RNAP initiated from either the PRM, the PR or the PRE promoter (see Fig. 4.4). A central promoter for the *cl* gene is PRM, whereas the promoter for the *cro* gene is PR. Both promoters are controlled through the operator OR ("operator right") located in the middle of the $\sim 48.5\,\mathrm{kbase}$ λ DNA in Fig. 4.4.

Figure 4.4 shows the DNA sequence for the operator OR. The DNA is a double helix with a period of 10.4 base pairs (bp). The operators consist of 17 bp, separated by 6 or 7 base pairs. Thus the center-to-center distances are about 23 base pairs, and therefore the binding of CI dimers to two consecutive operators will place the CI on nearly the same

---

[5] Homologous recombination with the help of host proteins can allow the new phage to recombine with the resident prophage and thereby make a double lysogen.

3′ end   5′ end

TATCACCGCAAGGGATAAATATCTAACACCGTGCGTGTTGACTATTTTACCTCTGGCGGTGATA

ATAGTGGCGTTCCCTATTTATAGATTGTGGCACGCACAACTGATAAAATGGAGACCGCCACTAT

5′ end   3′ end

OR3   OR2   OR1

Figure 4.4 Detailed view of the DNA sequence of the OR operator. Each operator covers 17 base pairs, and are in the figure indicated by bold letters. Further, each operator is close to palindromic. For example in *OR*3 the upper 3′ end is initiated by TATC, which is also found at the lower 3′ end of *OR*3 (on the opposite strand). This means that the protein binding is mirror symmetric, which is natural, as both Cro and CI bind as dimers, with one dimer on each of the three operators.

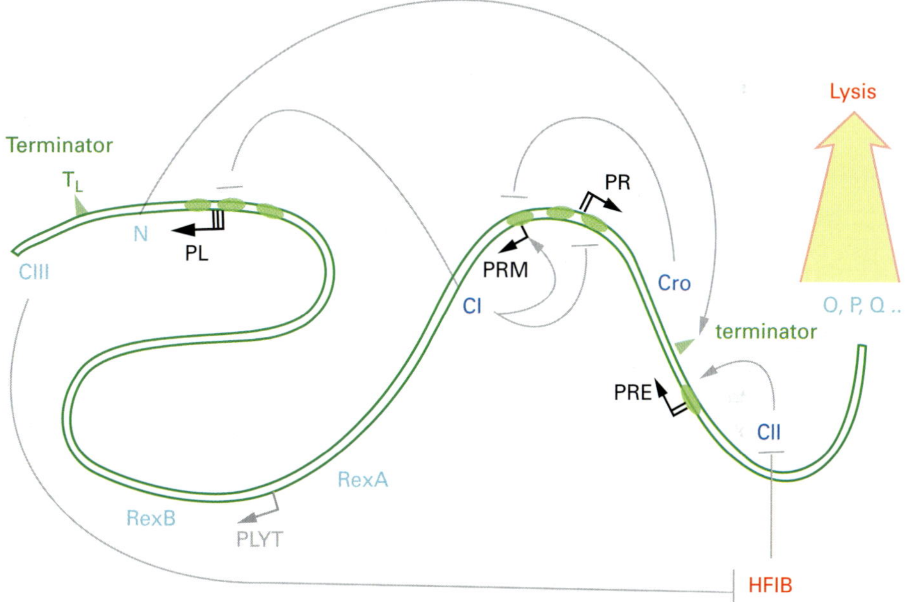

Figure 4.5 "λ-phage switch" and its surroundings in the 48502 bp long genome. Left of OR, one finds genes for CI, RexA and RexB, as well as the operator OL, which controls the promoter PL. PL initiates production of anti-terminator *N*, as well as other proteins. *N* allows the RNAP transcription initiated from PR to continue past a terminator (indicated by the small triangle) for transcription of CII. Transcription factors are indicated with blue letters, whereas regulatory links are shown in light gray. Right of OR, after the cro gene, the protein CII activates the promoter PRE for left-directed transcription of CI. Production of CI from PRE occurs in the early phases of infection, where it favors establishment of repression (PRE = Promotor for Repressor Establishment).

side of the DNA. This feature facilitates interactions between the CI dimers [164]. The OR1, OR2 and OR3 operator sequences are similar, and CI and Cro in fact bind to each operator with dissociation constants K that vary between 1 nM and 100 nM, whereas the mutual interaction between two CI dimers modulates the binding by about a factor of 100 [165].

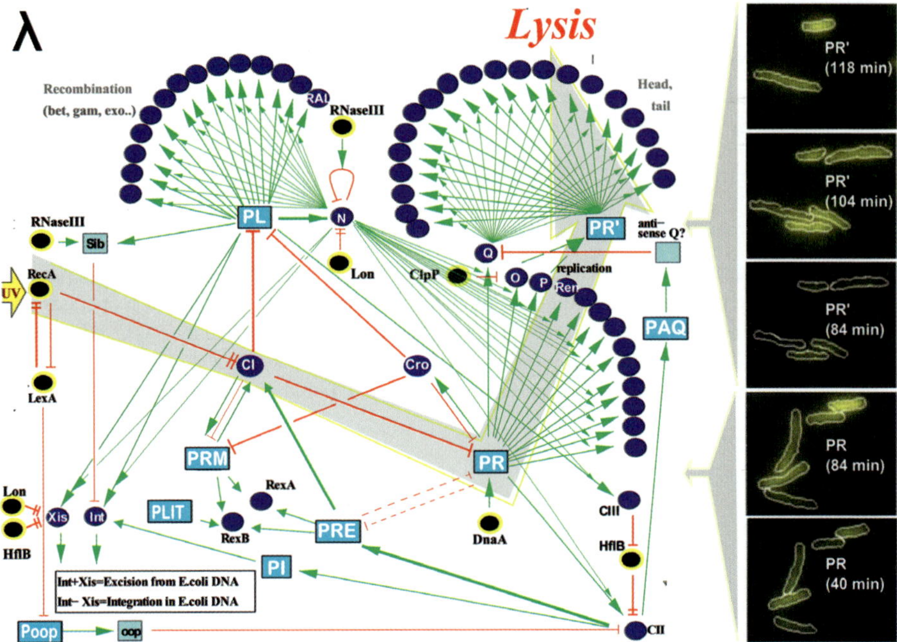

Figure 4.6 Regulatory network of phage λ [157] with highlighted induction path that is initiated by activated RecA. Input from *E. coli* is marked by black ovals, and includes the protease Lon [158] and the DNA replication initiation protein DnaA [159]. Right panels show induction of fluorescent proteins[6] inserted after promoter PR and PR′ [161]. PR′ is a late lytic promoter that initiates transcription of head and tail genes, provided that the anti-terminator Q allows the transcript to pass a terminator. Experiments were done in minimal media with moderately long cell generations. Notice that the induction signals are also transmitted to the upper-left hub, where "kill" and "bin" stop replication of the *E. coli* chromosome [162]. Lysogeny is favored (+), disfavored (−) by promoters +PRM −PR +PRE −PR′ +PAQ.

When CI binds to the right part of this operator, OR1 or OR2 in Fig. 4.3, it represses the *cro* gene and thereby prevents entry into lytic development. The cell is then persistently maintained in the lysogenic state, where CI is produced due to the ongoing activity of the promoter PRM. CI is called the repressor.

Cro is produced from PR provided that this is not represssed by CI. Cro, if present, represses production of CI by binding strongly to OR3. If Cro is present in substantial amounts, the cell is most likely on the way to lysis. This way to lysis is not because Cro in itself does anything apart from preventing CI from stopping PR. Rather, an active PR also implies transcription of genes downstream of *cro* that initiate lysis.

When CI is effectively inactivated by RecA [163], transcripts from the promoter PR initiate the lytic pathway. The regulatory sequence from RecA activation to lysis

---

[6] GFP = Green fluorescent protein, which is a protein extracted from jellyfish that turns green when irradiated by UV light. GFP allows direct visualization of transcription dynamics [160]. Typical problems of GFP are its maturation time of at least 5-10 minutes, and the fact that one needs many GFP to get a substantial signal.

is marked with grey in Fig. 4.5. The timing of the subsequent cascade was quantified by [161], who inserted reporter genes at different positions in the genome (see Fig. 4.5). Also consult [224]. The CI:Cro switch ensures that the turning-on of lysis is non-reversible, provided that CI has declined to below about 10% of its normal lysogenic concentration.

4.1.7 Questions

4.1.1 Why is it a good strategy for the λ phage to favor lysogeny when a bacterium is infected simultaneously by many λs?

4.1.2 One theorem in game theory states that for a single player there always exists a deterministic strategy that is no worse than any random strategy. This is apparently not true for λ. Argue how a random decision could be good, as an alternative to deterministically going lytic or lysogenic as a function of external conditions.

4.1.3 A λ lysogen is superinfected with another λ phage. Consider the situation where there is already a phage in a lysogenic state inside the bacterium and a second new λ phage infects. Which of the genes in the new λ is transcribed, and which are not? Why does the new phage not induce lysis?

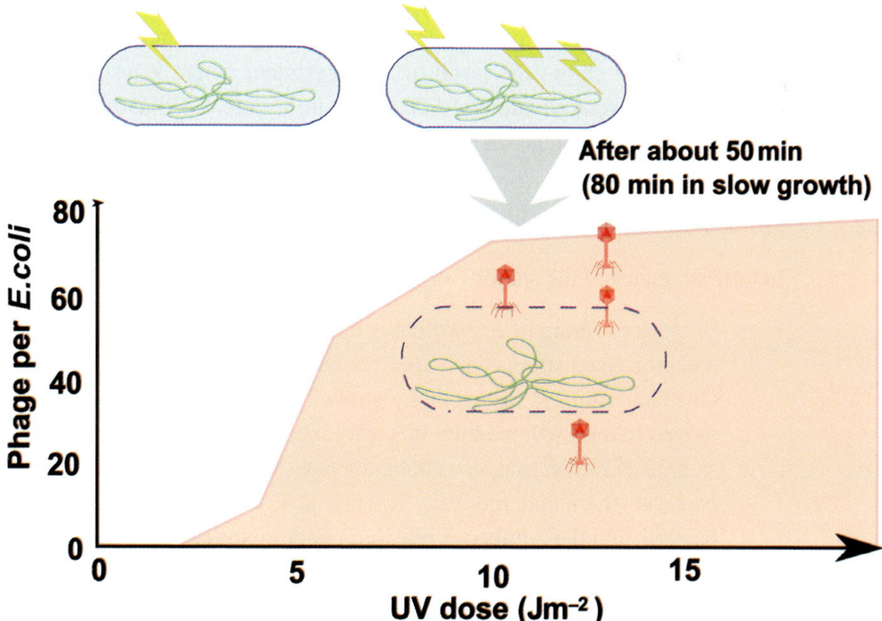

Figure 4.7 When exposing a λ lysogen to UV light, some phages may choose to lyse the host. The lysis frequency depends on the UV dosage in a threshold-like way, as here recorded by Little *et al.* [166].

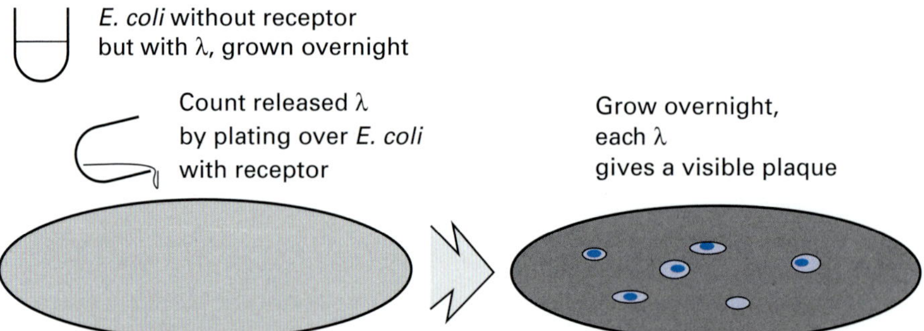

Figure 4.8 Measurement of phage release, here illustrated for an experiment to count spontaneous phage release. Similar counting is done for quantifying UV-induced release, see Fig. 4.7 (in fact 1/10 000 is clear). The infection including subsequent phage development can be carried out in a test tube, whereas the counting is done by plating the contents of the test tube on a growing strain of bacteria that is ready for λ infection. Each λ will infect and kill its host, leading to a plaque which is visible by the naked eye after one night's incubation at 37°C.

4.2   Bacterial growth and counting

4.2.1   Using exponential growth as a "microscope"

One illuminating feature of molecular biology is that it facilitates its own exploration. Often one can use the control of some parts to explore others. One example is the counting of phage release, using exponential bacterial growth and death as a "magnifying tool." Figure 4.7 shows the result of an experiment where λ lysogens (in *E. coli*) are exposed to different amounts of UV radiation. Figure 4.8 illustrates how to count the number of released phages at each exposed sample.

4.2.2   Protocol to count phage per bacteria

In order to measure the stability of the lysogenic state one could proceed as follows:

- First take a strain of *E. coli* that is infected with λ phage, but does not have the λ receptor. Such strains have been prepared.
- One bacterium (in fact a colony grown from one bacterium) from this strain is moved to a growth medium in a test tube. This is grown overnight into a large sample of *E. coli* that also contain phages that eventually are released from bacteria which undergo lysis. The aim is now to count both the number of bacteria and the number of released phages. Notice that because none of the bacteria have the maltose receptor, the released phages cannot enter another bacterium. Therefore no λ phages are lost by absorption into immune bacteria.[7]

---

[7] If a phage infects a bacteria with another λ in the lysogenic state, the development of the infecting phage is repressed, and the phage is subsequently lost.

(a)　　　　　　　　　　　　　　　　　　　(b)

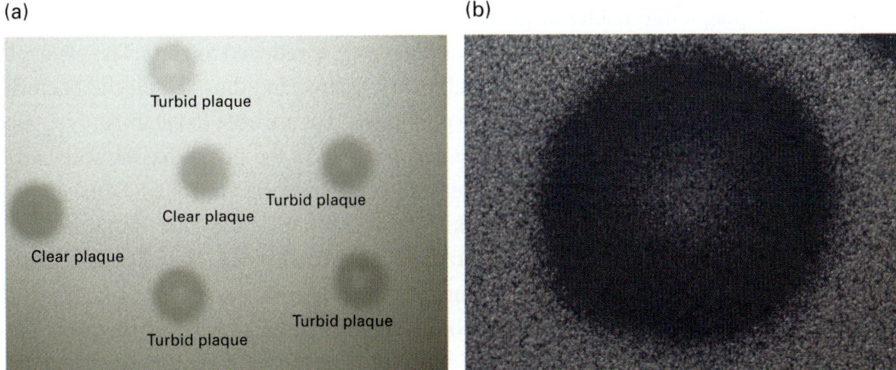

Figure 4.9 (a) Clear and turbid plaques formed by λ phages on a bacterial lawn that is formed by growing *E. coli* microcolonies that are susceptible to the λ phage. (b) Close up of a turbid plaque that measures about 3 mm across. Notice the granular structure of the bacterial lawn, which in fact consists of microcolonies that each contain of the order of $\sim 10^5$ bacteria. The resistant colonies in the center of the plaque contain bacteria that are all lysogens. Micrographs are by the author.

- To count bacteria, one makes an appropriate dilution, spreads it (distributes it homogeneously) on a culture dish, and lets it grow overnight. Each bacterium then gives rise to one colony, and if the solution of bacteria is sufficiently dilute, the growing colonies do not overlap and can therefore be counted. By counting the number of colonies on the dish, one can calculate the bacterial density.

- To count phages, one first centrifuges (or filters) the *E. coli* out of the solution, leaving a medium that contains only the released phages. An appropriate dilution of this solution is poured over a detector strain of *E. coli*, which contains the receptors that allow λ to enter. This plate is grown overnight, allowing each phage to initiate a local plaque.[8] The reason is as follows: The phage will infect a bacterium and nearly always ($\sim 99\%$) go lytic. Then new phages are produced that infect and kill the nearby bacteria. This is seen as a small circular area where bacterial growth is reduced: a plaque. By counting the number of plaques one can calculate the original number of phages in the full culture.

If there are many phages one dilutes the fraction poured over the plate accordingly. Thereby one can measure phage release in a range from 100 phages per *E. coli* to one phage per 10^7 *E. coli*.

4.2.3　Clear and turbid plaques

There is more information one can extract from phage-release experiments. It turns out that there are two types of plaques, clear and turbid, as shown in Fig. 4.9. For both types

---

[8] The plaque size depends on details of phage properties, see model in [167, 168].

of plaque the bacterial growth is reduced compared to the growth in the surrounding phage-free area, but for the clear plaques it is reduced to zero. The clear plaques appear because the phages cannot establish lysogeny, and thus kill all the bacteria they infect. The turbid plaques originate from phages that are able to establish lysogeny (in a small percentage of their hosts). The bacteria where the phage has established lysogeny are the only ones to survive, whereas all other bacteria are killed. This survival is because lysogenic bacteria cannot be killed by another λ: they are immune. In these regions many bacteria are killed and overall growth delayed. However, as soon as lysogenic bacteria appear their descendants grow at a normal rate to form a colony. This produces a turbid plaque that can easily be distinguished from the clear plaques.

Consider finally a culture of lysogens, i.e. *E. coli* that all contain a dormant λ phage. Let this culture be exposed to UV radiation. The released phages will then consist of normal healthy phages, and the amount of mutants will be negligible. Accordingly one only observes turbid plaques. One can further quantify the effect of UV radiation by counting the number of plaques per bacteria exposed to a given dose of UV. This was exactly what was done to obtain Fig. 4.7. On the other hand, if one oblates the DNA damage-response pathway,[9] the rare transition to lysis is dominated by mutated phages. Accordingly, clear plaques will dominate. However, among phages released from a wild-type λ lysogen in a wild-type *E. coli*, clear plaques are rare.

4.2.4   Questions

4.2.1   There exist virulent phages that always lyse bacteria after infections. An example is the phage T4 which infects *E. coli*. After infection, the phage copies its genome through a rolling circle [169, 170],[10] where one copy is generated about every 10 s, giving a phage production rate $\gamma \approx 6\,\mathrm{min}^{-1}$. If the bacteria lyse at a latent time τ, the number of phage progeny would be $\beta = \gamma(\tau - \tau_0)$, where $\tau_0 \approx 5$ min is the time needed before phage production starts. If the time needed to find and infect a new *E. coli* is $1/(\alpha\rho)$, where $\alpha \approx 10^{-9}\,min^{-1}\,ml$ is the infection rate per phage per bacterium and ρ is the bacteria density. Write an expression for the exponential growth rate, g. Show that the latent time (τ_{opt}) that maximizes the phage growth rate (g) satisfies [172]:

$$\frac{1}{\alpha \cdot \rho} + \tau_{\mathrm{opt}} = (\tau_{\mathrm{opt}} - \tau_0) \cdot \log[\gamma \cdot (\tau_{\mathrm{opt}} - \tau_0)] \tag{4.1}$$

Investigate the solution graphically.

---

[9] Using a RecA-strain of *E. coli*.

[10] Here simplified, as the latter stage of phage growth often occurs through exponential θ replication. The final burst time depends on produced holin proteins [171], which puncture the *E. coli* cell when their number reaches a critical threshold.

4.3 Chemical binding and counting

One molecule in one *E. coli* cell volume V corresponds to a concentration $1/V \sim 1$ nM, as derived in Chapter 2. With that in mind, this section introduces elementary chemistry inside such a volume in a way that emphasizes the relation between counting and chemical equilibrium. We derived some of the equations in section 3.1 using kinetic equations of reactions. Following [13] this section re-derives the equations from a statistical-physics point of view, with the aim of making equations that are easily generalized to the more complex situations studied in the next section.

Consider, for example, binding of a regulator protein R to a specific operator on the DNA. The probability that a given molecule is bound depends on the (Gibbs) free energy difference between bound and unbound. Call this difference $\Delta G' = G(\text{bound}) - G(\text{unbound})$. When $\Delta G'$ is negative, the "on" state has lower free energy than the "off" state.

4.3.1 Counting all the molecular positions in a cell

The statistical weight (non-normalized probability) of a state is given by the number of ways the state can be realized multiplied by $\exp(-\text{energy of state}/k_{B}T)$. For the case of one protein in a cell, the counting is outlined in Fig. 4.10. For the case of N proteins the two states are an "off state," where all N molecules are free, and an "on state," with $N - 1$ free molecules and one R bound to the DNA.

The statistical weight of the case with one molecule bound is:

$$Z(\text{on}) = \frac{1}{(N-1)!} \left(\int_V \int \frac{d^3 r\, d^3 p}{h^3} e^{-p^2/(2mk_B T)} \right)^{N-1} e^{-\Delta G'/k_B T} \tag{4.2}$$

The statistical weight of having no molecules bound is:

$$Z(\text{off}) = \frac{1}{N!} \left(\int_V \int \frac{d^3 r\, d^3 p}{h^3} e^{-p^2/(2mk_B T)} \right)^{N} \tag{4.3}$$

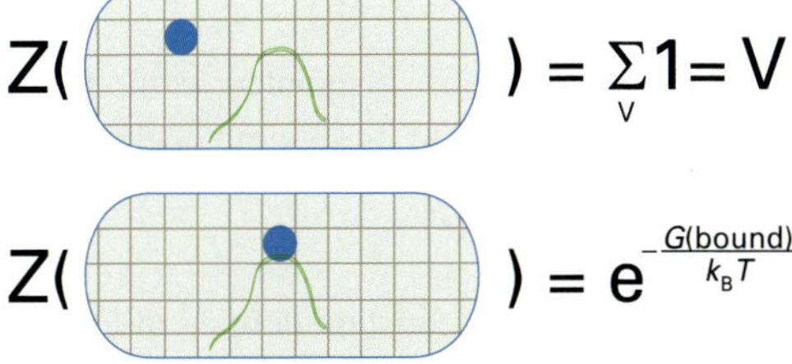

Figure 4.10 Counting statistical weight of unbound and bound configurations of a protein relative to its operator on the DNA. In cases where the protein is unbound there are many possible positions, effectively given by the volume of the cell (as indicated by squares), which by convention is set to a unit volume $v = 1/M \sim 1\,\text{nm}^3$. A bound protein is only allocated one square, where it is located due to a (negative) binding energy $G(\text{bound}) = \Delta G$.

Here the integrals count all the possible positions r and momenta p of all the N molecules in the cell. The division by Planck's constant h takes into account the discreteness of phase space imposed by quantum mechanics. The division by $(N-1)!$ or $N!$ counts all permutations of molecules in the volume V, and is included because these molecules are indistinguishable. This is necessary in order to avoid double counting of any of the identical free molecules in the cell volume V.

In principle, there are many other parts in the cell volume V, including water. Here we assume that everything else besides the N regulator molecules is independent of the state of the system (whether "on" or "off"), and therefore contributes equally to all weights.

Integrating, (and using $\int_{-\infty}^{\infty} dy e^{-y^2} = \sqrt{\pi}$), one obtains the statistical weights of the "on"-state:

$$Z(\text{on}) = \frac{\left(V[2mk_{\mathrm{B}}T\pi/h^2]^{3/2}\right)^{N-1}}{(N-1)!} \cdot \exp\left(-\frac{\Delta G'}{k_{\mathrm{B}}T}\right)$$

$$\propto c^{N-1}\rho^{-(N-1)} \exp\left(-\frac{\Delta G'}{k_{\mathrm{B}}T}\right). \tag{4.4}$$

The statistical weight of the "off" state, where all N particles are free, is

$$Z(\text{off}) = \frac{\left(V[2mk_{\mathrm{B}}T\pi/h^2]^{3/2}\right)^{N}}{N!} \propto c^{N}\rho^{-N} \tag{4.5}$$

where $\rho = N/V$ and $c = (2mk_{\mathrm{B}}T\pi/h^2)^{3/2}$. The last step uses Sterling's equation for factorials $x! \approx (x/e)^x\sqrt{2\pi x}$ and the approximation $(N-1)/N \approx 1$. This assumes that the density of free particles ρ does not change significantly when one molecule binds.

4.3.2 Stat-mec counting reduced to standard chemistry

The system of N particles in V with either one bound (on) or none bound (off) now has a total statistical weight (partition sum):

$$Z = Z(\text{on}) + Z(\text{off}) \tag{4.6}$$

and the probability of the on state is:

$$P_{\text{on}} = \frac{Z(\text{on})}{Z} \quad \text{and the ratio} \quad \frac{P_{\text{on}}}{P_{\text{off}}} = \frac{Z(\text{on})}{Z(\text{off})} = \frac{\rho}{c}e^{-\Delta G'/k_{\mathrm{B}}T} \tag{4.7}$$

The counting determines the statistical weights associated with, for example, the chemical reaction of the DNA-binding protein R to its specific operator O:

$$R + O \leftrightarrow RO \tag{4.8}$$

Usually this is characterized by the dissociation constant defined in Eq. (3.4):

$$K = \frac{[R][O]}{[RO]} \quad \text{where} \quad [O] + [RO] = [O_{\mathrm{T}}]$$

where $[O]$ is the free operator concentration, $[RO]$ is the bound operator concentration and $[O_{\mathrm{T}}] = 1/V$ is the total operator concentration corresponding to its presence in the bacterial volume. Since the operator O can only be "on" or "off", the probability

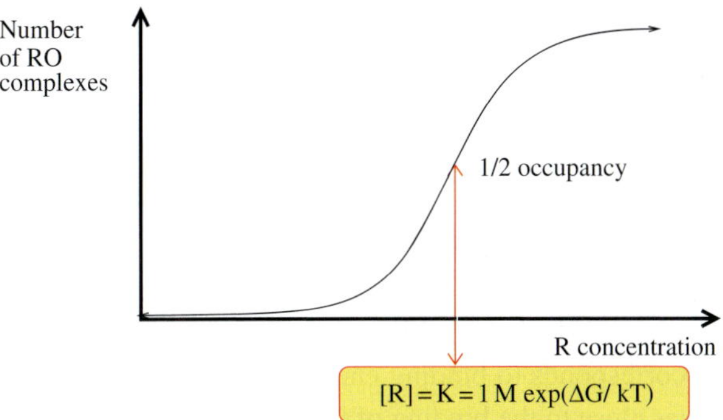

Figure 4.11 How to measure ΔG for the binding of R to an operator. The dissociation constant is identified from the concentration where there is half occupation of the operator. Thus $K = [R_{half}]$ $= 1\,M \cdot \exp(\Delta G/k_B T) \rightarrow \Delta G = k_B T \ln([R_{half}]/[M])$. If for example $[R_{half}] = 100\,nM$, the binding energy $\Delta G = 0.62\,kcal/mol \cdot \ln(10^{-7}) = -10\,kcal\,mol^{-1}$.

of being on is $P_{on} = [RO]/[O_T]$, and of being off is $P_{off} = [O]/[O_T]$. Using standard chemistry from Eq. (3.4) implies that:

$$\frac{P_{on}}{P_{off}} = \frac{[RO]}{[O]} = \frac{[R]}{K} \tag{4.9}$$

This is consistent with Eq. (4.7) when we set $[R] = \rho$ and identify

$$K = c \cdot e^{\Delta G'/k_B T} = [1M] \cdot e^{\Delta G/k_B T} \tag{4.10}$$

Here the last equality sign adopts the usual convention for ΔG by measuring the prefactor in moles per liter (M).[11] In the following we will not write the [1 M] in our equations, and always implicitly count concentrations in moles per liter. The approach to identifying ΔG from a standard chemical titration experiment is illustrated in Fig. 4.11.

4.3.3   Normalized counting

As a standard in the following we count all statistical weights normalized by Z_{off}. Equation (4.4) is then:

$$Z(\text{on}) = [R]e^{-\Delta G/k_B T} = [R]/K \quad \text{and} \quad Z(\text{off}) = 1 \tag{4.11}$$

Let us consider the range of possible values of K and associated ΔG values. From Eq. (3.4) we obtain a bound fraction:

$$\frac{[RO]}{[O_T]} = \frac{[R]}{[R] + K} = \frac{Z_{on}}{Z_{off} + Z_{on}} = \frac{[R]/K}{1 + [R]/K} \tag{4.12}$$

---

[11] Notice that one sometimes uses the association constant $K_A = 1/K$, instead of the dissociation constant K. Also notice that $\Delta G = \Delta G' + k_B T \cdot \ln(c/1\,M) = \Delta G' + k_B T \cdot \ln((2mk_B T\pi/h^2)^{3/2}/(6 \cdot 10^{23}\,/liters))$.

The occupancy therefore increases beyond 1/2 occupancy when the R level is increased beyond the half occupation point:

$$[R_{half}] = K = e^{\Delta G/k_B T} \qquad (4.13)$$

A "living" system should be able to address both sides of the regulation with a number of molecules of the order of $\sim 1-1000$. Because one molecule in a bacterial volume $V = 1\,\mu^3$ corresponds to a concentration of $1\,nM$, one expects K to be of the order of $10 \rightarrow 1000\,nM$ or $K \sim 10^{-8}$–$10^{-6}\,M$. Using $K = [1\,M] \cdot e^{\Delta G/k_B T}$ with $k_B T \approx 0.62\,kcal\,mol^{-1}$, this corresponds to binding energies of $\Delta G \sim -11.5$–$8.5\,kcal\,mol^{-1}$, which indeed are values close to the binding of CI to its operator sites in phage λ.[12]

4.3.4   Questions

4.3.1   Consider a titration experiment where one varies CI concentration for a fixed operator concentration in order to probe the first-order chemical reaction $CI + O \leftrightarrow CIO$. Compare the fraction of bound [CIO] to the operator as a function of $[CI_T]$ for the case where $[O_T] = 10 \cdot K$, and a case where $[O_T] = K/10$.

4.3.2   Consider one DNA-binding protein that may bind to $L = 5\,000\,000$ possible positions on the *E. coli* DNA with a binding free energy $\Delta G = -3\,kcal\,mol\,bp^{-1}$ (counted as the energy for each of the $L = 5\,000\,000$ positions). Alternately the protein may be anywhere in the volume $V = 1\,\mu m^3$. Express the statistical weight for being in each of these two states, and calculate which is the most likely.

4.3.3   If there are N of the above DNA-binding proteins, write the equation for the probability of possible partitions of the proteins in the DNA-bound state.

4.4.4   If a living system demanded 10^{12} different proteins to work, instead of a diversity of $\sim 10^3$, a cell would have a volume $1\,mm^3$. With about one regulator of each type per cell, what would then be the expected binding energy between the proteins and an operator? How much would the time for finding a particular operator be changed?

4.4   Chemistry and co-operativity as statistical mechanics

4.4.1   Shea–Ackers formalism

Let us consider more complex situations, where several regulators can be bound to nearby operators. The counting then involves a number of different states s, each

---

[12] ΔG counts the free energy difference between the bound and the free states when $[R] = 1\,M$. If $[R] < 1\,M$, then the entropy gain by going from "on" to "off" is larger. The free energy difference between a bound and a free state at density $[R]$ is deduced from Eq. (4.11):

$$\Delta G^* = k_B T\,(\ln(Z_{off}) - \ln(Z_{on})) = \Delta G - k_B T \ln([R]/1\,M)$$

reflecting that smaller $[R]$ makes ΔG^* more positive and thereby lowers the binding probability.

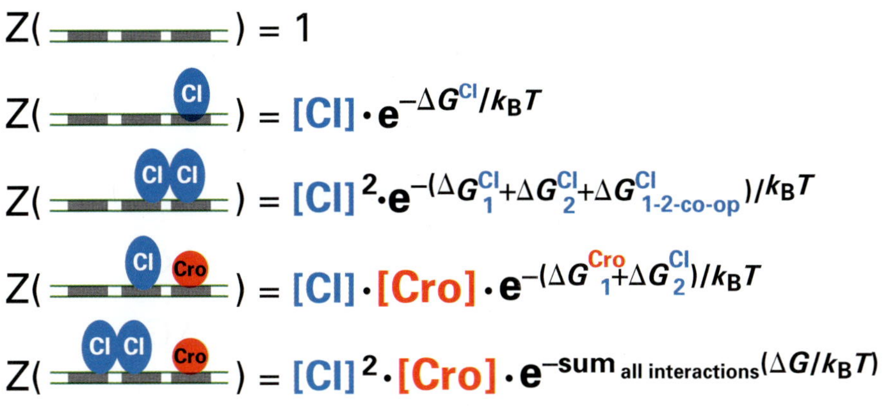

$$Z(\underline{\quad\quad\quad}) = 1$$

$$Z(\underline{\quad\text{CI}\quad}) = [\text{CI}]\cdot e^{-\Delta G^{\text{CI}}/k_{\text{B}}T}$$

$$Z(\underline{\quad\text{CI}\;\text{CI}\quad}) = [\text{CI}]^2\cdot e^{-(\Delta G^{\text{CI}}_1+\Delta G^{\text{CI}}_2+\Delta G^{\text{CI}}_{\text{1-2-co-op}})/k_{\text{B}}T}$$

$$Z(\underline{\quad\text{CI}\;\text{Cro}\quad}) = [\text{CI}]\cdot[\text{Cro}]\cdot e^{-(\Delta G^{\text{Cro}}_1+\Delta G^{\text{CI}}_2)/k_{\text{B}}T}$$

$$Z(\underline{\quad\text{CI}\;\text{CI}\;\text{Cro}\quad}) = [\text{CI}]^2\cdot[\text{Cro}]\cdot e^{-\text{sum}\;\text{all interactions}(\Delta G/k_{\text{B}}T)}$$

Figure 4.12 Statistical mechanics of genetic regulation [13, 173, 174, 175, 176, 177]. For any state the operators can be assigned a statistical weight that is proportional to the free concentrations of each of the bound molecules multiplied by an affinity that is quantified in terms of a binding (free) energy ΔG. In phage λ the relevant regulators are CI dimers and Cro dimers. One may also include a closed complex of RNAP as a variable.

characterized by a number $i(s)$ of bound regulators. For a state with $i(s)$ bound regulators, the exponent in Eq. (4.2) would be $N - i(s)$ instead of $N - 1$. The corresponding concentration-dependent factor in Eq. (4.4) would be $\rho^{-(N-i(s))}$, and in Eq. (4.11) it is $[\text{R}]^{i(s)}$ instead of $[\text{R}]$.

The counting of statistical weights of each state requires a calculation of R (which in the case of λ would be the dimer concentration [CI] given by total CI concentration from Eq. (3.6)). The statistical weights for any of the possible s states is then given by, in terms of [R]:

$$Z(s) = [\text{R}]^{i(s)}e^{-\Delta G(s)/k_{\text{B}}T} \tag{4.14}$$

which is normalized such that the statistical weight of the state where nothing is bound $Z(s = 0) = 1$. The concentration-dependent factor [R] reflects the entropy loss in going from a freely moving regulator to a bound regulator. Notice that Z has no dimensions, it is just a number, as all concentrations are counted in units of M. The probability of a state s is then given by the appropriately normalized ratio [173, 174]:

$$P(s) = \frac{Z(s)}{\sum_{s'} Z(s')} \tag{4.15}$$

Figure 4.12 shows the statistical weight Z for a number of operator states exposed to binding by one or two types of regulatory proteins.

4.4.2 λ regulation revisited

Consider the case of phage λ where both CI dimers [CI] and RNAP can be bound to the operators. The equation for Z is given by Eq. (4.15) with weights of individual states s of the form:

$$Z(s) = [\text{RNAP}]^{j(s)}[\text{CI}]^{i(s)}e^{-\Delta G(s)/k_{\text{B}}T} \tag{4.16}$$

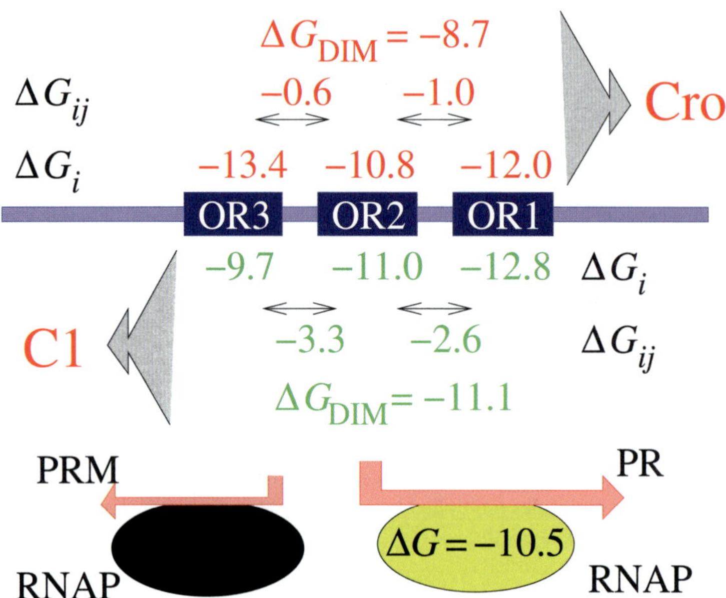

Figure 4.13 ΔGs (*in kcal mol*$^{-1}$) for OR regulation in phage λ [165]. Top numbers refer to Cro, with G_i and ΔG_{ij} referring to individual respective co-operative bindings. Bottom part is corresponding bindings for CI and RNAP. The dimerization energies are both denoted by ΔG_{Dim}. The co-operative bindings act such that only one counts if there is CI on all 3 operator sites [112]. *RNAP* binding to PR is prevented when CI or Cro binds to OR1 or OR2. RNAP binding to PRM is excluded when Cro or CI bind to OR3. Activation rate of RNAP from PRM increased by a factor of 9 when CI are bound to OR2. RNAP binding is here calibrated to a free RNAP concentration that was, perhaps wrongly [76], assumed to be 30nM [78, 79] All numbers are from *in vitro* measurements, and will also depend on the salt inside the cell, which in turn depends on external salt [178].

where $j(s)$ is the number of bound RNAP molecules in state s, and i is the number of bound CI dimers. Weightings for CI and Cro binding to OR are outlined in Fig. 4.12, with chemical binding for CI and Cro to OR listed in Fig. 4.13. Accordingly, Eq. (4.14), or rather a generalization of Eq. (4.16), allows all equilibrium properties to be calculated. In particular, one can calculate the probability for any subset of states as a function of the total amount of CI in the cell.

As an example, let us calculate the promoter activity of the two promoters for CI (PRM) and Cro (PR) in an *E. coli* cell, where there is only CI and no Cro. CI binds to the DNA in the form of a dimer with dimerization constant $K_{\mathrm{CI}} = e^{\Delta G_{\mathrm{DIM}}/k_{\mathrm{B}}T}$ $\mathrm{M} = e^{-11.1/0.62}\mathrm{M}$. Subsequently the CI dimers bound to neighboring operators also bind to each other [138, 164], which is why one seldom sees an isolated dimer on the operator OR.

OR in phage λ has three nearby operator sites, OR1, OR2 and OR3, as outlined in Figs. 4.12 and 4.13. Therefore, there are $2^3 = 8$ possible states of CI binding to these operators, and detailed modeling requires ΔG for all occupations of these states (plus some configurations associated to Cro and RNAP bindings that is ignored here). First consider the *cro* promoter PR. This promoter can only be accessed by RNAP when both OR2 and OR1 are free. Additionally, in order for the promoter to be active, RNAP must first be bound. The probability of the corresponding state where RNAP is bound to PR is:

$$P(0, \text{RNAP}) + P(1, \text{RNAP}) = \frac{Z(0, \text{RNAP}) + Z(1, \text{RNAP})}{\sum_s Z(s)} \tag{4.17}$$

where "0" refers to the absence of CI on OR3, whereas "1" implies that CI is bound to OR3 [165, 173, 174, 175, 179].

To calculate the statistical weights one makes use of Eq. (4.16) with the values of ΔG given in Fig. 4.13. For example:

$$
\begin{aligned}
Z(1, \text{RNAP}) &= [\text{RNAP}] \cdot [\text{CI}] \cdot \exp\left(-\frac{\Delta G(1, \text{RNAP})}{k_\text{B} T}\right) \\[2mm]
&= [\text{RNAP}] \cdot [\text{CI}] \cdot \exp\left(\frac{(9.7 + 12.5)\,\text{kcal mol}^{-1}}{0.62\,\text{kcal mol}^{-1}}\right) \\[2mm]
&= [\text{RNAP}] \cdot [\text{CI}] \cdot \exp\left(\frac{22.2\,\text{kcal mol}^{-1}}{0.62\,\text{kcal mol}^{-1}}\right)
\end{aligned}
\tag{4.18}
$$

where the appropriate binding energies of CI and RNAP are taken from Fig. 4.13. The concentrations should be in molar (M) which translates into a CI dimer concentration of $100\,\text{nM} = 10^{-7}\,\text{M}$ that contributes to the factor $[\text{CI}] = 0.000\,000\,1$.

The right-hand panel in Fig. 4.14 shows PR activity as a function of the amount of CI. As CI bound to either OR1 or OR2 prevents RNAP from binding to PR, the probability of initiating PR decreases strongly with CI concentration.

The similarly calculated activity of PRM is shown in the left-hand panel of Fig. 4.14. The CI promoter PRM is weak in itself, but is activated by a factor of ~ 9 when a CI dimer binds to OR2 while OR3 is free [112]. Therefore the states with CI on OR2 have a more active PRM than states without CI on OR2. The modeling accordingly reflects an increase in PRM activity when increased CI allows substantial binding to OR2.

The binding of CI dimers to both OR1 and OR2 is strengthened by additional binding between the two CI dimers, as seen from the additional binding of the (011) state compared to the sum of the (010) and the (001) states in Fig. 4.13. Therefore when reaching CI concentrations where CI binds to OR1, another CI also binds to OR2. This co-operativity is responsible for the strong increase in PRM activity for small CI concentrations in Fig. 4.14. This is in itself a small positive feedback circuit. The sharpness of the response of PRM to CI is important for the stability of the CI:Cro switch.

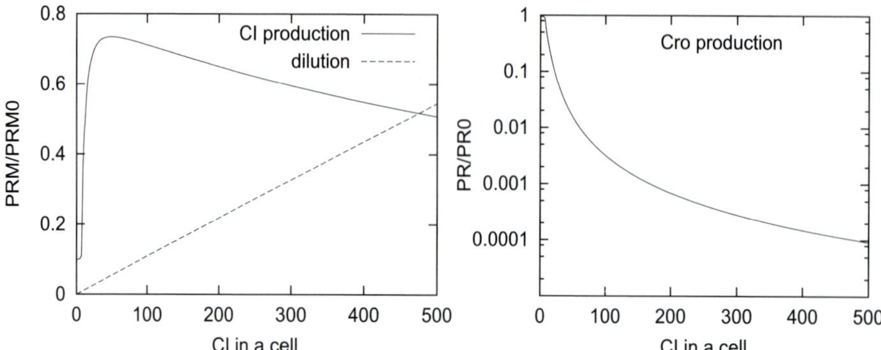

Figure 4.14 PRM and PR activity as a function of number of CI molecules in an *E. coli* cell that is assumed to have zero Cro. PRM activity is calculated to be proportional to RNAP polymerase bound to OR3 (= $\sum_{i=0}^{1} \sum_{j=0}^{1} P(RNAP, i, j)$), whereas PR activity is proportional to one RNAP molecule occupying PR that covers both OR1 and OR2 (= $\sum_{i=0}^{1} P(i, RNAP)$). RNAP binding to PRM has an energy of $-11.5\,kcal\,mol^{-1}$, and to PR is $-12.5\,kcal\,mol^{-1}$, whereas CI is here assumed to bind only to operator DNA. Furthermore an RNAP at PRM contributes to an activity of 1 if OR2 is occupied by CI and to an activity of 0.1 when OR2 is not occupied by CI.

When OR3 is occupied by either CI or Cro, PRM is blocked: an excessive amount of CI is prevented by negative auto-regulation of CI, which accordingly maintains the lysogen in a state that is easily modified by external signals. The *in vivo* activity of PRM has been measured, both with and without Cro [180], and in fact Fig. 4.14 differs from these experiments in several ways, mainly due to the influence of DNA that is far away from the operator OR.

4.4.3   Operators and the lytic patway

PR and PRM are not only a function of CI, but also of Cro in the cell. When Cro dimers are present, they bind first and most easily to OR3 (with $\Delta G(2, 0, 0) = -13.4\,kcal\,mol^{-1}$), thereby blocking transcription of *cI*. This action of Cro secures the stability of the lytic pathway when the switch is thrown. Furthermore, Cro binds to OR1 and OR2 and thereby represses itself. This negative auto-regulation prevents excessive amounts of both Cro and CII during the early decision process, where multiple copies of the phage genome may otherwise give too much CII. Also, when entering the lytic pathway, Cro self-repression limits the expression of early lytic genes.

The CI-dominated state is identical to the lysogenic state and is characterized by 150–300 CI monomers in the *E. coli* cell [181].[13] Conversely, the Cro-dominated state

---

[13] The estimate was production of 3.75 CI monomers per minute in a lysogen growing with a doubling time of 40 minutes. On the other hand, in a lysogen, CI amounts to a fraction (0.000080%) of total protein mass, corresponding to about 300 CI per cell (there are 3.5 million proteins of average size 330 aa in *E. coli*). During infection, the production of CI is estimated to be 29 per minute per genome, mainly governed by PRE.

can also be realized for a phage genome inserted into the *E. coli* chromosome, and this anti-immune state is estimated to contain about 300 Cro proteins [151, 179].

4.4.4 Questions

4.4.1 Develop a computer program to calculate the activity of PRM and PR as a function of CI concentration, given the tabulated free energies for CI to OR sites in Fig. 4.13. To simplify the problem disregard the RNAP in the partition sum, and include only the eight states that bind CI dimers in various ways. Activity is then given by the states that allow for RNAP binding to respective promoters, i.e. PRM with OR3 free and PR with both OR1 and OR2 free. Remember that the PRM activity of states with CI on OR2 is ten times larger than that of states without CI on OR2. Also remember that co-operativity between OR1–OR2 and OR2–OR3 is mutually exclusive. First, calculate PR and PRM activity as a function of CI, assuming that CI dimerization energy is $-\infty$ (all CI in dimers). Second, calculate the same activity using a dimerization constant $K_{\mathrm{Dim}} = [1\,\mathrm{M}] \cdot e^{-11.1/0.62}$. Third, in the last calculation investigate the effect of removing the co-operative binding between CI bound to OR1 and CI bound to OR2.

4.4.2 If CI is cleaved by RecA, its dimerization is prevented. What concentration of CI monomers is needed to maintain a similar probability of having OR1 occupied as 100 nM of CI dimers does. Assume that monomer–OR1 binding is half of the dimer CI–OR1 binding ($= -12.8\,\mathrm{kcal\,mol}^{-1}$), and also simplify the problem by assuming that only one CI monomer can bind to the operator.

4.4.3 Often there are very few molecules of a given transcription factor inside a cell. Accordingly, it may be useful to consider the partition function for the case where occupancy of the single operator depletes the amount of free CI substantially. Argue that for a state characterized by n_{M} free monomers, n_{D} free dimers and an operator state s where $i = i(s)$ dimers are bound to operators, the exact statistical weight is [175]:

$$Z(s, n_{\mathrm{M}}) = \frac{V^{n_{\mathrm{D}}}}{n_{\mathrm{D}}!} \cdot \frac{V^{n_{\mathrm{M}}}}{n_{\mathrm{M}}!} \cdot K_{\mathrm{D}}^{n_{\mathrm{D}}+i(s)} \cdot e^{G(s)/k_{\mathrm{B}}T} \tag{4.19}$$

where $n_{\mathrm{D}} = (N - n_{\mathrm{M}} - 2i)/2$, since the number of free dimers in the cell is fixed by the conservation requirement: $N = n_{\mathrm{M}} + 2n_{\mathrm{D}} + 2i$, with N being the total numbers of repressors in the cell. Hint: each of the $n_{\mathrm{D}} + i$ dimers contribute by their dimerization binding free energy through $K_{\mathrm{D}} = e^{\Delta G_{\mathrm{D}}/k_{\mathrm{B}}T}$, whereas only the dimers bound to operators contribute to $-\Delta G(s)$. Again the total partition function can be written as a sum over all the states s, n_{M} that the N molecules can be in: $Z = \sum Z(s, n_{\mathrm{M}})$.

4.5   Distant DNA in gene regulation

4.5.1   Non-specific binding to DNA

4.5.1.1   Dilution of free regulators

Gene regulation often involves transcription factors that bind to specific DNA sites, as well as to the many non-specific DNA sites there are on the *E. coli* chromosome. The non-specific binding can be taken into account by using Eq. (4.14) for $Z(s)$, where the free dimer concentration [CI] is adjusted by the loss to the many non-specific sites. The probability for a given dimer to be bound to a given non-specific site can be calculated through the statistical weightings:

$$Z(\text{unbound}) = 1$$

$$Z(\text{bound}) = \frac{1}{V} \cdot e^{-\Delta G_u/k_B T} \tag{4.20}$$

where $1/V$ represents the concentration of one dimer in volume V. If instead of one site, the dimer could be bound to L sites along the chromosomal DNA, then the corresponding statistical weightings would be:

$$Z(\text{unbound}) = 1 \tag{4.21}$$

$$Z(\text{bound}) = \frac{L}{V} \cdot e^{-\Delta G_u/k_B T} \tag{4.22}$$

representing counting over all the different DNA sites that the dimer could bind.

The probability of being unbound is:

$$P(\text{unbound}) = \frac{1}{1 + \frac{L}{V} \cdot e^{-\Delta G_u/k_B T}} \tag{4.23}$$

If there are a total of N_D dimers in the cell, the average number of unbound dimers will be:

$$n_D = \frac{N_D}{1 + \frac{L}{V} \cdot e^{-\Delta G_u/k_B T}} \tag{4.24}$$

Here $1/V$ is the concentration, measured in molar, in *E. coli*, with $1/V = 10^{-9}$ and non-specific binding becomes substantial when

$$e^{-\Delta G_u/k_B T} \cdot \frac{L}{V} \sim 1 \tag{4.25}$$

where L is the number of DNA base pairs. When $L \sim 5 \cdot 10^6$, this balance happens when $\Delta G_u \sim -3\,\text{kcal mol}^{-1}$: When non-specific binding is stronger, the $L = 5 \cdot 10^6$ binding sites on the DNA "win" over the $V \sim 10^9$ free positions in the cell.

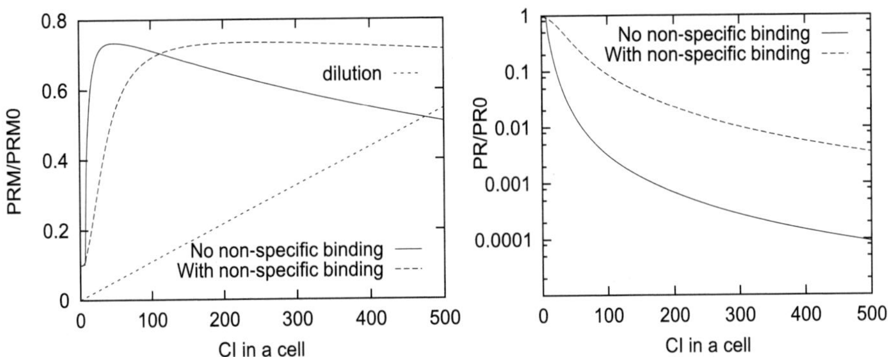

Figure 4.15 PRM and PR activity as functions of CI molecules in a bacterial cell, that have zero Cro (PRM0 parameterizes the maximum strength of PRM, and PR0 the maximum of PR). The figure illustrates the effect of a non-specific CI–DNA binding of $\Delta G_u = -4\,kcal\,mol^{-1}\cdot bp^{-1}$, in comparison to no non-specific binding. Non-specific binding is found to shift the amount of CI needed to influence the promoters. The dashed line shows the rate of CI depletion due to cell growth and division.

4.5.1.2 Non-specific influence on occupation of specific sites

The statistical weight for a given specific state s of operator occupation is given by:

$$Z(s) = \left(\frac{n_D}{V}\right)^{i(s)} e^{-\frac{\Delta G(s)}{k_B T}} = \left(\frac{N_D}{V + L \cdot e^{-\Delta G_u/k_B T}}\right)^{i(s)} \tag{4.26}$$

where $i(s)$ is the number of dimers bound to operators. Here V counts the statistical weight of a free CI, whereas $L \cdot e^{\Delta G_u/k_B T}$ counts the statistical weight of a CI being bound to the L non-specific DNA sites. The above equation reduces the available dimers because a fraction of them bind to non-specific DNA. Assuming a $\Delta G_u = -4\,kcal\,mol^{-1}$, the predicted effect of non-specific CI–DNA binding on PRM and PR activity is illustrated in Fig. 4.15. These profiles roughly resemble the *in vivo* promoter dependences with CI concentration measured by [180].

4.5.1.3 Buffering of DNA binding

Non-specific binding is not only a cost that has to be accounted for. In fact, if nearly all transcription regulators bind mostly non-specifically to the DNA, then an overall change in protein–DNA binding constant by a factor of 100 will not change the relative strength of operator to non-specific binding. This buffering follows from:

$$Z(s) = \left(\frac{N_D}{V + L \cdot e^{-\Delta G_u/k_B T}}\right)^{i(s)} e^{-\frac{\Delta G(s)}{k_B T}} \approx \left(\frac{N_D}{L}\right)^{i(s)} e^{-\frac{(\Delta G(s) - i(s)\Delta G_u)}{k_B T}} \tag{4.27}$$

where the last approximation uses the strong non-specific binding limit where the volume factor is relatively small and therefore all regulators remain on the DNA.[14]

---

[14] When non-specific binding is sufficiently strong, $L \cdot e^{-\Delta G_u/k_B T} \gg V$.

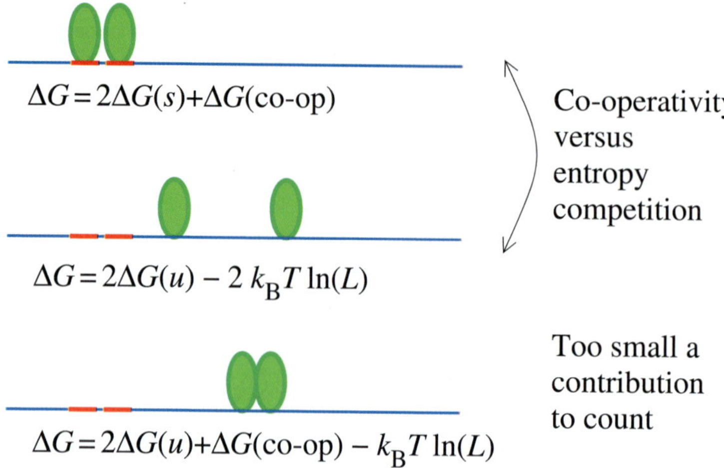

Figure 4.16 Co-operativity as a mechanism to counteract non-specific binding, here examined in the limit where all proteins are confined to the DNA. When co-operative binding between two transcription factors $\Delta G(co - op)$ contributes less than the free energy associated with all L non-specific binding sites, then the non-specific binding in the lower panel is disfavored. For the remaining two upper scenarios the co-operative binding favors specific binding. Notice that a moderate cooperative energy similarly favors specific binding in cases where the protein was free within the cell volume.

When protein–DNA binding is changed by, say, $3\,\mathrm{kcal\,mol^{-1}}$, then both the specific $G(s)$ and the unspecific binding ΔG_{u} change. This makes the difference in the exponential in Eq. (4.27) independent of the change, and non-specific binding therefore effectively buffers the specific binding. This buffering may well be important when considering the pronounced effect that changes in salt concentration have on protein–DNA binding [182].

4.5.1.4   Co-operativity and non-specific DNA binding

Given the large number of non-specific DNA sites in a cell, they may out-compete the specific DNA binding to a particular operator. One biological remedy to this is to increase co-operativity, and design operator sites that demand simultaneous binding of several proteins. Consider two proteins that each bind with energy ΔG_{s} to nearby specific sites, an energy $\Delta G(\text{co-op})$ to each other and energy ΔG_{u} to any non-specific site. The competition between configurations is illustrated in Fig. 4.16, where the $\ln(L)$ term represents the entropy of the L non-specific binding positions.

The lower configuration in Fig. 4.16 can be ignored for realistic co-operativities, $\Delta G(\text{co-op}) \sim -3\,\mathrm{kcal\,mol^{-1}}$, since $\Delta G(\text{co-op}) \gg -k_{\mathrm{B}}T \cdot \ln(L) \sim 0.6\ln(10^{7})$ $\mathrm{kcal\,mol^{-1}} \sim -10\mathrm{kcal\,mol^{-1}}$ and therefore the loss in entropy relative to the middle situation is too large. However, if the co-operativity energy becomes very strong, the last situation will dominate, and the specificity associated with co-operativity vanishes.

The effect of co-operative binding is investigated by comparing the specific and non-specific binding scenarios in the top and middle configurations of Fig. 4.16, respectively:

$$\Delta G_{\text{specific}} - \Delta G_{\text{non-specific}} = 2\Delta G_s + \Delta G(\text{co-op}) - 2\Delta G_u + 2k_B T \cdot \ln(L) \qquad (4.28)$$

Overall, we conclude that moderate co-operative binding $\Delta G(\text{co-op}) \in [-k_B T \cdot \ln(L), 0]$ favors specificity. For example, when the energy of co-operative binding is of the order of $-3\,\text{kcal mol}^{-1}$, as found between CI on OR sites in λ, these CI dimers cannot form tetramers without the help of specific binding.

Finally, one should be aware that especially non-specific binding of transcription factors is in competition with abundant DNA-binding proteins HNS, HU, FIS, IHF, HFQ, and DPS, each with copy numbers so high that they cover a substantial fraction of the *E. coli* genome,[15] and thereby influence the available space on the DNA. These major classes of DNA-binding proteins are involved in the organization of DNA, with HU and IHF bending the DNA, whereas HNS can bridge distant sites on the DNA and thereby sustain large loops.

4.5.2 DNA-looping and the λ-switch

4.5.2.1 The probability of forming a loop between OL and OR

Co-operativity has multiple layers: CI dimerization, co-operative binding between CI dimers bound to OR1 and CI bound to OR2, and finally a "layer" of cooperativity that involves DNA loop formation. As indicated in Fig. 4.10, λ phage has an operator left (OL), $\sim 2.4\,\text{kbase}$ left of OR, with almost identical sequences. OR and OL can interact through DNA looping, as demonstrated by the experiments described in Fig. 4.17. The effect of this loop is to repress PR [187] and modulate PRM activity [180, 188]. The looping is also seen directly in *in vitro* experiments using single-molecule techniques [189]. Figure 4.17 shows that PRM with OL is about 2.5 times less active than without. PR is repressed by an additional factor of 4 by the loop [187], when compared to the wt system at wt CI levels.

The interaction between OL and OR can be quantified in terms of effective binding energies associated with states where the two operators interact (closed loop). In order to introduce the reader to considerations concerning organization of DNA and the role of DNA-binding proteins, we now want to estimate the binding energy associated with loop formation.

Assume for simplicity that only one combination of CI–operator binding contributes to both the open and closed states. The statistical weight of a given binding pattern s is then:

$$Z(\text{open}) = f(CI) \cdot e^{-\Delta G_R(s)/k_B T} \cdot e^{-\Delta G_L(s)/k_B T} \qquad (4.29)$$

---

[15] Refs. [183, 184] list the following number for abundant DNA-binding proteins per *E. coli* cell: 20 000 HNS, 50 000 FIS (in log growth, only 100 FIS in stationary phase), 30 000 HU, 12 000-55 000 IHF, 30 000-60 000 HFQ and between 6000 and 180 000 DPS (largest in the late stationary phase). The reader may also consult [185], and for the integrase host factor IHF [186].

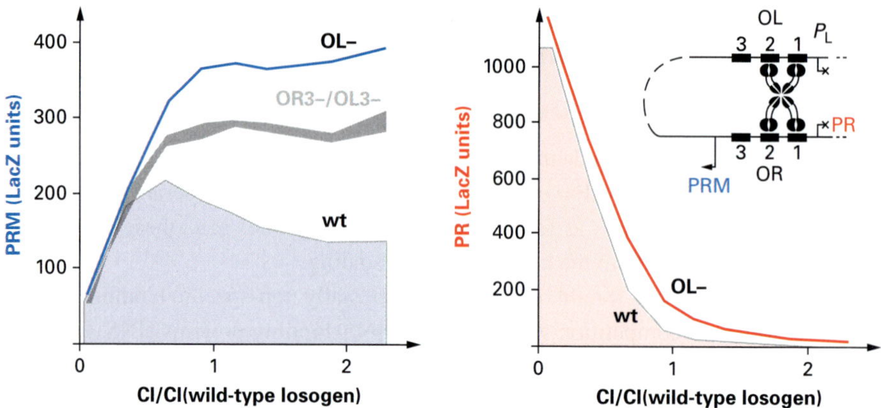

Figure 4.17 The activity of PRM and PR measured as a function of an externally regulated CI, as reported by [180]. The system is engineered such that CI is not produced from PRM. The activity is measured through a reporter gene (LacZ) that is inserted after the respective promoters. The CI level is controlled indirectly, through a seperate *cI* gene that is under the control of the lac represssor. The figure shows wt (wild-type construct) and a mutant where OL is removed (OL-), as well as mutants where CI does not bind to OR3. Notice that OL represses PRM by about a factor of 2.5–3, and PR by a factor of 4 at CI levels found in normal lysogens. The OL3-, or the OR3- mutant activities of PR resemble that of the wt.

and

$$Z(\text{closed}) = Z(\text{open}) \cdot e^{-\Delta G(\text{complex},s)/T} \tag{4.30}$$

depending on whether OL and OR interact (Eq. 4.30) or not (Eq. 4.29). Here, $f(CI)$ is the concentration-dependent factor that depends on CI binding to both OR and OL.

Figure 4.17 demonstrated that a strain with OL+ (wt) has a PR activity level that is a factor of 4 more repressed than an OL− strain, at CI levels of a normal lysogen. Assume now that PR can only be active during periods where the loop is open. In these time intervals, the PR activity equals the activity of the OL− strain, and therefore the probability of non-looping must be:

$$\frac{Z(\text{open})}{Z(\text{open}) + Z(\text{open}) \cdot e^{-\Delta G(\text{complex},s)/T}} \sim 0.25$$

implying that $e^{-\Delta G(\text{complex},s)/k_{\mathrm{B}}T} = 3$, or $\Delta G(\text{complex}) = -0.7\,\text{kcal}\,\text{mol}^{-1}$. This complex bridges a connection between OL and OR, and can be facilitated by two dimers on OR that bind to two dimers on OL, thereby forming a 8-mer closing of a loop with total net energy ΔG (8-mer loop). In addition, the activity of PRM is also found to depend on the ability of CI to bind to OR3 and all the OL sites [180], implying that at high CI, dimers on all three OR sites bind to dimers at all three OL sites, forming a 12-mer loop.

Figure 4.18 shows the predicted activity of PRM and PR that reproduces a known repression of CI in a wt lysogen containing an assumed number of 250 CI proteins in the *E. coli* cell. The key parameters of the fit are a non-specific CI binding energy of

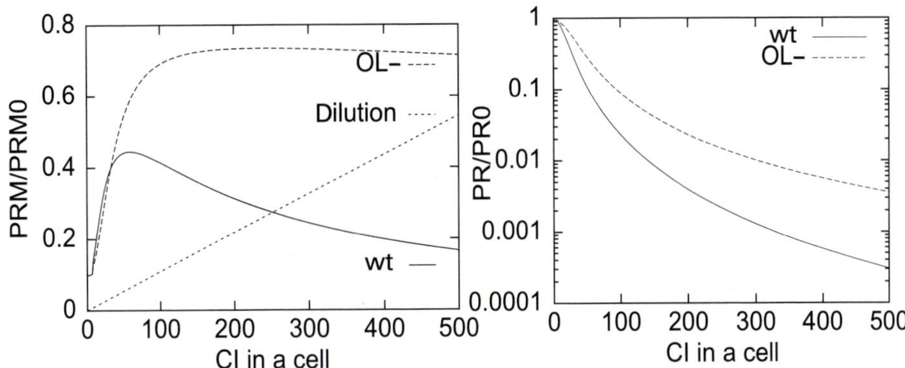

Figure 4.18 PRM and PR activity as functions of CI level, with non-specific DNA binding and loop-formation energies adjusted to fit the experiments of Fig. 4.17. In particular, the loop energy is made to fit the PRM auto-repression of about a factor of 3, measured by [180].

$-3.5\,\text{kcal mol bp}^{-1}$, a complex-forming energy

$$\Delta G(\text{8-}mer\ loop) = -1.0\,\text{kcal mol}^{-1} \tag{4.31}$$

and a full 12-mer energy of $\Delta G(\text{12-}mer\ loop) = -3.0\,\text{kcal mol}^{-1}$.

4.5.2.2   DNA trajectories and loop entropies

The L–R free energy $\Delta G(\text{8-mer loop})$ is given by the binding energy and the entropy cost associated with the loop:

$$\Delta G(\text{8-mer loop}) = \Delta G_8 - k_\text{B}T\Delta S_\text{loop} \tag{4.32}$$

Here ΔG_8 is the energy of CI on OL interacting with CI on OR, and ΔS_loop is the entropic cost associated with closing a loop between OL and OR, separated by $l \sim 2.4\,\text{kb}$.

The entropy cost of loop closure $\Delta S_\text{loop} = S(\text{bound}) - S(\text{open})$ declines with the number of ways that the two loop ends can be separated from each other (free volume $V_{(\text{OL}-\text{OR})}$):

$$e^{S_\text{loop}/k_\text{B}} = \frac{1}{V_\text{OL:OR}} = J_\text{loop} \tag{4.33}$$

where we also introduce the often used J-factor [190], which counts the effective density of one loop end at the other loop end [191]. The J-factor is measured in M, and in *E. coli* a J-factor of about 1 nM corresponds to a loop that is so long that the two ends meet as often as if they was freely diffusing inside the *E. coli* cell. From the above equation the J-factor can be converted into a free energy cost of forming the loop:

$$\Delta G_\text{loop} = -T \cdot S_\text{loop} = -k_\text{B}T \cdot \ln(J_\text{loop}) \tag{4.34}$$

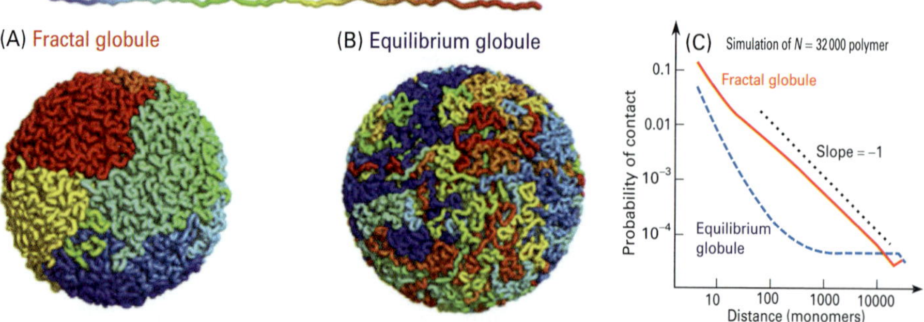

Figure 4.19 Comparison of a fractal globule with an equilibrium polymer, as envisioned by Mirny [194]. The fractal globule is the compact state of the polymer as it appears just after collapse and before random movements have allowed the polymer to entangle into a random structure with knots. There is increasing evidence that chromosomes in both prokaryotes and eukaryotes compare most accurately to fractal globules, and not to the equilibrium structure [193, 194]. Panel C shows the probability of contact between two monomers as a function of their distance along a backbone of 32 000 monomers, as simulated by [194].

Here we estimate $V_{\text{OL:OR}}$ using two alternative assumptions about the chromosome of *E. coli*: one is that the nucleoid is randomly organized around the OR–OL complex, the other assumption assumes that the nucleoid is a fractal globule [192, 193, 194]. These two widely different types of chromosomal organizations are outlined in Fig. 4.19. In any case, one copy of the bacterial chromosome consists of about 5 megabases (Mbp) of DNA that are believed to occupy a volume of 0.3 μm^3 [195].

Using the random-coil model for a DNA segment with length l between the operators, the distance, *dist*, between them is given by a random walk that changes direction for each Kuhn length (l_k)[16]: $dist \sim l_k(l/l_k)^{0.5}$. The corresponding volume $V_{\text{OL:OR}} = \frac{4\pi}{3} \cdot dist^3$, giving a entropy cost of:

$$J_{\text{loop}} = e^{\Delta S(\text{loop})/k_B} = \frac{1}{V_{\text{OL:OR}}} = \frac{1}{(4\pi/3)\left(\left(\frac{l}{l_k}\right)^{0.5} \cdot l_k\right)^3} \tag{4.35}$$

$$\sim \frac{1}{4(8^{0.5} \cdot 100\,\text{nm})^3} \sim 18\,\text{nM}$$

when using a Kuhn length $l_k \sim 300\,\text{bp} \sim 100\,\text{nm}$. Notice that the OL–OR segment then should stay within a volume $V \sim 0.09\,\mu\text{m}^3$ that is only moderately smaller than the total volume of the nucleoid volume of 0.3 μm^3 [195]. The corresponding $\Delta S/k_B \sim -18$ corresponds to a looping cost of $\Delta G_{\text{loop}} = 11\,\text{kcal mol}^{-1}$ and

$$\Delta G(\text{8-mer loop}) = \Delta G_8 + 11\,\frac{\text{kcal}^{-1}}{\text{mol}} = -1\,\frac{\text{kcal}^{-1}}{\text{mol}} \Rightarrow \Delta G_8 = -12\,\frac{\text{kcal}^{-1}}{\text{mol}}$$

[16] The Kuhn length is two times the persistence length of the DNA, and is defined as the length that should be used if simulating the DNA in steps where it typically changes 90°. Persistence length, on the other hand, refers to the length scale for an exponentially decaying correlation of the direction. The persistence length of DNA outside the cell is typically measured to $\sim 200\,\text{bp}$, but depends on salt concentration. Inside the cell it may well be much smaller due to abundant DNA-binding proteins.

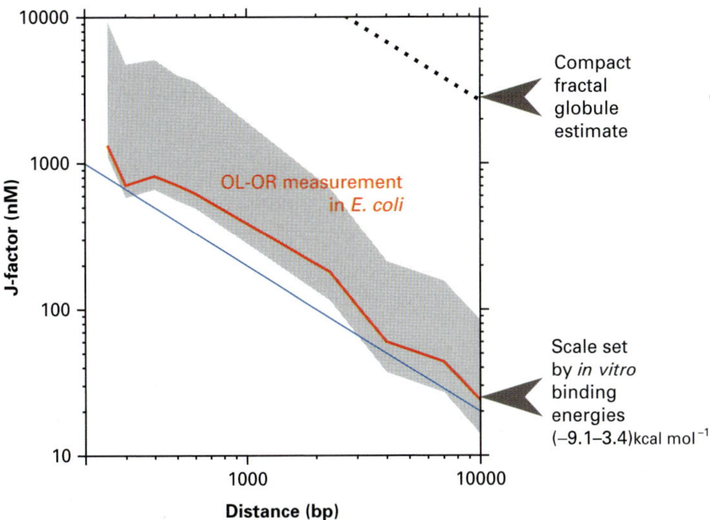

Figure 4.20 Measurement of OL–OR looping as their mutual distance is varied in the *E. coli* chromosome [198]. The gray area marks the uncertainty in the experiments. The blue line shows the 1/*distance* scaling of a fractal globule. The pointer with a subsequent dashed line marks the prediction assuming that DNA is packed such that the volume between any two points on the DNA only contains DNA from between these two points.

when using $\Delta G(\text{8-mer loop}) = -1\,\text{kcal mol}^{-1}$ from Eq. (4.31). This estimate is substantially stronger than the $-9.1\,\text{kcal mol}^{-1}$ octamerization energy measured *in vitro* [196, 197], reflecting that a random polymer is not a convincing model for chromosonal DNA. The Shearvin lab recently reported that the OL–OR looping probability scales as 1/distance, and not with the steeper relation implied by the random polymer assumption, see Fig. 4.20.

There is another and more restricted way to estimate the cost of loop formation, assuming that the chromosome of *E. coli* is a fractal globule [194] (see Fig. 4.19). This type of organization resembles the expected organization of the DNA under the constraint that it does not have time to randomize itself to the knotted equilibrium configuration that was implicitly assumed in the previous estimate. In the fractal globule a DNA segment of length l will occupy a space:

$$V(l) = \sigma \cdot l \tag{4.36}$$

where σ parameterizes an area perpendicular to the DNA strand inside the nucleoid. For simplicity we assume a compact organization, where all DNA within any two points on the DNA is contained within the volume spanned by these two points. σ can then be estimated from the measured size of the volume $0.3\,\mu\text{m}^3$ for one chromosome in the nucleoid

$$0.3\,\mu\text{m}^3 = L\sigma \Rightarrow \sigma = \frac{0.3\,\mu\text{m}^3}{5\,000\,000\,\text{bp} \cdot 0.3\,\text{nm bp}^{-1}} \sim 200\,\text{nm}^2$$

Thus the OL–OR piece of DNA should occupy (at least) a volume:

$$V_{\text{OL:OR}} = L \cdot \sigma = \left(2400\,\text{bp} \cdot 0.3\,\text{nm bp}^{-1}\right) \cdot 200\,\text{nm}^2 = 1.4 \cdot 10^{-4}\,\mu\text{m}^3$$

a volume that is indeed much smaller than the $900 \cdot 10^{-4}\,\mu\text{m}^3$ associated with DNA that was in a random coil (from Eq. 4.35). This volume corresponds to a J-factor:

$$J_{\text{loop}} = \frac{1}{V_{\text{OL:OR}}} = \frac{1}{1.4 \cdot 10^{-4}\,\mu\text{m}^3} = 10^{-5}\text{M}$$

With this extreme assumption, then the entropy cost for closing the loop:

$$S(\text{loop})/k_\text{B} = \ln(J_{\text{loop}}) = -11.5$$

with $\Delta G_{\text{loop}} = 7.1\,\text{kcal mol}^{-1}$. Thereby the estimated octamerization energy $\Delta G_8 = -8.1\,\text{kcal mol}^{-1}$ becomes close to the *in vitro* data [196, 197].

Maintaining the $-9.1\,\text{kcal mol}^{-1}$ octamerization energy and using an additional energy of $-3.4\,\text{kcal mol}^{-1}$ when CI on all ORs is bound to CI on all OLs, the Shearwin lab in Adelaide [198] obtained the J-factors shown in Fig. 4.20. A calibration that is also supported by the fact that similar J-factors are obtained from use of the Lac system, with independently measured *in vitro* binding energies [198]. Their scaling is consistent with the fractal globule idea, but with an astounding factor of 50 lower concentration than the compact estimate above. This disfavors the idea of a fractal globule, and leaves a third option: super-coiling, which is found to favor OL–OR binding [199], as envisioned in Fig. 4.21. Supercoiled DNA may be twisted around itself, forming

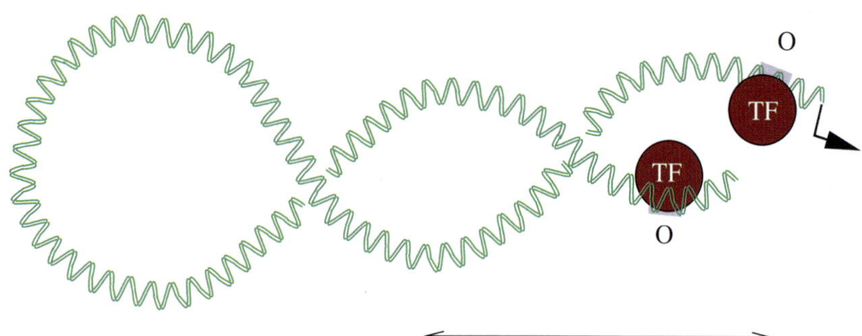

Figure 4.21 A supercoil of DNA tends to twist the DNA around itself and thereby makes it easier for transcription factors to act on promoters that are separated by long distances along the DNA. In the figure *TF* refers to transcription factor, and the regions on the DNA marked *O* to operators.

occasional side branches that may isolate OL–ORs from each other (Ian Dodd, private communication).

4.5.3   Questions

4.5.1   The linear dimensions of a human cell are about 10 times that of a bacterium. Human DNA consists of $3 \cdot 10^9$ base pairs. If one (wrongly!) assumes that the human cell can be viewed as one big "bag" of proteins and DNA, what would be the non-specific binding energy that makes a protein as equally likely to be on the DNA as in the cell volume?

4.5.2   Consider a DNA-binding protein in an *E. coli* cell that binds to 90% of the its DNA with $\Delta G = -3\,\mathrm{kcal\,mol^{-1}}$ and to 10% of its DNA with $\Delta G = -5\,\mathrm{kcal\,mol^{-1}}$ What is the probability that such a protein will be free in the cell? Hint: The statistical weight for one repressor to be bound to a sample of binding sites with energies $\Delta G_i, i = 1, \ldots$ is $Z = \sum_i e^{-\Delta G_i/k_\mathrm{B}T}$.

4.5.3   The interaction between a DNA-binding protein and the DNA can be written as a sum of individual interactions between amino acids and the base pairs at the corresponding positions [200, 201]. Assume, for a repressor in *E. coli*, that each of the $5\,000\,000$ non-specific bindings are drawn from a Gaussian distribution with mean $-3\,\mathrm{kcal\,mol^{-1}}$ and standard deviation $2\,\mathrm{kcal\,mol^{-1}}$. What must the binding energy to the specific operator site O be in order that a protein should spend at least half its time at O?

4.5.4   Repeat the above question if the typical non-specific binding is $+3\,\mathrm{kcal\,mol^{-1}}$. (Trick question, remember that the protein may also be free).

4.5.5   Simulate a lattice polymer on a 3-dimensional cubic lattice. Let it start at position (0,0,0) and make 12 steps, each in a random, uncorrelated direction compared to its previous step. What is the likelihood that it returns to the origin after exactly 12 steps and exactly 20 steps. Repeat the above questions for a polymer where we do not allow self-interactions, apart from the last point.

4.5.6   Other temperate phages are also governed by two antagonistic promoters, with one promoter directing the production of a lysogen-maintenance protein that one may call CI. In [202] the CI promoter *PL* in phage 186 shows an activation curve with respect to its product CI, which resembles that of PRM in phage λ in Fig. 4.17. In 186 the self-activation comes from an indirect effect of CI that represses a PR promoter that would otherwise initiate RNAP, which removes RNAP that is bound to the promoter for CI. Assume that this interference is of the form $PL = PL_0/(PR + PL_0)$ with $PR = PR_0/(CI + 1)$ and deduce the activity of *PL* as a function of CI. Set the base promoter activities to $PR_0 = 10 \cdot PL_0$.

4.6   **Summary**

- The statistical weight of a configuration of molecules is proportional to the product of their free concentration:

$$Z(A, B, C) = [A] \cdot [B] \cdot [B] \cdot \exp\left(-\sum \Delta G / k_B T\right) \tag{4.37}$$

where $\sum \Delta G$ includes the free energies of all binding that keeps the molecules from being free.
- Non-specific bindings weaken the effect of transcription factors, but buffer against changes in their DNA binding that, for example, can be caused by changes in salt.
- DNA looping allows genes to be regulated from distant operator sites.

5 Diffusion and randomness in transcription

5.1 Random walks and diffusion

Transcription, as well as its regulation, is a dynamic process. Previous chapters focused on cases where transcription factors and RNAP binding was so fast that their activity was simply proportional to an effective binding affinity. This chapter explores dynamic aspects of transcription regulation on a shorter timescale, emphasizing the time needed for transcription initiation, and the time needed to adjust binding patterns of regulatory proteins.

To put protein motility and transcription-factor binding in perspective, we introduce a few equations for diffusion as applied to conditions inside the bacteria. Although this is a very crowded solution of macromolecules, it will be assumed that protein motion can be characterized by a single diffusion constant D:

$$D = \frac{1}{2} v \cdot \ell \tag{5.1}$$

Here v is a characteristic velocity for the random thermal motion of the particle and l a characteristic length (the mean distance that the particle has to travel in order to randomize its direction due to collisions with solvent molecules). D may also be expressed as $\ell^2/(2\tau)$ where $\tau = \ell/v$ is the time between collisions, where the diffusing particle changes velocity. Therefore, the dimensions of D are length^2/time. For the protein albumin in water, the measured value is $D = 60\,\mu\mathrm{m}^2\,\mathrm{s}^{-1}$.

Inside a living cell diffusion is much slower than in pure water, with reported values of D varying between $0.1\,\mu\mathrm{m}^2\,\mathrm{s}^{-1}$ and $10\,\mu\mathrm{m}^2\,\mathrm{s}^{-1}$ (see Table 5.1). Inside an *E. coli* cell GFP[1] has been measured to diffuse with $D = 3\text{–}7\,\mu\mathrm{m}^2\,\mathrm{s}^{-1}$ [203]. D depends on the size of the diffusing protein, with $D = 7\,\mu\mathrm{m}^2/\mathrm{s}$ for 25 kDa proteins, $D = 1\,\mu\mathrm{m}^2/\mathrm{s}$ for 100 kDa proteins, and $D = 0.5\,\mu\mathrm{m}^2/\mathrm{s}$ for a 250 kDa protein [7] D may, in addition, depend on the transient binding time that the protein may have with the less mobile part of the interior of the cell. In particular, proteins may bind to chromosomal DNA, which, because of its size, diffuses much slower than proteins. For example, telomeric DNA in eukaryotic cells has a value of D between $0.0002\,\mu\mathrm{m}^2\,\mathrm{s}^{-1}$ and $0.0006\,\mu\mathrm{m}^2\,\mathrm{s}^{-1}$ [205].[2]

---

[1] GFP is a protein from jelly-fish that emits green light when exposed to UV, and therefore allows visual measurements using microscopy.

[2] Noticeably [205] the movement of telomeric DNA in mammalian cells is restricted to a volume of about $0.5\,\mu\mathrm{m}$ in radius, a small sub-section of the cell nucleus.

Responders	
ERCC1-XPC,PCNA, RAD51,RAD52, RAD54	$\sim 10\,\mu m^2\,s^{-1}$
NBS1, MDC1	$2\text{–}3\,\mu m^2\,s^{-1}$
Scanners	
Ku70, Ku 86, WRN, BLM	$0.1\text{–}0.4\,\mu m^2\,s^{-1}$
Chromatin binders	
Fibrallarin, HMG-17,SF2-AFS	$0.2\text{–}0.5\,\mu m^2\,s^{-1}$
UBF, Nucleolin, B53, Rpp29	$\sim 0.1\,\mu m^2\,s^{-1}$

Table 5.1 Diffusion constants for a sample of proteins in the cell nucleus of eukaryotic cells (measured on hamster cells or "Hela" cancer cells). From [204], which suggested a subdivision into freely diffusing "responders," and more constrained "scanners" with a diffusion that is reduced by transient bindings to DNA or chromatin.

In the equation

$$D = \frac{1}{2}v \cdot \ell = \frac{1}{2}v^2\tau \tag{5.2}$$

τ is given by the time it takes for the particle to randomize its movement entirely by collisions with other particles. This time is a factor of two larger than the decay time for the momentum p obtained from the last kick,[3] a time $(\tau/2)$ which sets the friction in the corresponding Langevin equation:

$$\frac{dp}{dt} = -\frac{2}{\tau}\cdot p - \frac{dV}{dx} + \text{noise} \tag{5.3}$$

For fast relaxation, the term $p/(\tau/2)$ dominates the dynamics and the inertia can be ignored ($dp/dt = 0$). Inserting $p = m \cdot v = m \cdot dx/dt$:

$$\frac{dx}{dt} = -\left(\frac{\tau}{2 \cdot m}\right)\frac{dV}{dx} = -\mu\frac{dV}{dx} \tag{5.4}$$

where $\mu = \tau/(2m)$ is the mobility. In these terms, D can be given an alternative interpretation: because of equipartition of energy associated with thermal noise, $\frac{1}{2}\langle mv^2\rangle = \frac{1}{2}k_BT$ and

$$D = \frac{1}{2}v^2\tau = \frac{1}{2}v^2 \cdot \mu \cdot 2 \cdot m = v^2 \cdot m \cdot \mu = (k_BT) \cdot \mu \tag{5.5}$$

The pre-factor, k_BT reflects the fact that high thermal noise gives faster speed between collisions and thereby faster diffusion. The proportionality between diffusion and mobility, Eq. (5.5), is called the fluctuation-dissipation theorem [207].

To put the above equations in perspective, consider a single amino acid, say, glycine, in water, with a diffusion constant $D = 1000\,\mu m^2\,s^{-1}$. According to the fluctuation-dissipation theorem, the corresponding timescale for dissipation is:

---

[3] To prove this, we follow [206] and consider the motion as a sequence of random steps of duration τ. Within each step the motion is collision-free and thus with constant acceleration $mdx^2/dt^2 = -(dV/dx)$ $\Rightarrow \Delta x = -(1/2m)\cdot(dV/dx)\tau^2 \Rightarrow \Delta x/\tau = -(\tau/2m)\cdot(dV/dx)$ which matches the equation for overdamped motion with timesteps $dt = \tau$.

$$\tau = \frac{2mD}{k_{\mathrm{B}}T} = 0.06 \cdot 10^{-12}\ \mathrm{s}$$

Accordingly, after 0.06 ps (picoseconds) a single amino acid will have forgotten its momentum, and its dynamics degenerate into first-order Langevin dynamics, where velocity is proportional to the force. Similarly, for a protein with $D = 60\ \mu\mathrm{m}^2\ \mathrm{s}^{-1}$ then $\tau \sim 1$ ps.

Diffusion is closely associated with random walks. Imagine a point particle on a one-dimensional lattice with lattice spacing ℓ. If at each time step τ we move one step to the right or left with equal probability, then the position of the particle at time t is:

$$x = \sum_{i=1}^{t/\tau} \eta(i) \tag{5.6}$$

where $\eta(i)$ takes values $\pm \ell$ randomly. Considering the ensemble-averaged square of the position (averaged over many copies of a random walker starting at $x = 0$ at time $t = 0$):

$$\langle x^2 \rangle = \left\langle \sum_{i=1}^{t/\tau} \eta(i) \sum_{j=1}^{t/\tau} \eta(j) \right\rangle$$

$$= \sum_{i,j=1}^{t/\tau} \langle \eta(i)\eta(j) \rangle$$

$$= \sum_{i=1}^{t/\tau} \langle \eta(i)\eta(i) \rangle$$

$$= \sum_{i=1}^{t/\tau} \ell^2 = (t/\tau)\ell^2 = \left(\ell^2/\tau\right) t = 2 \cdot D \cdot t \tag{5.7}$$

where we assume that steps at different times are uncorrelated (for example, $\langle \eta(1)\eta(2) \rangle = ((+1) \cdot (+1) + (+1) \cdot (-1) + (-1) \cdot (+1) + (-1) \cdot (-1))/4 = 0)$. Again D represents the diffusion constant, equal to the step size squared, divided with the time τ that it takes to move one step.

To describe the average behavior of a diffusing particle, one sometimes introduces the density $\rho(x,t)$ where $\rho(x,t)\mathrm{d}x$ is the probability that the particle is between x and $x + \mathrm{d}x$ at time t. Consider the current $J(x)$, counted in units of particles per second. The current is given by particles at positions $x - \ell/2$ and $x + \ell/2$ that move across position x during the time τ. The current of particles across position x is:

$$J = \frac{1}{2}\left(\rho(x - \ell/2)\frac{\ell}{\tau} - \rho(x + \ell/2)\frac{\ell}{\tau} \right)$$

$$= \frac{\ell}{2\tau}\left(\left(\rho(x) - \frac{\ell}{2}\cdot\frac{\rho(x)}{\mathrm{d}x} \right) - \left(\rho(x) + \frac{\ell}{2}\frac{\mathrm{d}\rho(x)}{\mathrm{d}x} \right) \right)$$

$$= -D \cdot \frac{\mathrm{d}\rho(x)}{\mathrm{d}x}$$

with $D = \ell^2/(2\tau)$. Here the factor $\frac{1}{2}$ is because only half of the particles at position $x - \ell/2$ move forward (and only half of those at position $x + \ell/2$ move backward). The

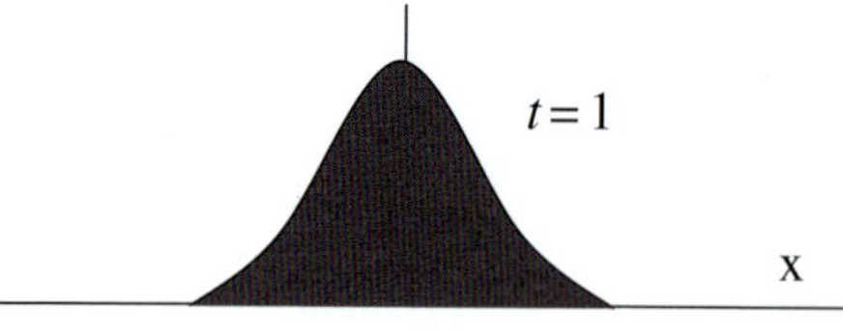

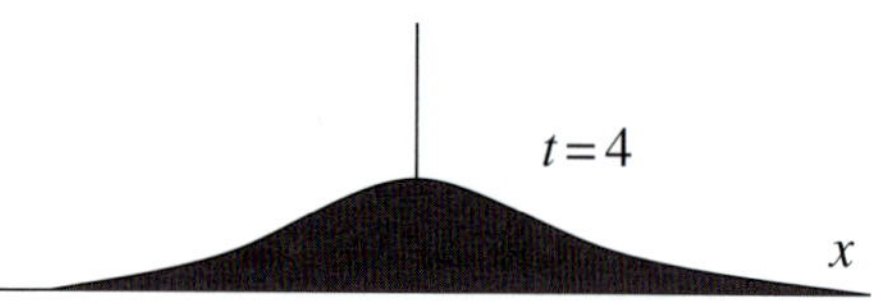

Figure 5.1 Diffusion of particles at position $x = 0$ when $t = 0$. When time $t \to 4t$, the width of the Gaussian curve is doubled compared to the distribution at time $t = 1$.

change in density is subsequently given by the difference between what moves in, and what moves out:

$$\frac{d\rho}{dt} = -\frac{dJ}{dx} = \frac{d}{dx}\left(D\frac{d\rho}{dx}\right) \tag{5.8}$$

In three dimensions the diffusion equation reads[4] (when D is independent of x):

$$\frac{d}{dt}\rho(\mathbf{r}, t) = D \cdot \left(\frac{d^2\rho}{dx^2} + \frac{d^2\rho}{dy^2} + \frac{d^2\rho}{dz^2}\right) \tag{5.9}$$

where $\rho(\mathbf{r}, t)$ is the density of particles at position $\mathbf{r}$ at time t. A particle that starts at $\mathbf{r} = \mathbf{0}$ at time $t = 0$, will at time t be found at position $\mathbf{r}$ with probability density

$$\rho(\mathbf{r}, t) = \left(\frac{1}{\sqrt{4\pi t D}}\right)^3 \cdot \exp\left(-\frac{\mathbf{r}^2}{4Dt}\right) \tag{5.10}$$

Technically, this equation is the solution of Eq. (5.9) with a very narrow Gaussian distribution as the initial condition for ρ. In other words the diffusion equation has the property that a Gaussian curve stays as a Gaussian curve, but with an ever-increasing width (see Fig. 5.1). Equation (5.10) propagates a δ-function at time $t = 0$ to a Gaussian distribution at time t.[5]

Equation (5.10) provides a typical time for diffusion across a distance d in the *E. coli* cell:

$$t_{\text{diffusion}} = \frac{d^2}{2D} \tag{5.11}$$

---

[4] Where the corresponding random walk takes one random step along each of the three dimensions at each time step.

[5] Diffusion can occasionally be modified by coupling its motion to other degrees of freedom, potentially causing anomalous diffusion (see review [208]).

This estimate is the time it takes before the density at position $x = d$ in Eq. (5.10) becomes substantial. Thus for a bacterial cell of length $d = 1\,\mu$m, a protein will typically have reached the opposite end after a time $t_{\text{diffusion}} = d^2/(2D) \sim 0.1$ s. Thus, for timescales larger than 0.1 s, all freely diffusing proteins should be homogeneously distributed inside the *E. coli* cell.

5.1.1 Questions

5.1.1 What is the timescale for homogenizing protein concentrations in a eukaryotic cell with diameter $20\,\mu$m. Assume that $D = 5\,\mu\text{m}^2\,\text{s}^{-1}$, or $D = 0.2\,\mu\text{m}^2\,\text{s}^{-1}$.

5.1.2 What time does it take to homogenize protein concentrations in a neuron by use of diffusion, connecting the brain with tissues that are 1 m away? Assume that $D = 5\,\mu\text{m}^2\,\text{s}^{-1}$.

5.2 Timescales for target location in a cell

5.2.1 On rates

Let us now consider the time it takes a molecule to find a binding site in a volume V if the only process is diffusion. This is defined as the time it takes the molecule to visit a specific reaction center, which we model as a small sphere of radius $\epsilon << r/2$, where r is the much larger radius of the cell volume $V = \frac{4\pi}{3} \cdot r^3$. Before calculating the binding time in detail, consider a simple dimensional argument (m is meters, s seconds):

- D with dimension: m^2/s^{-1}
- ϵ with dimension: m
- V with dimension m^3

The "on-time" τ_{on} should grow proportionally to the space that has to be searched (V). Preserving this proportionality, the only combination of the three quantities that has the right dimension is:

$$\tau_{\text{on}} = \frac{V}{D\epsilon} \tag{5.12}$$

This estimate is correct apart from a factor 4π that we will now argue for [33, 209]. Far away from the reaction volume there is a concentration of particles N/V, which is held constant. The reaction volume is a sphere of radius ϵ that is perfectly absorbing for the particles (see Fig. 5.2).

Consider the density ρ that in fact will increase with distance r from the reaction volume in the center of the cell. Increasing ρ corresponds to a negative flux, which again reflects a flux directed towards the absorbing reaction volume at small $r \sim \epsilon$. For any distance r, far away from the reaction volume, $r > \epsilon$, the steady-state flux away

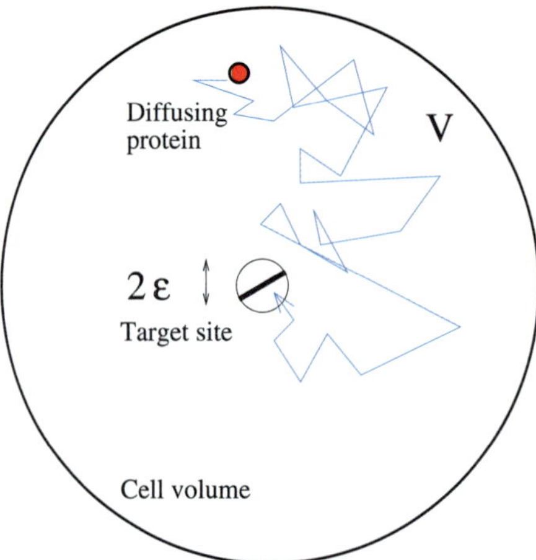

Figure 5.2 A diffusing particle in a volume V will at some time enter a smaller subvolume of radius ϵ. In the text we estimate the time it takes for the first encounter with this smaller volume. This time sets the diffusion-limited time for reaction between the particles in a cell volume, where ϵ is the radius of the reaction volume.

from the absorbing reaction volume is given by[6]:

$$J = -D\, 4\pi\, r^2\, \frac{d\rho}{dr} \tag{5.13}$$

Here, J is the current integrated over a spherical shell of radius r, hence the factor $4\pi r^2$ where the units of J are particles s^{-1} (see Fig. 5.2 for an illustration of the idealized geometry). The simplest integration of this equation around a target of size $r = \epsilon$ with an estimated gradient given by $d\rho/dr \sim (N/V)/\epsilon$ gives:

$$J = \frac{1}{\tau_{\text{on}}} = 4\pi D\epsilon \cdot \frac{N}{V} \tag{5.14}$$

This expression is called the Smoluchowski equation (for a formal proof, see Question 4.2.1). For a cell of volume $V = 1\,\mu m^3$, the diffusion-limited encounter time with a region of radius $\epsilon \sim 1 \rightarrow 10$ nm is about:

$$\tau_{\text{on}} \sim (2\text{--}20)s/N. \tag{5.15}$$

where the number N of an given protein type in an *E. coli* cell is between 1 and 10^4.

For RNAP from *E. coli* the measured *in vitro* on-rates to a typical promoter is consistent with $\epsilon = 0.75$ nm [69]

---

[6] The diffusion equation 5.10 in spherical coordinates is given by:

$$\frac{d\rho}{dt} = \frac{1}{r^2} D \frac{d}{dr}\left(r^2 \frac{d\rho}{dr}\right)$$

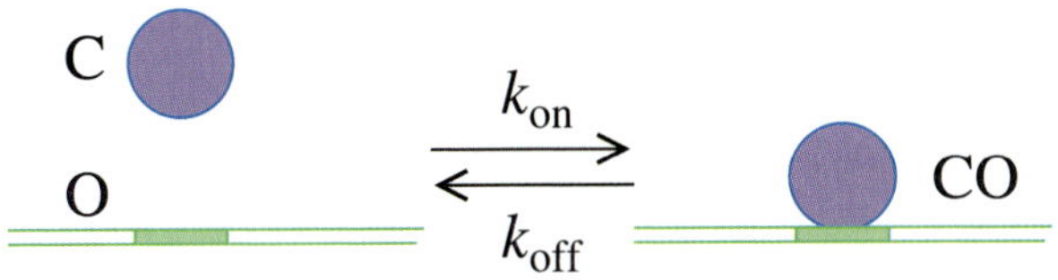

$$k_{on}[C] = \frac{1}{\tau_{on}} = \frac{4\pi D\varepsilon N}{V}$$

$$\frac{k_{off}}{k_{on}} = K = [1\,M]\,\exp(\Delta G/k_B T)$$

Figure 5.3 The on rate from chemistry and its idealized relation to the diffusion-limited on rate. Notice the concentration $[\rho] = N/V$.

5.2.2 Off rates

We can now address the off rate $k_{off} = 1/\tau_{off}$ associated with establishing an equilibrium distribution. The equilibrium chemistry is obtained by balancing the two rate constants k_{on} and k_{off} where $K = k_{off}/k_{on}$ is the dissociation constant that is measured in molar (M). Notice that the on-rate constant in chemistry is measured in numbers per molar per second! Thus, the on rate from chemistry has to be multiplied by the concentration of the target, in order to count the on rate to a single binding site. That is, $k_{on} \cdot$ concentration is the real on rate $\left(\frac{1}{\tau_{on}}\right)$ for a given binding site (see Fig. 5.3). Accordingly, the relationship between k_{on} and the above-mentioned τ_{on} is:

$$\frac{1}{\tau_{on}} = k_{on} \cdot \frac{N}{V} \tag{5.16}$$

From the assumption of a diffusion-limited on rate we derive:

$$\frac{1}{\tau_{on}} = \frac{4\pi D\epsilon N}{V} \tag{5.17}$$

and thus obtain an estimate of the chemical constant:

$$k_{on} = 4\pi \cdot D \cdot \epsilon \tag{5.18}$$

Using equilibrium chemistry:

$$\frac{k_{off}}{k_{on}} = K = [1\,M] \cdot \exp(\Delta G/k_B T) \tag{5.19}$$

where $[1\,M]$ again refers to the unit of molar that is in front of any binding constant. The off rate is then:

$$\frac{1}{\tau_{off}} = k_{off} = k_{on}[1\,M]\exp(\Delta G/k_B T)$$
$$= 4\pi D\epsilon\,[M]\exp(\Delta G/k_B T) \tag{5.20}$$

provided that the on rate is diffusion limited, with ϵ quantifying the size of the binding site.

We also introduced [1 M] to emphasize that $D\epsilon$, with formal unit length^3/s, where volume is counted in M^{-1} (= 1.6 nm^3).[7] If one assumes $\epsilon = 5$ nm, and $D = 5\ \mu m^2\ s^{-1}$, then:

$$\frac{1}{\tau_{\text{off}}} \sim 4\pi \cdot \left(5\ \mu m^2\ s^{-1}\right) \cdot (0.005\ \mu m) \cdot \frac{6 \cdot 10^{23}}{(10^{-3}\ m^3)} \cdot \exp\left(\frac{\Delta G}{k_B T}\right)$$

$$\sim 2 \cdot 10^8 s^{-1} \cdot \exp\left(\frac{\Delta G}{k_B T}\right) \tag{5.21}$$

where the pre-factor corresponds to one escape attempt per 5 ns. Thus, without binding the off rates are very fast.

For a binding energy $\Delta G = -12\ \text{kcal mol}^{-1}$, the expected time until a protein leaves is:

$$\tau_{\text{off}} \sim 2\ s$$

reflecting the typical residence time of one molecule on a medium-strong binding site on the DNA.

5.2.3   Facilitated target location

The time a transcription factor uses to finds its specific operator site on the DNA may well be inhibited by non-specific binding, which also allows the protein to diffuse *along* the DNA for some time [210]. Remarkably, the measured "on rates" for specific sites on DNA in test tubes can increase by up to a factor of 20 when the length of the flanking DNA varies from 20 to about 800 base pairs [211] (see Fig. 5.4). In contrast, locating a specific target inside the living cell presents the challenge of searching very long DNA.

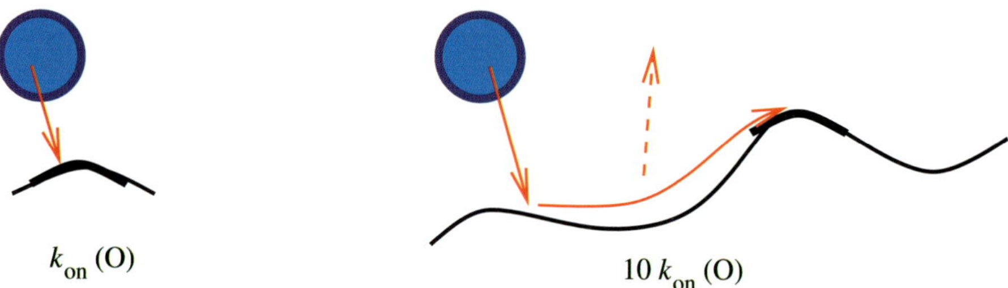

Figure 5.4 Facilitated target search *in vitro*. A protein searching for a specific operator site (O) is "helped" by flanking DNA, as the original three-dimensional search on the left is replaced by an easier search for the larger DNA, followed by a one-dimensional search along the flanking DNA. One typically observes increased on rates, k_{on} by a factor of up to 20. This observed upper limit will in part be set by the non-specific binding strength, which naturally limits the time the protein can stay on the flanking DNA.

---

[7] If each particle occupies a volume of 1.6 nm^3 then we have 1 M concentration of the particles. For comparison, one protein with a diameter of 5 nm occludes a volume of ~ 100 nm^3.

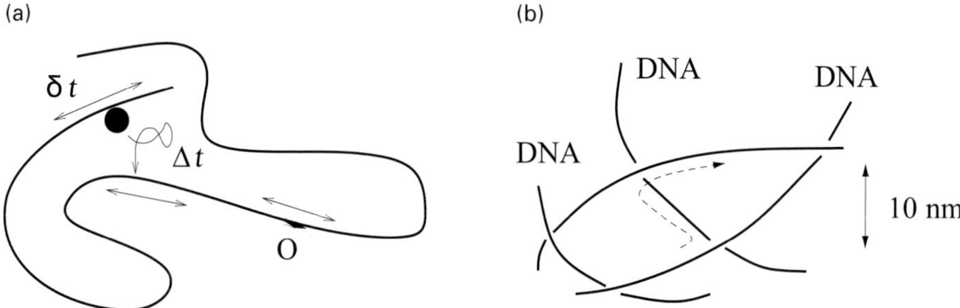

Figure 5.5 A protein searching for a specific operator site inside a cell. (a) The protein both diffuses along the DNA, and diffuses in three-dimensions. The ratio between the time used in these states is given by non-specific binding. When this binding is small, the overall search time approaches the free search time. When non-specific binding is large, the search is slowed down because most time is spent on repeatedly searching along the same part of the chromosome. (b) *In vivo* facilitated search, where no time is spent in three-dimensions, and the one-dimensional search only involves small segments. With typical *in vivo* DNA density, parts of the DNA that are separated by hundreds of thousands of base pairs may be close in three dimensions, allowing for direct and fast jumps between different DNA regions.

As a consequence, a process that saves time under *in vitro* conditions can become time-consuming inside the cell.

Let Δt denote the typical time for one transient visit to the DNA. For $\Delta t \to \infty$ the protein is always bound to the DNA and the search is purely one-dimensional with a search time:

$$t_{\mathrm{on}}(1\mathrm{D}) \sim \frac{L^2}{D_1} \tag{5.22}$$

where D_1 is the diffusion constant for the protein along the DNA. One would expect $D_1 < D \sim 5\,\mu\mathrm{m}^2\,\mathrm{s}^{-1}$,[8] but even in the case where diffusion along the DNA is fast, the time to locate the specific site (in *E. coli* with 1.5 mm of DNA) will be of the order of $t_{\mathrm{on}}(1\mathrm{D}) = (1.500\,\mu\mathrm{m})^2/(5\,\mu\mathrm{m}^2\,\mathrm{s}^{-1}) = 400.000\,\mathrm{s} \sim 5$ days. This is a very long time, leaving plenty of time for jumps. In fact, the time it takes to find a specific site depends on the time spent serching along one stretch of DNA, before leaving it to be recaptured by another region of the DNA (see Fig. 5.5).

An estimate for the size of the facilitating DNA region is the $\Delta x \sim 800$ bp $\sim 0.2\,\mu\mathrm{m}$ DNA that was the maximum flanking region found to increase the on rate [211]. To scan this segment by diffusion takes time $\Delta t \sim (0.25\,\mu\mathrm{m})^2/D_1 \sim 0.01$ s if $D_1 = D \sim 5\,\mu\mathrm{m}^2\,\mathrm{s}^{-1}$. To scan the entire $L = 5 \cdot 10^6$ bp *E. coli* genome, the protein needs of the order of $L/\Delta x$ scans of size Δx. This will take a "facilitated" search time of about:

$$t_{\mathrm{on}} = \left(\frac{L}{\Delta x}\right) \cdot \Delta t \sim 70\,\mathrm{s} \tag{5.23}$$

---

[8] For MutS, a mismatch repair protein that dimerizes to form a ring around the DNA, and then diffuses along the DNA. *In vitro* the corresponding diffusion constant is measured to be $D_{\mathrm{muts}} = 0.1\,\mu\mathrm{m}^2\,\mathrm{s}^{-1}$, which allows MutS to facilitate information transfer across several thousands of base pairs along the DNA within a few minutes.

plus the insignificant time[9] spent in jumping between the $L/\Delta x$ different segments. Therefore the pool of transcription factors is fastly reshuffled between being free floating and binding to non-specific DNA.

If, for example, 90% of the transcription factor is bound to non-specific DNA, a total on rate $= 10^5 \, \text{s}^{-1}$ must be accompanied by an off rate that is about $= 10^4 \, \text{s}^{-1}$. This rate is in itself not enough to secure scanning of the $\Delta x \sim 800 \, \text{bp}$ assumed in Eq. (5.23). In fact, $\delta t = 10^{-4} \, \text{s}$ only allows us to scan a segment of length $\Delta x \sim 80 \, \text{bp}$, giving a total search time:

$$t_{\text{on}} = \left(\frac{L}{\Delta x} \right) \cdot \frac{\Delta x^2}{D_1} \sim \frac{L \Delta x}{D_1} \sim 7 \, \text{s} \tag{5.24}$$

if we again assume that $D_1 = D$.

Overall, "facilitated" search times from Eq. (5.23) in principle may be comparable to three-dimensional searching without non-specific binding. Also one can see that an eventual slowing down will be of the order of $\sim D_1/D$, implying that a copy number of about 100 proteins may still give an on rate for a specific operator of about one per second. At least this will be the case if diffusion along the crowded chromosomal DNA[10] is not several orders of magnitude slower than diffusion in the cytoplasm [212].

5.2.4 Questions

5.2.1 Follow and supplement the formal proof below for diffusion-limited target location as illustrated in Fig. 5.2. In a steady state the current J is independent of r. Use Eq. (5.13) to argue:

$$\rho(r) = \frac{J}{D 4 \pi r} + \rho(\infty)$$

Using the concentration at infinite $\rho(\infty) = N/V$ and $\rho(\epsilon) = 0$ (why are these sensible limits?) one may obtain:

$$\frac{-J}{D 4 \pi \epsilon} = \frac{N}{V} \Rightarrow |J| = 4 \pi \epsilon D \frac{N}{V} \Rightarrow \tau_{\text{on}} = \frac{1}{|J|} = \frac{V}{4 \pi D \epsilon N}$$

5.2.2 Derive the Smoluchowsky equation in two dimensions (use the two-dimensional analog $J = -2\pi r d\rho/dr$ with $\rho(\infty) = N/A$ counting particles per area).

---

[9] The time to jump from one part of the DNA to another depends on density of the DNA. As any part of the chromosomal DNA is a target, the rate to reach the total length L of DNA can thus be estimated by summing up the rates to reach all parts:

$$\text{d}(rate) \sim \frac{4\pi DdL}{V} \Rightarrow \text{rate} \sim \frac{4\pi DL}{V} \Rightarrow \delta t = 1/\text{rate} \sim \frac{V}{4\pi DL}$$

giving $\delta t \sim 10^{-5} \, \text{s}$ for $L = 1.5 \, \text{mm}$ and $V = 1 \, \mu\text{m}^3$.

[10] As many proteins bind weakly to DNA, the diffusion of any particular protein is slowed down due to crowding. Assuming fast reshuffling of these other proteins, the crowding may be taken into account by replacing $D_1 \to D_1 \cdot (1 - \rho)$, where ρ is the density of other DNA-binding proteins in units of maximum possible coverage of the DNA.

5.2.3 Plot τ_{off} as a function of binding energy ΔG, and find ΔG where τ_{off} equals the *E. coli* cell generation time (say, 1 hour).

5.2.4 Simulate a random walker on a lattice, confined to a box represented by integer positions between -5 and 5 (included), in all six directions: i.e. the particle at (x,y,z) can move one step in any one of six directions, except when this move brings it outside the box. Compare the time to find point $(0,0,0)$, with the time to find any point in the cube $[-1:1]^3$ and also to find the $(0,0,0)$ point in $[-2:2]^2$. First use an initial position at the border and investigate all three system sizes. Then simulate the largest system when starting at $(0,0,0)$.

5.2.5 Argue that the escape rate to a distance R away from a one-dimensional DNA strand before recapturing scales as: $r \propto \frac{2\pi Dl}{ln(R/b)}$, where b is the diameter of the DNA, l is the length of the DNA and where one assumes that the touching distance $= b$ always leads to absorption (this is the diffusion-limited case, reaction is instant when possible).

5.2.6 DNA mismatch repair generally repairs about 99% of single nucleotide mismatches after DNA replication. It involves the protein MutS which dimerizes around the mismatch and thereafter, in *E. coli*, diffuses along the DNA until a hemi-methylated GATC site is located. This "half-methylated" site serves as a detector for new DNA (which initially is non-methylated). Estimate the minimum diffusion constant that one MutS needs to have in order to reach a GATC site within a characteristic time interval of 2 minutes (the time it takes the protein Dam to methylate the GATC site). Allow for a maximum of 1% failure rate and assume that it starts at a position that is 128 base pairs from a GATC site. Hint: with a characteristic time window of 2 min there will be 1% probability that the GATC site is methylated within 1 s.[11]

5.2.7 Simulate the problem in question 5.2.3, starting a random walk at position $x = 0$ and let it walk until it first passes position $+128$ or position -128. Use the simulation to deduce the diffusion constant when one assumes a 99% chance of reaching ± 128 before they are invisible, when methylation of each of these positions occurs at a rate of 0.01s^{-1}.

5.3 Traffic on DNA

DNA is not only the object for binding/unbinding events and simple diffusion. It is also a one-dimensional highway with substantial directed traffic. In fast-growing *E. coli*

---

[11] In *E. coli*, the hemi-methylated GATC site is identified by the MutL–MutH complex, which binds to the DNA-diffusing MutS dimer and "nicks" (= cuts) one DNA strand at the hemi-methylated site. Subsequently the protein UvrD unwinds the DNA, thereby allowing synthesis of new DNA around the GATC site and including the mismatch site. Interestingly, the loss of any of MutS, MutL or MutH allows a mutator mutant of the bacteria, which can eventually be reverted by horizontal transfer of genes from other bacteria.

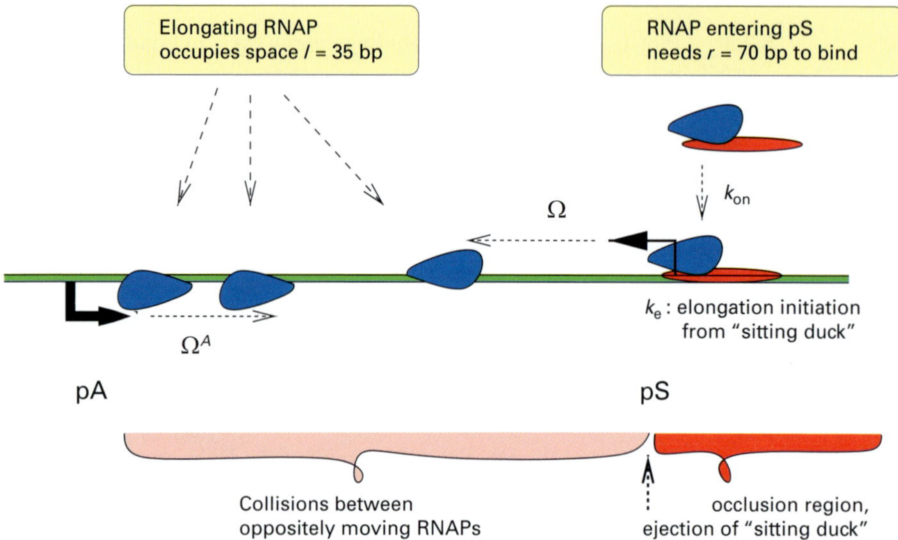

Figure 5.6 Promoter interference between an aggressive promoter pA that interferes with the activity of promoter pS by "firing" RNAP though it. There are several mechanisms for such interference: occlusion, collisions between moving RNAPs and collisions between an RNAP from pA and an RNAP bound to the pS. For small distances L between the convergent promoters, the last interference mechanism denoted by "sitting duck," becomes largest.

a DNA polymeraze (DNAP) moves across any point every 25 minutes, disrupting any protein–DNA complex, and splitting the DNA into separate strands. Other molecules are in constant action, including gyrases and topo-isomerases, that maintain the topological properties of DNA. Finally DNA is constantly transcribed by RNAP that moves along the DNA, while separating the DNA strands locally and transcribing them into RNA. This last traffic aspect of transcription allows regulation [213, 214], where one promoter on one strand can initiate an RNAP that eventually is bound to promoters on the opposite strand (see Fig. 5.6); thereby opposing promoters can repress each other's activity.

To model promoter interference, consider first an isolated promoter, with properties determined through the "on rate," k_{on} and the "firing rate" k_e, from a so-called "sitting duck" complex [215]. The "sitting duck" complex is similar to the open complex described in Chapter 2, as it represents a state of the RNAP that can only leave the promoter by either transcription initiation or by active direct ejection by another RNAP.

The two-step transcription initiation model includes: (1) a one-way formation of the sitting duck complex with rate k_{on},[12] and (2) a one-way initiation of elongation from the sitting duck complex with rate k_e ("e" for elongation).

The average occupancy of this complex θ (dimensionless, just counting the probability of being there) and the total activity Ω (in units of RNAP leaving the promoter per

---

[12] In terms of the three-step model presented in Chapter 2, k_e is the elongation initiation from the open complex, whereas $k_{on} = 1/k_o + (1 + (k_u/k_b))/(k_b[\text{RNAP}])$ represents the total on rate to the open complex, given the free [RNAP] concentration in the cell.

second) of the undisturbed promoter can be calculated from conservation of occupation at steady state:

$$\text{What comes in to a state} = \text{What goes out from this state}$$

giving the steady state for RNAP on the undisturbed promoter:

$$k_{\mathrm{on}}(1 - \theta) = k_{\mathrm{e}}\theta \quad \text{with} \quad \Omega = k_{\mathrm{e}}\theta \tag{5.25}$$

The $(1 - \theta)$ term takes into account that RNAP can only bind to pS if there is no "sitting duck." Notice that θ is a probability (dimensionless), counting the fraction of an ensemble of promoters that will have RNAP sitting on them. Similarly $1 - \theta$ is the fraction of the promoters that are free. Obviously these fractions depend hugely on the on and elongation rates of RNAP on the promoters.

Equation (5.25) predicts the average occupancy of an undisturbed promoter:

$$\theta = \frac{k_{\mathrm{on}}}{k_{\mathrm{e}} + k_{\mathrm{on}}} \quad \text{and} \quad \Omega = k_{\mathrm{e}}\theta = \frac{k_{\mathrm{on}}\,k_{\mathrm{e}}}{k_{\mathrm{e}} + k_{\mathrm{on}}} \tag{5.26}$$

This teaches us that a strong promoter needs to have both a large on rate, k_{on}, and a large firing rate k_{e} once RNAP is on the promoter. The above equation can also be expressed in terms of times:

$$\frac{1}{\Omega} = \frac{1}{k_{\mathrm{on}}} + \frac{1}{k_{\mathrm{e}}} \tag{5.27}$$

stating that the average time between promoter initiations is the sum of the time it takes to form an open complex, plus the time it takes to initiate elongation.

In the above equations we have ignored that RNAP has a finite length, and thus that it also occupies a promoter for some time ($\sim \Omega l/v$, $l \approx 35\,\mathrm{bp}$ is the length of elongating complex, $v \sim 40\,\mathrm{bp\,s^{-1}}$, its velocity) after the elongating complex has left.

Consider now an antagonistic promoter pA firing into the above-modeled promoter (see Fig. 5.6). Any RNAP sitting on pS can be destroyed with a probability given by the total firing rate of this antagonistic promoter Ω^{A}. We here further assume that $\Omega^{\mathrm{A}} >> \Omega$, such that we can ignore changes in effective firing of pA due to the activity of pS. Equation (5.25) is therefore changed to:

$$k_{\mathrm{on}}(1 - \theta) = (k_{\mathrm{e}} + \Omega^{\mathrm{A}})\theta \quad \text{with} \quad \Omega = k_{\mathrm{e}}\theta \tag{5.28}$$

This equation again uses steady state, where the flux of RNAP into the promoter region is equal to the flux away from the promoter. Here, however, there are two ways to leave the promoter, either as a transcribing RNAP or as an RNAP that is pushed away by a collision with an RNAP from promoter pA. Notice that both of these processes occur with a rate proportional to θ. The average occupation of the promoter is:

$$\theta = \frac{k_{\mathrm{on}}}{k_{\mathrm{e}} + k_{\mathrm{on}} + \Omega^{\mathrm{A}}} \tag{5.29}$$

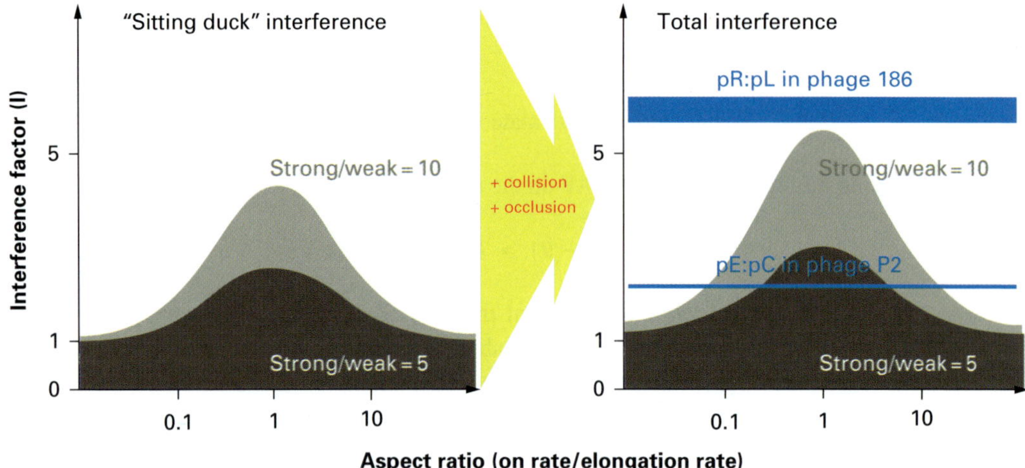

Figure 5.7 Transcription interference as a function of relative promoter strength Ω^A/Ω and aspect ratio $\alpha = k_{on}/k_e$ of the sensitive promoter. The right panel includes corrections from occlusion and collision. The horizontal cyan areas show two promoter combinations from 186 and P2 phage respectively: $\Omega_{PR} = 1/18$ s, $\Omega_{PL} = 1/180$ s for 186, and $\Omega_{PE} = 1/36$ s $\Omega_{PC} = 1/360$ s for phage P2. Fitting the PR/PL promoters requires maximum interference, thus limiting the aspect ratio to $\alpha = 1$. Fitting the much weaker interference for the *PE:PC* data implies that the aspect ratio has to be either small or large, $\alpha = 1/25$ or $\alpha = 25$.

with an activity:

$$\Omega(\text{with } pA) = k_e\theta = \frac{k_{on}\, k_e}{k_e + k_{on} + \Omega^A} \tag{5.30}$$

This equation describes the main promoter interference, where a stronger promoter pA will reduce the activity of a given promoter with a factor (see Fig. 5.7):

$$\mathcal{I} = \frac{\Omega(\text{without } pA)}{\Omega(\text{with } pA)} = \frac{k_{on} + k_e + \Omega^A}{k_e + k_{on}} \tag{5.31}$$

The interference for promoters in phage 186 is found to be $\mathcal{I} = 5.6$ [214], where the interfered promoter is a factor of 10 weaker than the aggressive promoter pA, $\Omega = \Omega^A/10$. Equation (5.31) predicts somewhat less interference, but it also ignores effects associated with occlusion, due to collisions of RNAP from pA with RNAP that have left the sitting duck complex, and also corrections associated with the correlations between subsequent firings of RNAP from pA [215]. Occlusion may also be increased by the positive supercoil that is induced ahead of the transcribing polymerase [216].

The derived interference factor can also be rewritten as:

$$\mathcal{I} = 1 + \frac{\Omega^A}{k_e + k_{on}} = 1 + \left(\frac{\Omega^A}{\Omega}\right) \cdot \frac{\alpha}{(1+\alpha)^2} \tag{5.32}$$

where the promoter aspect ratio $\alpha = k_{on}/k_e$, (see Fig. 5.7). The remarkable symmetry $\mathcal{I}(\alpha) = \mathcal{I}(1/\alpha)$ means that interference is weak, both in the case where the RNAP spends most time in binding to the promoter, and in the opposing case where the RNAP spends most of the time initiating transcription from an easily reached "sitting duck" state on the promoter.

Promoter interference is often found in the regulation of the repressor protein CI in all P2-like phages. In addition, it is also active in initial lysis–lysogeny decision establishment of lysogens in λ-like phages through PR–PRE interference, and thus in general it plays a role in all phages that have a CII type of protein to regulate entry into lysogens. In *E. coli* there are about 100 promoters that are known to be placed face to face at fairly close distances [215], making promoter interference part of the regulation of $\sim 5\%$ of the promoters. Finally, promoter interference may also act when promoters initiate transcription in parallel, as seen in the elegant demonstration of promoter occlusion by Adhya and Gottesman [213].

5.3.1   Questions

5.3.1   Reconsider Eq. (5.26), taking into account that RNAP needs time $\Omega l/v$ to leave the promoter before a new RNAP can bind to it. Let $l = 35$ bp be the length of an elongating RNAP and $v \sim 40\,\mathrm{bp\,s^{-1}}$ the velocity of the RNAP. At what promoter firing strength does this correction become more than a factor of two?

5.3.2   Show that $\mathcal{I} = 1 + (\Omega^{\mathrm{A}}/\Omega) \cdot \alpha/(1+\alpha)^2$, where $\alpha = k_{\mathrm{on}}/k_{\mathrm{e}}$ is the aspect ratio of the promoter. Which value of α gives maximal interference? Discuss why interference decreases for both very small and very large α.

5.3.3   Consider occlusion, the fact that an entering RNAP from pA prevents an RNAP bind to pS for a time given by $l + r = 35 + 75$ bp (see Fig. 5.6). ($r = 75$ bp is the length an RNAP needs to bind to a promoter.) What is the interference factor $\mathcal{I}$ if one includes this occlusion effect? Calculate $\mathcal{I}$ for $\Omega = k_{\mathrm{on}}/2 = k_{\mathrm{e}}/2 = 0.01\ \mathrm{s^{-1}}$, $\Omega^{\mathrm{A}} = 0.1\ \mathrm{s^{-1}}$ and $v = 40\mathrm{bp\,s^{-1}}$. What is the interference if both promoters are four times stronger?

5.3.4   Assume that RNAP from pS (see Fig. 5.6) has to travel a distance $L - 40$ before it has escaped possible collision with RNAP from pA. How does $\mathcal{I}$ change with increasing L?

5.3.5   Implement a stochastic model for promoter interference on a computer, including only the collision effect. Thus, each promoter starts one RNAP at each of their respective initiation rates, and each RNAP moves one step at a time towards the right, or left respectively. RNAP are assumed to fall off when colliding. Use a promoter strength of pA $= 0.1\,\mathrm{s^{-1}}$ and pS $= 0.05\,\mathrm{s^{-1}}$ and an RNAP speed of $40\,\mathrm{bp\,s^{-1}}$. Calculate the promoter activity for distances $L = 100$ bp, 1000 bp and 2000 bp. Compare with the analytical predictions from the previous question. *Hint:* Use $dt = 0.01$ s as the time unit, and let each promoter initiate a new RNAP with probability $dt \cdot pA$, or $dt \cdot pS$. At each time step move the RNAPs in their specified direction with probability $v \cdot dt = 0.4$. One may speed up the simulation by implementing steps of 10 bp, thus using time steps $dt = 0.1$ and accordingly adjusted promoter initiation probabilities per time step.

5.4   Bursty transcription initiation

5.4.1   Correlations in transcription initiation

Noise in intracellular concentrations of various protein species has been highlighted in a number of natural and engineered genetic circuits, including the λ circuit. Gene expression is perhaps the most important stochastic process in the cell. Transcription involves the production of small numbers of mRNAs, which are then translated multiple times [217], creating and amplifying noise in protein concentrations.

If transcription initiations are uncorrelated, the noise in final expression will diminish proportionally to $1/\sqrt{n}$, where n is the average number of transcription events within the lifetime of the observed molecule (here mRNA) [218]. These natural bursts associated with the amplification step of translation is itself large [219] (see Fig. 2.10). However, if transcription events are correlated, the fluctuations can become even larger. Such a variable rate of initiation will produce substantial fluctuations that can lead to heterogeneous behavior across populations of genetically identical cells. It is widely speculated that this variability may allow populations of unicellular organisms to cope with variable and unpredictable environments.

It is in this context that a study by Golding *et al.* [220], shown in Fig. 5.8, is interesting. Golding and colleagues used a system where an mRNA wih multiple MS2 binding sites could be visualized by binding many copies of MS2-GFP fusion proteins to the mRNA. This allowed for detection and counting of individual mRNAs inside an *E. coli* cell. This detectable mRNA was subsequently placed under the control of the $P_{lac/ara}$ promoter. When this promoter was fully induced, the transcription of the mRNAs was observed to occur in an unexpected fashion, with very long periods of quiesence that separed bursts of transcription, where several mRNAs were produced in rapid succession. This phenomenon is called transcriptional bursting. The bursts of activity (on-periods) were exponentially distributed, with a mean of 6 minutes. During on-periods, an exponentially distributed number of transcripts are produced, characterized

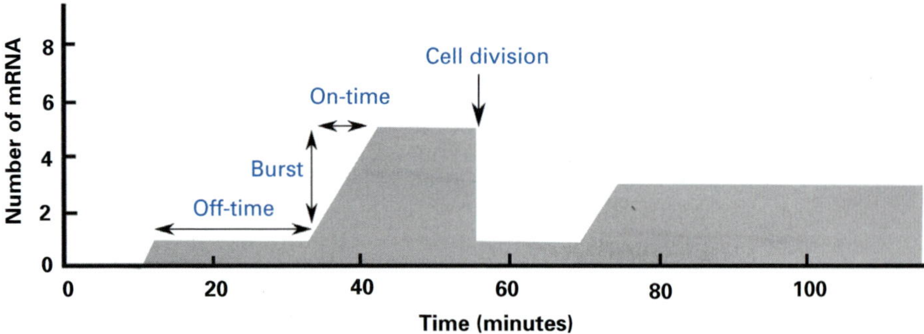

Figure 5.8 Activity from fully induced $P_{lac-ara}$, as reported by [220]. Bursty promoter activity is seen, even in the absence of any transcription factors. This suggests that an elongating RNAP leaves the vacated promoter in a transient state that is particularly open for recruitment of a new RNAP [221]. Figure redrawn from [221], printed with permission from Elsevier.

by a mean of 2.2 transcripts per on period. The long periods without transcription (off periods) were also exponentially distributed, with a mean period length of 37 minutes.

5.4.2 CV and the Fano factor

It is often difficult to measure directly the "bunchiness" of the transcription process. Instead one can quantify the level of noise in terms of two factors. First there is the spread to mean of a signal S, also called the coefficient of variation:

$$\text{CV} = \frac{\text{spread}}{\text{mean}} = \frac{\sqrt{\langle S^2 \rangle - \langle S \rangle^2}}{\langle S \rangle} \tag{5.33}$$

which can be used in all cases. Typically $\text{CV} \propto 1/\sqrt{n}$ for n independent events. Thereby the spread to mean is a measure for the effective number (n) of independent events that form the signal S. Examples of CVs for plasmid expression systems in *E. coli* are shown in Fig. 5.9.

When the signal S is a discrete number of molecules, the Fano factor:

$$\text{Fano} = \frac{\left(\langle S^2 \rangle - \langle S \rangle^2\right)}{\langle S \rangle} \tag{5.34}$$

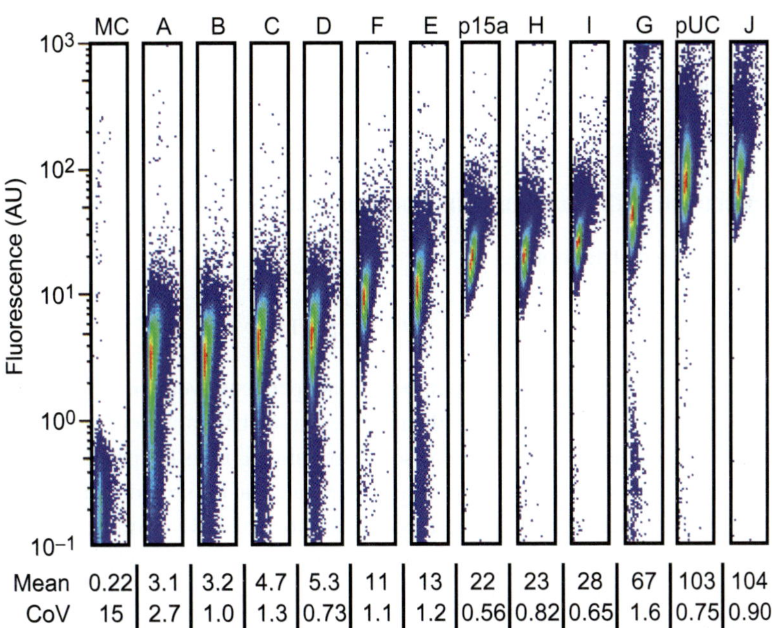

Figure 5.9 Distribution of fluorescent protein expressed from promoters on various plasmid systems in *E. coli*. Figure is taken from [222], which also reports the mean number of plasmids per *E. coli* genome for the different plasmid systems (see first row below figure). The second row shows the coefficient of variation (CV). Notice that some plasmids have huge expression variations, presumably reflecting a large variation in the number of plasmids between different cells.

measures the size of the single events that produced these molecules. For example, Fano = 1 for a simple Poisson process where one molecule is produced for each event. In contrast, Fano = $s > 1$ when $s > 1$ molecules are for produced each of $\langle S \rangle / s$ independent events.

5.4.3 Distribution of time to the next random event

This is relevant when, for example, one considers the number of proteins from individual mRNAs. The mRNA decays stochastically with a lifetime that is exponentially distributed. As translation initiation occurs at a nearly constant rate throughout the lifetime of the mRNA, the number of proteins per mRNA also becomes exponentially distributed. The exponential distribution has the form:

$$P(n) = \frac{1}{\lambda} \cdot \exp(-n/\lambda) \tag{5.35}$$

and is characterized by:

$$\langle n \rangle = \lambda$$
$$\langle n^2 \rangle - \langle n \rangle^2 = \lambda^2$$
$$\text{Fano factor} = \lambda \tag{5.36}$$
$$\text{spread/mean} = 1 \tag{5.37}$$

Thus an exponential has a Fano factor that grows with the scale λ of the exponential decay.

5.4.4 Distribution in the number of independent events

The Poisson distribution:

$$P(n) = \frac{\lambda^n}{n!} \exp(-\lambda) \tag{5.38}$$

describes the distribution of a number of independent events n, given that the average expectation is λ. Thus, whereas the exponential distribution concerns the waiting time between subsequent events, the Poisson distribution counts the probability that a number of these events occur within a given time interval. The Poisson distribution, for example, describes the distribution of the expected number of mRNA transcripts n where these transcripts are produced randomly and independently of each other [223].

The Poisson distribution with mean λ has the statistics:

$$\langle n \rangle = \lambda$$
$$\langle n^2 \rangle - \langle n \rangle^2 = \lambda$$
$$\text{Fano factor} = 1 \tag{5.39}$$
$$\text{spread/mean} = 1/\sqrt{\lambda} \tag{5.40}$$

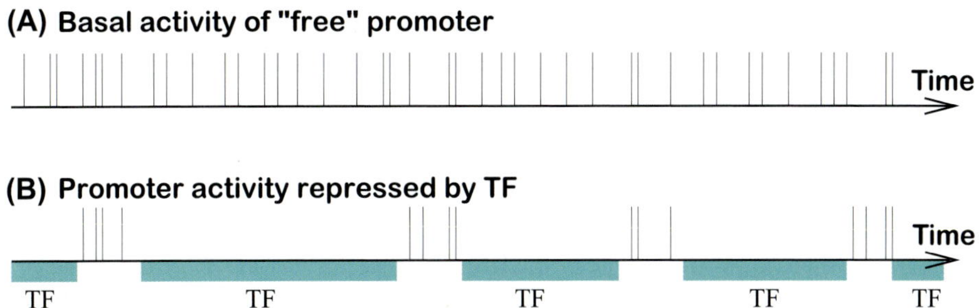

Figure 5.10 Schematic diagrams for the time sequence of promoter activity for a bare promoter (A) and a promoter with a transcription factor (TF) that represses gene expression (B). The vertical lines represent the times when transcriptions are initiated. The shaded(cyan) intervals in panel (B) mark when a TF is bound to the operator site, thus repressing its activity. The transcription bursts take place in the periods when the TF does not bind, provided that the on rate for the TF is smaller than the on rate for the RNAP.

Thus, effectively, the Poisson distribution gets narrower as λ increases. However, the variance remains constant in units of average expectation, making this measure particularly suited for characterizing transcriptional noise (under the condition where one can count individual messages).

Assume, for example, that mRNA has a Poisson distribution and that each transcript produces exactly s proteins. Thereby the distribution of number of proteins $S = x \cdot s$ will be determined by $P(x) = \frac{\lambda^x}{x!} \exp(-\lambda)$ with $x = 0, 1, 2.....$ Therefore there is, on average, $s \cdot \lambda$ proteins produced, whereas the variance in this number is $s^2\lambda$ and the Fano factor $f = s$.[13] Fano factors for individual mRNA levels are reported to be between 1 and 2 in the genome-scale experiment of [65].

Golding *et al.* [220] reported a Fano factor of about 4 for the number of mRNA transcripts from the isolated PRM promoter in phage λ. If each mRNA only generates one protein, the Fano factor for protein production would similarly be 4. If, on the other hand, each mRNA was translated into 10 proteins, then proteins would be produced in bursts of characteristic size $4 \times 10 = 40$, and the Fano factor for protein variations between different cells would then be correspondingly high.

5.4.5 Bursts caused by slow transcription factors

We remind the reader that a transcription factor is a protein that regulates RNAP initiation by binding to the DNA. One should be aware that bursty transcription could occur simply as a consequence of correlations imposed by a slowly binding transcription factor (see Fig. 5.10, adapted from [113]) That is, if the "off periods" of a transcription factor are larger than the time it takes an RNAP to bind to the promoter, then subsequent RNAP binding events will be correlated, as shown in Fig. 5.10B. In contrast, if

---

[13] The Fano factor would be smaller, because degradation of individual messages smear out the bursty production.

the transcription factor (TF) binds quickly, then there will be no burst. The Fano factor is approximately given by the number of RNAPs bound within the time taken to bind one repressor TF [113]:

$$\frac{\langle n^2 \rangle - \langle n \rangle^2}{\langle n \rangle} \approx 1 + \frac{k_{on}}{k_b^{TF}} \tag{5.41}$$

where k_{on} is the on rate of the RNAP and k_b^{TF} is the on rate of the TF.

The derivation of Eq. (5.41) comes from calculation of the number of subsequent RNAP initiations, after one successful RNAP initiation. In more detail: given that one RNAP has just left the promoter, one may calculate the distribution of number n subsequent RNAPs that bind to the promoter before the first TF binds. Each time an RNAP leaves the promoter, the likelihood that an RNAP binds before a TF is $p = k_{on}/(k_{on} + k_b^{TF})$. For strong TF binding, the cumulative distribution of n can be easily expressed in terms of the small number $q = 1 - p$:

$$P(> n) = (1 - q)^n \sim \exp(-q \cdot n) \tag{5.42}$$

a cumulative distribution that corresponds to the probability density $p(n) = -dP(> n)/dn = qe^{-q \cdot n}$. Thus to have at least n consecutive initiations of RNAP, then RNAP must be faster to bind than the TF at least n times in a row. This cumulative distribution represents an exponential distribution with mean $\lambda = 1/q$, and Fano factor:

$$\text{Fano factor} = \frac{1}{q} = \frac{1}{1 - p} = \frac{k_{on} + k_b^{TF}}{k_b^{TF}} \tag{5.43}$$

The Fano factor is therefore equal to one plus the number of times that RNAP can bind within the average time it takes one TF to bind: indeed, it counts the burst size before one TF binds and blocks transcription for a longer time interval.[14]

5.4.6   Bursts without transcription factors

Let us now return to the bursty promoter examined in Fig. 5.8, revisiting the explanation proposed in ref. [221]. Promoters can be very sensitive to supercoiling; for example, *in vitro* the activity of the LacP promoter increases by more than a factor of 10 when the super-coiling is changed from zero to the value -0.065, which represents the typical supercoil of DNA in *E. coli*. Furthermore, an elongating RNAP increases the negative supercoiling of the DNA behind it [224, 225]. Therefore bursts of transcription might

---

[14] In more elaborate modeling that takes into account that RNAP may fall off and rebind several times before forming an open complex, the on rate for RNAP should be redefined to

$$k_{on} = \frac{k_o}{1 + k_u/k_b} \tag{5.44}$$

where k_o is the rate for the isomerization step (see Chapter 2), and k_u/k_b is the binding affinity of RNAP to the open complex. k_{on} is the total on rate for open complex formation. The time used from opening to elongating initiation ($1/k_e$) does not matter because this is assumed to be independent of the TF.

be caused by a transcribing RNAP that assists the recruitment of a subsequent RNAP via the disturbed supercoiling it leaves behind when leaving the promoter.

Supercoiling-mediated burst may in particular occur when the open complex formation is a rate-limiting step that is sensitive to supercoiling [221]. In the simplest terms, this effect could be parameterized in terms of a probability q, that a prior RNAP allows a subsequent RNAP to rapidly form an open complex before normal supercoiling is reached. This then creates two possible states of the promoter:

(1) If the promoter is in the on state, open complex formation is enhanced and transcription events occur at the rate for elongation initiation k_e.

(2) If the promoter is relaxed, then open complex formation is much slower; transcriptional events are still exponentially distributed, but now with the much lower isomerization rate k_o.[15]

The distribution of time intervals between subsequent RNAP initiations is now:

$$P(\Delta t) = q \cdot \frac{1}{\tau_e} \exp\left(-\frac{\Delta t}{\tau_e}\right) + (1 - q) \cdot \frac{1}{\tau_o} \exp\left(-\frac{\Delta t}{\tau_o}\right) \tag{5.45}$$

with $\tau_e = 1/k_e$ and $\tau_o = 1/k_{on}$ reflecting the two timescales. This equation comes about from counting time just after an elongation event, and then partitioning the history into a recruiting event (probability q) and a restart event (probability $1 - q$). This recruitment mechanism can reproduce the observations shown in Fig. 5.8 on the $P_{lac-ara}$ promoter, when using $q = 0.545$ and an on rate k_{on} that is about a factor of 10 slower than the elongation initiation rate k_e [221].

5.4.7 Questions

5.4.1 Make a computer program to select random numbers from a Poisson disribution with average 3. Test it against the real distribution. Hint: calculate cumulative distribution, and select random numbers between [0; 1] to select numbers from the cumulative distribution.

5.4.2 Consider a simplified production of proteins, where each mRNA on average gives one protein, and where one mRNA is produced every 10 minutes, whereas the decay time for proteins is 30 minutes. First assume that there is always exactly one protein per mRNA. Second assume that the mRNA lifetime is exponentially distributed, and that proteins are produced at random intervals as long as the mRNA survives.

5.4.3 Repeat as in 5.4.2, but now with 10 proteins produced simultaneously at each event (as it would be if we looked at proteins). In addition compare with situations where each mRNA gives an exponentially distributed "'burst" of proteins.

---

[15] In fact more exactly with the rate $k_{on} = k_o/(1 + k_u/k_b)$, which approches k_o for fast on–off binding rates.

5.4.4 Show that the large N, small p limit of a binomial distribution is a Poisson distribution. Hint:

$$p(n) = \frac{N!}{(N-n)!n!} \cdot p^n \cdot (1-p)^{N-n} \tag{5.46}$$

and use $(1-\frac{\lambda}{N})^{-n} \sim e^{-n\lambda/N} \sim 1$, $(1-\frac{\lambda}{N})^N \approx e^{-\lambda}$ and $\frac{N!}{(N-n)!N^n} \sim 1$.[16]

5.4.5 Simulate a single promoter with $\alpha = 0.1$ and overall firing activity of one elongation initiation every 20 s. Assume an mRNA decay rate of $0.01\ \text{s}^{-1}$ (100 s lifetime). Plot mRNA as a function of time, and calculate the Fano factor for the number of mRNA. Repeat the simulation for $\alpha = 1$ and calculate the Fano factor. Why is the Fano factor smaller than 1? Finally, repeat the simulation for an mRNA decay rate of $0.001\ \text{s}^{-1}$ and $\alpha = 0.1$, and convince yourself that the Fano factor remains around 1.

5.4.6 Repeat the simulation above, with decay time 100 s $k_{\text{on}} = 0.00505\ \text{s}^{-1}$ and $k_e = 0.505\ \text{s}^{-1}$, but now allow the elongating RNAP to recruit a new RNAP to the open complex with 90% probability. Thus the average promoter activity remains close to that in the previous question, but the fluctuations become larger. Calculate the Fano factor for the number of mRNA produced.

5.4.7 Consider the promoter from 5.4.5, but now also consider that RNAP binds competitively with a repressor that has an on rate of $1\ \text{s}^{-1}$ and an off rate of $0.05\ \text{s}^{-1}$. Vary α and discuss why repression varies. Calculate the Fano factor for the number of mRNA in each case.

5.4.8 Consider the promoter from 5.4.5 and 5.4.7, but now also consider that RNAP binds competitively with a repressor that has an on rate of $1\ \text{s}^{-1}$ and an off rate of $0.01\ \text{s}^{-1}$. Set $\alpha = 10$ and plot the dynamics on mRNA and promoter status. The lifetime of mRNA can again be assumed to be 100 s.

5.5   Economy of operons: noise minimization

An operon [226] is a set of genes that is transcribed from one promoter [72], producing one mRNA that often contains several genes each with their own ribosome start site. Operons with many genes are ubiquitous in the prokaryotic world, but almost never found in larger organisms. What are the evolutionary forces that drive operon formation in prokaryotes, but not in eukaryotes?

One idea is that the high rate of horizontal gene transfer in prokaryotes provides an advantage for functionally related genes to be grouped together to increase their probability of co-transfer in evolutionary exchange of genes between different organisms. Another benefit of operon formation is that it decreases the fluctuations between the concentrations of the co-expressed proteins [227]. Fluctuations in relative protein concentrations can be wasteful, for example, when proteins form a tight heterocomplex in order to function.

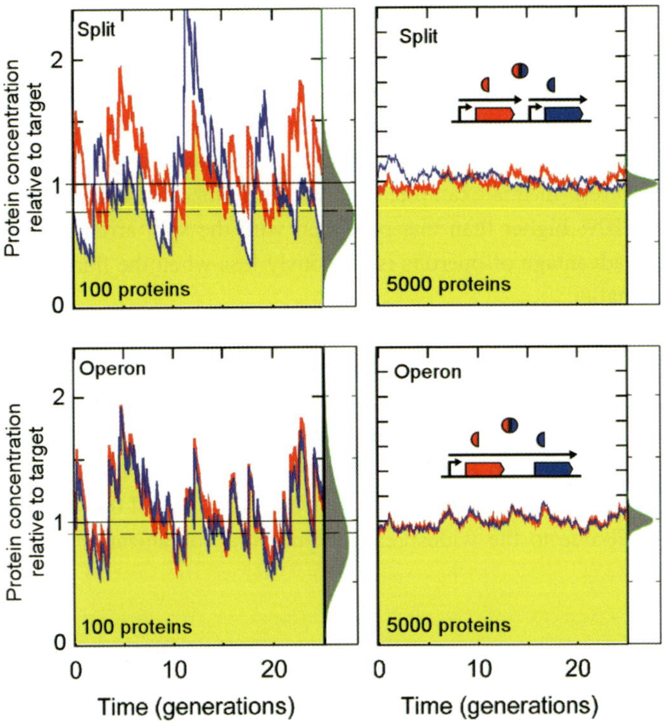

Figure 5.11 An abundance of a protein complex is sensitive to the least abundant protein [227]. The shortfall is larger when proteins are transcribed separately (monocistronic), whereas proteins produced from same polycistronic message have lower cost. The red and blue traces show fluctuations in the numbers of two proteins produced stochastically. The yellow areas show the concentration of a 1:1 complex of the two proteins, assuming that the complex level is the minimum of the two proteins. Individual RNA production, degradation and translation events are simulated as random Poisson processes, with an average of 5 in (A,B) and 250 in (C,D) mRNAs per cell generation. Simulations all assume an average of 20 proteins per mRNA. Proteins were stable and randomly distributed upon cell division. Figure reprinted with permission.

Translational coupling, in which ribosomes translating an upstream gene aid the translation of the downstream gene on the same mRNA molecule, is a way in which operon formation can reduce relative fluctuations. But strong translational coupling is not a general feature of operons. Rather, as we will stress here, co-transcription by itself can provide a substantial cost reduction in the production of protein complexes. This benefit decreases as the protein copy number increases, as it indeed will in the larger eukaryotic cells. Therefore the reduction in the shortfall of protein complexes provides an additional explanation for the abundance of operons in prokaryotes and archaea compared to eukaryotes.

Consider a functional 1:1 complex consisting of two different proteins. Now compare a system in which the two genes are co-transcribed, but not translationally coupled, to a system in which the two genes are transcribed independently from promoters of equal strength. If 100 copies of the complex are required, then in the absence of noise (and assuming a tight complex), it would be sufficient to produce 100 copies of each protein. In a living cell, both systems would fall short of the 100-complex target because the number of each protein will fluctuate around 100, and the number of complexes is determined by the minimum level of the two proteins. However, in the polycistronic operon arrangement, the levels of the two proteins tend to fluctuate in synchrony, and thus, the shortfall is lowered. In this example, the average number of complexes produced by the operon is 20% higher than that produced with the split arrangement (Fig. 5.11). The economic advantage of operons is obviously less when the fluctuations in protein numbers are smaller.

The size of fluctuations decreases when the number of mRNAs is larger (Fig. 5.11), which can be achieved by increasing the transcription rate of the mRNA. In *E. coli*, with its low number of transcripts, it is the variation in these numbers that makes the largest contribution to noise. Fig. 5.11 examines fluctuations when an average of 20 proteins are produced per mRNA. In *E. coli*, where about one-third of transcriptional units are polycistronic, it is estimated that there is an overall lowering of the cells metabolic cost by at least 0.2% due to the widespread polycistronic organization along its genome [227].

5.5.1   Questions

5.5.1   For a single promoter producing one mRNA (m), which encodes for one protein (P) the final protein concentration will have a distribution with width σ_P described by [223, 228]:

$$\frac{\sigma_P^2}{\langle P \rangle^2} = \frac{\gamma_P \cdot \gamma_m}{\Omega \cdot \omega} + \frac{\gamma_P \cdot \gamma_m}{\omega \cdot (\gamma_P + \gamma_m)}$$

where $\langle P \rangle$ is the average protein level, γ_P and γ_m are degradation rates of mRNA and its encoded protein, and Ω and ω are transcription and translation initiation rates, respectively. Reformulate this equation in terms of average proteins per message etc.

5.6   Summary

- Life on the molecular scale is dynamic, with molecules that bind to each other with on rates of the order of:

$$\text{rate} = \frac{D \cdot \epsilon}{V} \tag{5.47}$$

In *E. coli*, one expects on and off rates to be of the order of seconds to tenths of minutes.

- The fluctuation in protein numbers from cell to cell is especially sensitive to the number of transcription events within the transcribed proteins' lifetime.

- In prokaryotes, protein traffic on DNA plays a role in gene regulation. Transcription factors diffuse along the DNA, and RNAPs from different promoters collide and interfere with each other.
- Burstiness is ubiquitous in gene expression and results in large cell-to-cell variations.
- Operons in bacteria consist of several genes that are read from the same promoter. Thereby the relative variation in expression between these genes is minimized.

6 Stochastic genes and persistent decisions

6.1 Stochastic simulations

Consider the simple model for production and decay of a protein with N being its copy number in the cell:

$$\frac{\mathrm{d}N}{\mathrm{d}t} = \mathrm{Prom}(N) - \frac{N}{\tau} \tag{6.1}$$

Apart from an overall strength factor, the production $\mathrm{Prom}(N)$ can, in principle, be calculated from the average occupancy of the promoter by RNAP by use of the partition function from Fig. 4.12. Two examples of such promoter activity profiles are outlined in Fig. 6.1: on the left the promoter activity is independent of the number N of the investigated protein, as expected when the protein has no regulatory role. In contrast, the right panel investigates PRM-like activity as a function of the number of repressor CI molecules from phage λ. In steady state Eq. (6.1) often predicts a single stable steady-state number of proteins N, given by the value of N where $Prom(N) = N/\tau$.

6.1.1 Stochastic update using discrete time steps

In living cells the number of particular proteins and mRNAs is often low and cell-to-cell variation is substantial [35, 36, 37, 229]. Equation 6.1 in fact approximates dynamics where production takes place in discrete steps of a size that at least is on the scale of the number of proteins n_c per mRNA, i.e. per transcripion initiation event. Also, the decay is discrete, but typically takes place in small steps of one protein per event.[1] If proteins are produced with a rate $Prom(N)$, then doing this in discrete units of size n_c amounts to events occurring at a reduced rate $Prom(N)/n_c$. Similarly, degradation of one protein takes place with rate N/τ.

To implement the above dynamics in a stochastic manner, one may choose a small time interval for integration, Δt. For each timestep $[t, t + \Delta t]$ one selects two random numbers ran_1 and ran_2 uniformly over the interval $[0; 1]$: The number N is only changed:

---

[1] At cell division, proteins are presumably partitioned binomially at between the daughter cells [175, 230]. This result in a variance $= N/4$ and a spread to mean $CV =\sim 1/\sqrt{4N}$ which is on the same small scale as the contribution from independent degradation events.

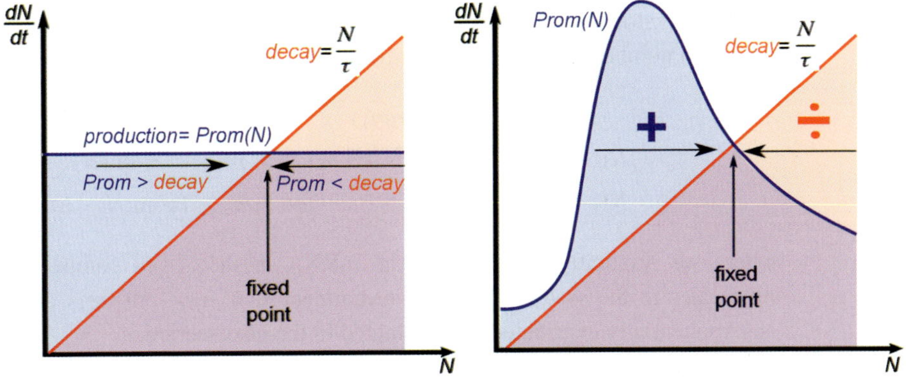

Figure 6.1 Dynamics of a protein N is influenced by production and decay/dilution. The black horizontal arrows indicate deterministic trajectories starting below and above the fixed point. In the left-hand panel the production $Prom(N)$ is independent of N, whereas the right-hand panel plots behavior that resembles that seen for CI self-regulation in phage λ. In both cases there is a stable fixed point, where production equals dilution. This will be the steady-state level of proteins in the cell. The decrease in $Prom(N)$ at large N in the right-hand panel mimics negative feedback that, in fact, will decrease the cell-to-cell variation of N [231].

$$\text{if } ran_1 < \Delta t \cdot \mathrm{Prom}(N)/n_\mathrm{c} \text{ then}$$

$$N(t) \to N(t + \Delta t) = N(t) + n_\mathrm{c} \tag{6.2}$$

$$\text{if } ran_2 < \Delta t \cdot N/\tau \text{ then}$$

$$N(t) \to N(t + \Delta t) = N(t) - 1 \tag{6.3}$$

Notice that the Δt interval should be chosen to be small enough that all probabilities are much smaller than 1. In particular the requirement $\Delta t \cdot N/\tau < 1$ slows down the simulation substantially, because one has to choose a very small value of Δt, and thus many integration steps are necessary. Alternatively one can approximate the time development by taking into account that degradation occurs one protein at a time and therefore contributes little to noise. Using this approximation, one may simulate the degradation by the approximate deterministic update:

$$N(t) \to N(t + \Delta t) = N(t) \cdot \exp(-\Delta t \cdot N/\tau) \tag{6.4}$$

such that the total update becomes:

$$N(t + \Delta t) = N(t) \cdot e^{-\Delta t \cdot N/\tau} + n_\mathrm{c} \quad \text{for } ran_1 < \mathrm{Prom}(N) \cdot \Delta t \tag{6.5}$$

$$N(t + \Delta t) = N(t) \cdot e^{-\Delta t \cdot N/\tau} \quad \text{for } ran_1 \geq \mathrm{Prom}(N) \cdot \Delta t \tag{6.6}$$

an approximation to true stochastic behavior that generates non-integer numbers of molecules in the cell.

The stochastic simulation can be extended to take into account the random decay of the mRNA, and thus the protein per message is exponentially distributed with mean n_c.

To take this stochastic behavior into account, the full update in time step Δt reads: select a random numbers ran_1 and ran_2 uniformly $\in [0; 1]$ and update:

$$x = -n_c \cdot \ln(ran_1) \tag{6.7}$$

$$N(t + \Delta t) = N(t) \cdot e^{-\Delta t \cdot N/\tau} + x \quad \text{for} \ \ ran_2 < \text{Prom}(N) \cdot \Delta t \tag{6.8}$$

$$N(t + \Delta t) = N(t) \cdot e^{-\Delta t \cdot N/\tau} \quad \text{for} \ \ ran_2 \geq \text{Prom}(N) \cdot \Delta t \tag{6.9}$$

where we implicitly assume that the mRNA is short-lived compared to the other timescales in the system. That the assignment $x = -n_c \cdot \ln(ran_1)$ does indeed give an exponentially distributed x is explained in the next section.

6.1.2   The Gillespie algorithm: an event-based simulation

Gillespie developed a stochastic approach to chemical kinetics, using an event-based algorithm to deal with many molecules that react with each other due to random encounters [232, 233, 234]. The main idea is to represent the dynamics in terms of discrete events, occurring at times determined stochastically from their respective rates. The simulation is event-driven, with changes occurring at discrete variable time steps and where nothing changes between these events.

Assuming that some event; occurs with rate r, it has the probability not to occur in the time interval $[t, t + \Delta t]$:

$$\left(1 - \frac{r\Delta t}{n}\right)^n = \exp(-r \cdot \Delta t) \ \ \text{for} \ \ n \to \infty \tag{6.10}$$

which is exponentially decaying with the duration of the time interval Δt, see Fig. 6.2. To select when it will actually occur, notice that this exponential decay specifies the cumulative probability that the event occurs at times larger than Δt. As explained in

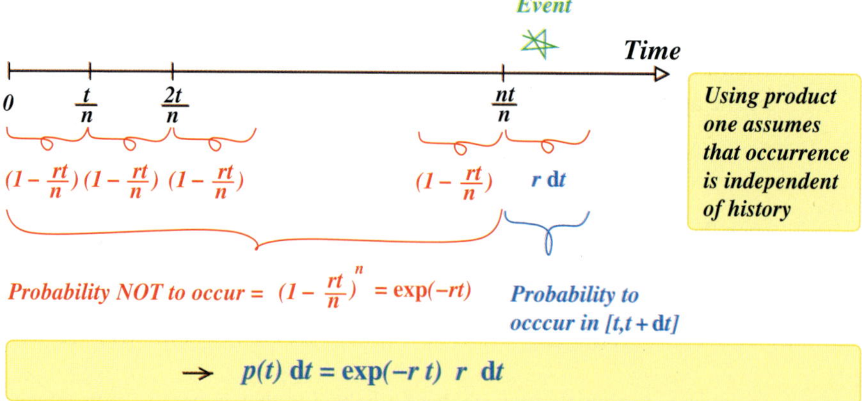

Figure 6.2 Occurrence of an event with constant rate r implies that the time of next occurrence is exponentially distributed, just like radioactive decay of an atomic nucleus. The main assumption is that there is no memory/aging in the process.

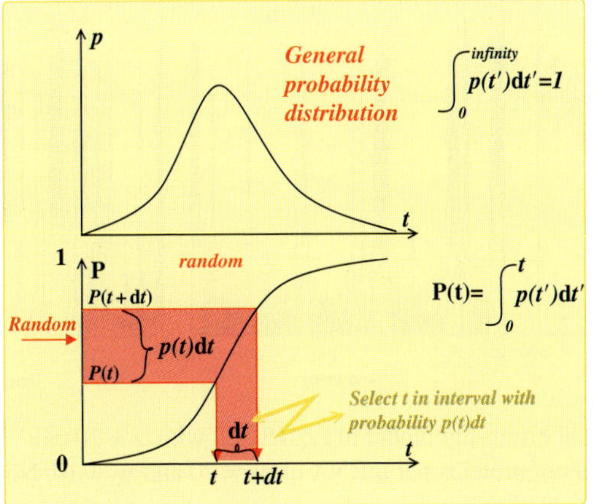
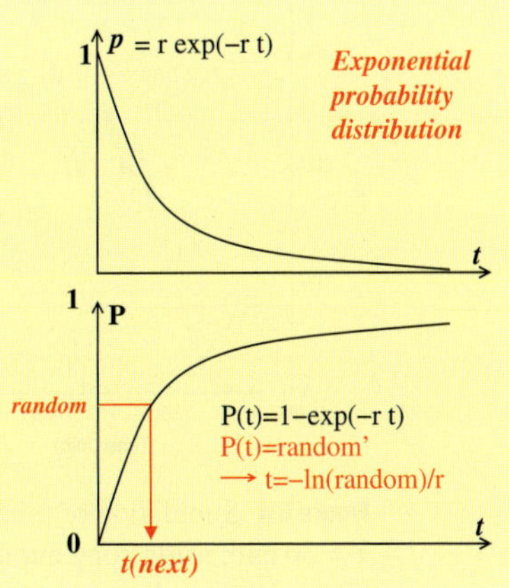

Figure 6.3 Selection of t according to some pre-defined distribution function $p(t)$. One constructs the cumulative distribution $P(t)$, selects a *random* number uniformly in $[0, 1]$, and finds the t that makes $P(t) = random$. Thereby the part of $p(t)$ that has high values will be more likely to be selected because there are more *random* values that correspond to values where $P(t)$ has a high slope. For example if p is doubled around some $t, t + dt$, the slope of P is doubled and the probability of selecting t is doubled. If $p(t)$ is exponential, then $P(t)$ is also exponential and the solution is analytical (as shown in the right-hand panels).

Fig. 6.3 one should select uniformly a random number $ran \in [0, 1]$ and find the δt that solves $\exp(-r\Delta t) = ran$. The selected Δt will be exponentially distributed. Accordingly, if the current time is t, then the next event should be assigned to occur at $t + \Delta t$ with:

$$\Delta t = -\frac{1}{r} \cdot \ln(ran) \tag{6.11}$$

Here $1/r$ is the average time until the next event.

If many competing events take place, at any one time, one needs to have a list of times for each of these events. One update consists of selecting the first occurring event in this list, and then assessing the consequences. Subsequent rates for many of the other events may have changed, and most simply one just reassigns new exponentially distributed event times to each event. One should also keep track of time during the simulation, always updating the time with a step given by the occurrence of the selected events.

After initialization with a start number for all relevant molecules an update step in the event driven algorithm reads:

(1)   Monte Carlo step: generate random numbers to determine the time step for all potential events, and select the first event for updating.

(2)   Update: increase the time by the generated time step (from Step 1) and update the molecule count based on the selected event.

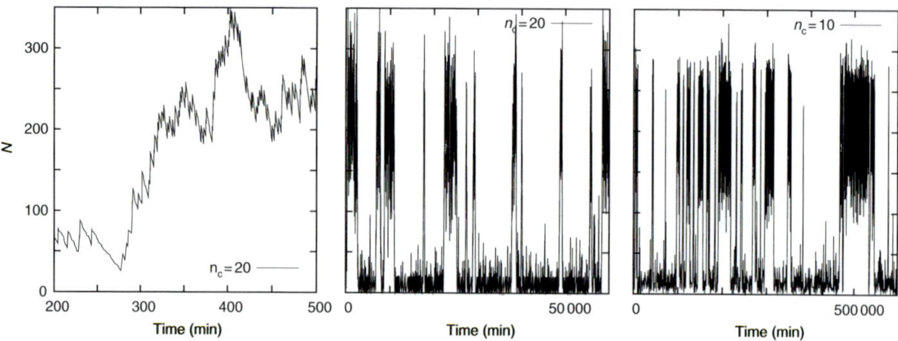

Figure 6.4 Simulation of minimal bistability model in Eq. (6.12), with $\alpha = 20\,\mathrm{min}^{-1}$ and $\tau = 30\,\mathrm{min}$, using copy numbers of proteins per mRNA of $n_c = 20$ and $n_c = 10$. Notice the time axis is 10 times longer in the $n_c = 10$ case.

The key assumption is that we consider systems without memory, which implies that any event depends only on the current state of the system, and thus is independent of how long has passed since the last changes in the system.

Gillespie orriginally formulated the algorithm to take care of chemical reactions between all molecules in a relatively small system [233]. However, the event-based simulation can also be used in coarse-grained models. For example, one may consider the simple production decay from Eq. (6.1) in terms of only two possible events, a production event and a degradation event.

As a case study, consider a minimal model of a lysis–lysogeny switch, where the mRNA for the repressor is produced with a rate:

$$\mathrm{Prom}(N) = \frac{\alpha}{n_c} \cdot \left(0.01 + \frac{(N/200)^2}{1 + (N/200)^2 + (N/200)^4} \right) \tag{6.12}$$

After production of the mRNA one may simplify the subsequent production of proteins by assuming that there are $x = -n_c \cdot \ln(ran_2)$ proteins produced per mRNA ($ran_2 \in [0, 1]$). Thus, a production event is a change in protein number, $N \rightarrow N + x$ after a time interval $\Delta t_p = -\ln(ran_1)/\mathrm{Prom}(N)$.

The other event maintaining the repressor concentration is dilution occurring with a rate:

$$d(N) = \frac{N}{\tau} \tag{6.13}$$

The dilution/degradation is a change $N \rightarrow N - 1$ that occurs after time $\Delta t_d = -(\tau/N) \cdot \ln(ran_3)$, $ran_3 \in [0, 1]$. For $\tau = 30$ min, $\alpha = 20\,\mathrm{min}^{-1}$ and $n_c = 20$, one obtains bistable dynamics that are characterized by long periods in one state, separated by fast jumps between the two states (see Fig. 6.4).

Notice that in the above example, either of the two events will influence the probability of the other event. Accordingly one needs to recalculate new times for each of these events after any update. For more efficient sampling methods of rare events the reader may consult [235, 236, 237].

6.1.3   Questions

6.1.1   Collect 10 000 numbers x, each selected by $x = -n_{\mathrm{c}} \cdot \ln(ran_2)$ where ran_2 is another random number selected in $[0; 1]$ and $n_{\mathrm{c}} = 10$. Plot the histogram $h(x)$ in a graph where the x axis is linear and the y axis is logarithmic.

6.1.2   Repeat the Gillespie simulation of Eqs. (6.12) and (6.13), including that the mRNA lifetime is exponentially distributed with a mean lifetime that gives $n_{\mathrm{c}} = 20$ proteins per mRNA. Compare with a simulation where we set $n_{\mathrm{c}} = 10$, but maintain the mean production level of proteins.

6.1.3   Consider protein and associated mRNA production and decay, through rate Ω for $M \to M + 1$, rate ω for $M \to M + P$ for each M, rate γ_m for $M \to M - 1$ for each M and rate γ_P for $P \to 0$ for each protein. What is the average number of proteins per mRNA? Simulate the process using the Gillespie algorithm and the rates $\Omega = 0.02$, $\omega = 0.1$ (protein per second per mRNA), $\gamma = 0.01$ (mRNA lifetime of 100 s) and $\gamma_p = 0.002$ (protein lifetime of 500 s).

6.1.4   Repeat the investigation of the toggle switch [238] shown in Fig. 6.5:

$$\frac{\mathrm{d}u}{\mathrm{d}t} = \frac{a}{1 + v^2} - u \quad \text{and} \quad \frac{\mathrm{d}v}{\mathrm{d}t} = \frac{b}{1 + u^2} - v \qquad (6.14)$$

for $a = b = 10$, by determining fixed points for both equations, as well as their crossing points. Hint: plot the right-hand side of the equations for u and v in $[0, 10]$. Simulate the equations using the Gillespie algorithm by assuming that both u and v are produced and decay in discrete units of size 1.

6.2   Maintenance of lysogeny in the λ system

Upon infection of an *E. coli* cell, the λ phage enters either a pathway leading to lysis, or to lysogeny, in which it can be passively replicated for very long periods. Indeed, the wild-type rate of spontaneous loss of λ lysogeny is only about 10^{-5} per cell generation, a lifetime of the order of 5 years. Moreover, this number is mainly due to random activation of another part of the genetic system (the SOS response, involving RecA), whereas the intrinsic loss rate has, in several independent experiments, been found to be much smaller [166, 240].

6.2.1   Stochastic simulations of the lysogenic state

Figure 6.6 outlines results of an experiment by [142]. The genetic construct uses a "defective" λ phage, which cannot lyse. The lytic pathway is instead detected by a reporter gene *gal*, downstream of *cro*. If *cro* is transcribed, then *gal* is also transcribed, and the corresponding enzyme β-galactoxidase catalyzes a reaction in the added growth

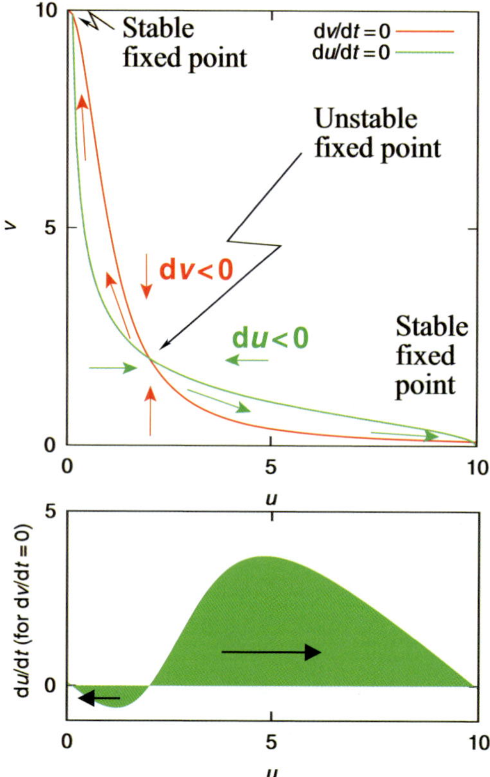

Figure 6.5 Toggle switch [238] as described in Question 6.1.4. The upper panel shows the predicted deterministic value of du/dt and dv/dt as a function of two protein concentrations u and v. The lower panel presents a projection to a trajectory where v is always assumed to be in a steady state with respect to the given concentration of its repressor u. Such plots have been measured for promoter PRM as a function of CI (see [239]).

medium that makes a red dye. Thus colonies in the lysogenic state are white, while colonies in the lytic state become red.

The experiment illustrates that both lysogen and lytic (anti-immune) states are stable, thus forming a truly bistable epigenetic system. Figure 6.6 also indicates the stability of each of the two epigenetic states.

We now describe a quantitative model for relating the stability of the λ phage switch to its underlying promoter activities. The stability of the λ phage depends in particular on the CI concentration, as even a relatively small amount of CI will be sufficient to repress PR. The stability is quantified by the spontaneous rate of escape from lysogeny to the lytic state. The dynamics of the lysogen–lytic states for a cell with a number of CI and Cro molecules (N_{CI}, N_{Cro}), can in principle be directly simulated by use of the Gillespie algorithm. However, because the stability of the lysogen state is very high, the simulation becomes very tedious.

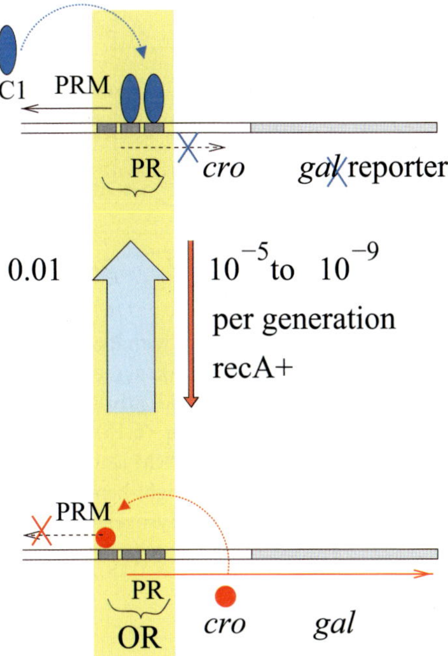

Figure 6.6 Switching between a sustained lysis state (anti-immune state) and the lysogen for a defective λ phage that cannot escape the *E. coli* chromosome [142]. The state was recorded through the color exhibited when the gene *gal* was expressed from PR (red). The *in vivo* stability of the lysogen of wild-type λ in a RecA+ strain is about 10^{-5} per cell generation [166, 229]. In a RecA− strain, phage escape is dominated by mutants, either by loss of CI (at a rate of $\sim 10^{-6}$) or by PRM mutants (at a rate of $2 \cdot 10^{-8}$ [166]), whereas real escapes without mutations may occur at rates lower than 10^{-9} per generation [240].

The deterministic equations for the development of CI and Cro are:

$$\frac{d[CI]}{dt} = PRM(CI, Cro) - [CI]/\tau_{CI}$$
$$\frac{d[Cro]}{dt} = PR(CI, Cro) - [Cro]/\tau_{Cro} \tag{6.15}$$

where PRM and PR0 are given in Fig. 6.7. The stochastic simulation shown in Figure 6.8 simulates dynamics in large, discrete timesteps, taking into account genome replication, growth in cell volume, decay of Cro between cell divisions [151] and reduction of protein number at cell divisions by random partitioning between daughter cells. The main contribution to the noise comes particularly from the number of Cro proteins per Cro mRNA, whereas the number of CI proteins per mRNA is expected to be small because this gene has almost no Shine–Delgarno sequence. The noise at cell division or from Cro decay is of no consequence.

Figure 6.7 in fact allows for two stable fixed points, one at high [CI] (= 250 molecules in a cell) and low [Cro], and a competing one at low [CI] and high [Cro]. The fixed points correspond to solutions of Eq. (6.15) where d[*CI*]/d*t* = 0 and d[*Cro*]/d*t* = 0. The simulation predicts possible fluctuations around the two fixed points, predictions that

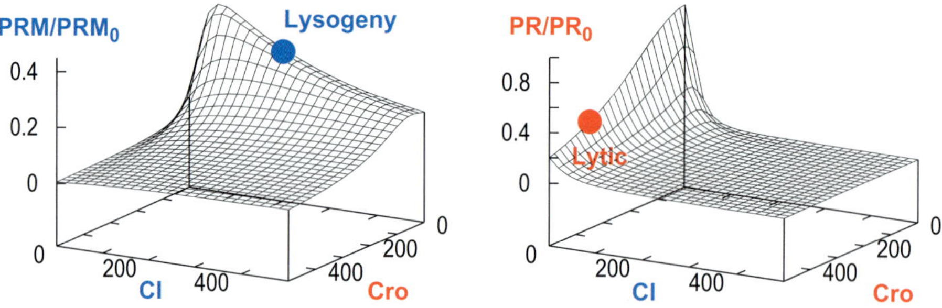

Figure 6.7 PRM and PR activity as a function of both CI and Cro with the same parameters as Fig. 4.18, supplemented with a Cro non-specific binding of $-3.5\,\mathrm{kcal\,mol^{-1}\,bp^{-1}}$. The relative strength of the two promoters is set by $\mathrm{PR}_0 = 10\cdot\mathrm{PRM}_0$ (I. Dodd, private communication, see also [239]). The promoter activity sets the production terms in Eq. (6.15) that include eventual looping between OL and OR. The left-hand plot illustrates that PRM activity decreases monotonically with [Cro], but is non-monotonic with [CI]. The right-hand plot shows that PR decreases rapidly especially with [CI], thereby securing a stable lysogenic state.

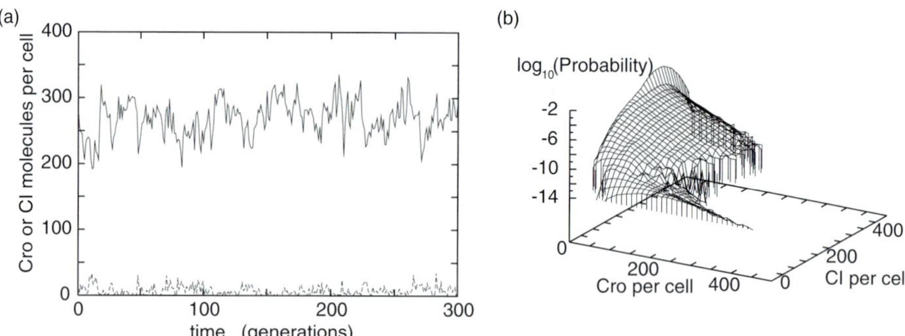

Figure 6.8 Left: Stochastic simulation of the number of CI and Cro inside an *E. coli* lineage [175]. The upper curve shows the CI level from generation to generation, and the lower curve the corresponding Cro level. Notice that CI and Cro are anti-correlated: the presence of Cro represses CI, and decreases the amount of CI in a given cell. Right: Histogram of predicted distribution of CI and Cro in a long simulation [175]. The tail at low CI concentrations corresponds to spontaneous transitions to lysis. Rare transitions are reflected by low probability. Notice that the z-axis is logarithmic and spans over a factor of more than billion. The assumed noise levels are consistent with a lysogen stability of 10^{-9} per cell generation.

can be compared to experiments. That is, the cell-to-cell variation in [CI] in lysogeny has been addressed by single cell measurements using GFP placed after the CI gene in the phage (see Fig. 6.9).

Remarkably, the measured cell-to-cell variation of [CI] is found to have a spread to mean variation CV = 0.3, a coefffficient of variation that corresponds to a burstiness of the PRM promoter, where effectively all CI in the cell is produced in less than 10 independent events. In fact, such a high level of noise predicts a high frequency of events where CI drops below the, perhaps, 10% that is needed to sustain the lysogenic state.

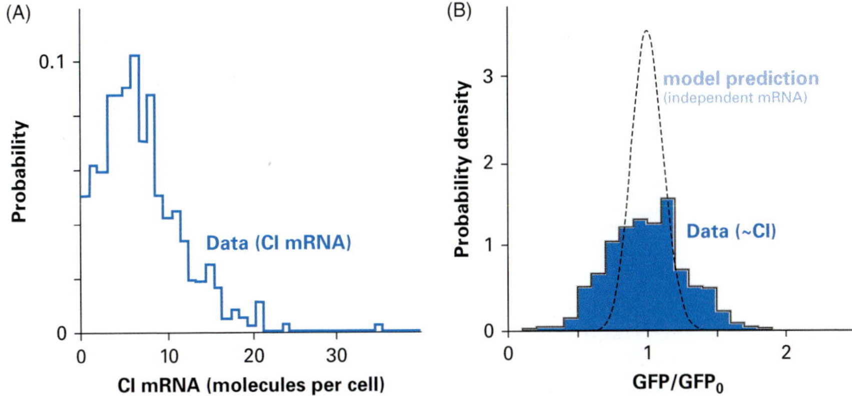

Figure 6.9 Histogram of cell-to-cell fluctuations in (A) mRNA for CI [241] and (B) a GFP reporter for CI protein, both expressed from PRM in phage λ, from [229]. The dashed line shows the prediction from the model in Fig. 6.8, where we assume that there are about 100 independently produced mRNA for CI in each cell generation, giving $\sigma/\text{mean} = 1/\sqrt{N_{\text{mRNA}}} \approx 0.1$. The observed fluctuation with $\sigma/\text{mean} \approx 0.3$, is much larger than predicted. With such high noise, the lysogen state should be more unstable than experimental limits allow for [166, 240].

Accordingly the lysogeny→ lysis frequency becomes much larger than experiments have constrained it to be [240]. This challenges the presented standard model for maintenance of the lysogenic state, a challenge that is supported by the existence of lysogens for radical operator mutants as well [166].

6.2.2   Dynamics in epigenetic landscapes

Stochastic dynamics may also be formulated in terms of Langevin equations [242] of the form [243]:

$$\frac{dN}{dt} = \text{production} - \text{decay} + \text{noise} \tag{6.16}$$

for both CI and Cro. The noise in the above equation represents the fact that both production and decay/diluton involve random events in a system of a few molecules.

The strength of the noise term can be quantified by its variance, $\sigma^2(N)$, which in fact will depend on the number of molecules N and in how big a burst these molecules are produced. The size of this variance is given by the sum of the variances of the independent processes, here the sum of the variance for production and the variance for decay:

$$\sigma^2(N) = \sigma^2_{\text{production}}(N) + \sigma^2_{\text{decay}}(N) \tag{6.17}$$

A simple generic equation that just states that the variance of a sum or difference of independent processes is the sum of the variances of each process.

Consider the one-dimensional graphic illustration of the system in Fig. 6.10. This may approximately correspond to the CI production rate along the most probable trajectory

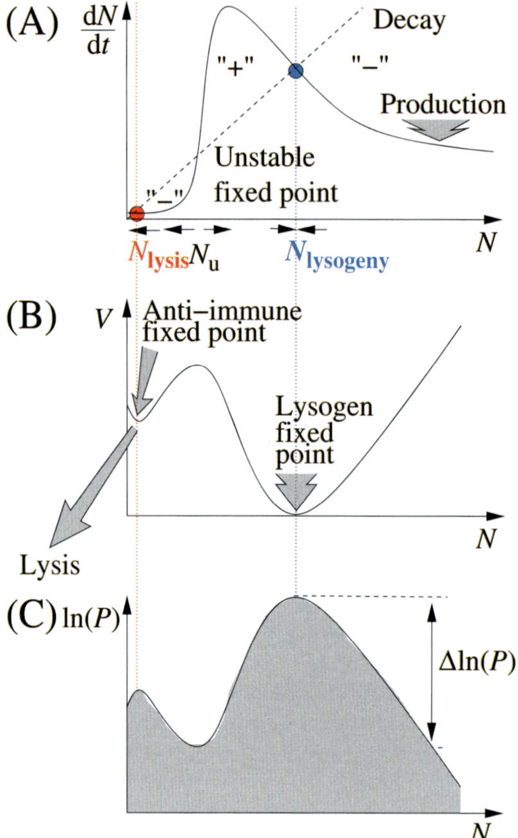

Figure 6.10 (A) Production of a protein that both activates and represses its own promoter. The dashed line shows dilution due to cell growth. The intersections between production and dilution define fixed points that are either stable (N_{lysis}, N_{lysogeny}) or unstable (N_{u}). (B) Effective potential V obtained by integrating the production–dilution from the top curve. The transition from lysogeny to lysis can be treated as a Kramer's escape problem with copy number N as the "reaction-like" coordinate. (C) Translation of the probability distribution $P(N)$ for N, in terms of an effective epigenetic landscape $V = -\log(P)$.

that connects lysogeny to lysis, in the two-dimensional CI–Cro phase plane describing the state of an infected *E. coli* cell. Figure 6.10A shows the production rate of the protein as a function of its copy number N in the cell, a system that, in principle, could also be constructed synthetically, provided that one had access to a transcriptional repressor that only binds to its operator in a highly co-operative form.

The points where the production and decay are equal determines the fixed point ($dN/dt = 0$), which can be either stable or unstable. As long as N is larger than the unstable fixed point $N = N_{\text{u}}$, the high N fixed point corresponding to lysogeny tends to be restored. For phage λ with $N \sim \text{CI}$ number, the unstable fixed point is at $N_{\text{u}} = CI_{\text{u}} \sim 0.1 \cdot CI_{\text{lysogen}}$. The relative distance between lysogen concentration and CI_{u} supports the large stability for the lysogenic state. The smaller difference between

0 and CI_u is consistent with the smaller stability of the anti-immune state, a state that in a more detailed description is maintained by high levels of Cro.

The stability and dynamics of the system can also be illustrated in terms of a potential energy landscape, as seen in Fig. 6.10B. This potential is in fact our first example of a so-called epigenetic landscape, a landscape with valleys and "mountain passes" that illustrates the available states and potential transitions for the system. The "epigenetic landscape" can be formally obtained if we assume that there is only one variable and then integrate the drift term f = production − decay, as it acts as an effective "force" on the variable N:

$$V(N) = -\int_0^N f(N')\mathrm{d}N' \tag{6.18}$$

The transition from the state N_{lysogeny} over the unstable (saddle) point N_u to the lysis state $N_{\text{lysis}} \sim 0$ is a first exit problem. Following Kramer's equation for escape over potential barriers, one obtains an estimate for the transition rate from lysogeny to lysis [243]:

$$r \propto r_0 \cdot \exp\left(-\int_{N_s}^{N_u} \frac{2f(N)\mathrm{d}N}{\sigma^2(N)}\right) \sim r_0 \cdot e^{-\Delta V/(0.5\sigma^2)} \tag{6.19}$$

with a noise σ^2 that parameterizes the cell-to-cell variation in N. The pre-factor r_0 sets the attempt rate for independently fluctuating states of the cell. For lysogeny in phage λ that is governed by long-lived CI and Cro molecules, the characteristic time will be set by cell generation. Thus as an order of magnitue estimate one should use r_0 = once per cell generation.

In Eq. (6.19), the generation-to-generation variation of production − decay plays the role of an effective temperature, whereas production − decay in itself plays the role of a potential. The exponential of the potential implies that stability can become huge, provided that the "barrier" is much higher than the variation in the production over a cell generation. Intuitively, using the lysogenic state of phage λ, the stability is given by

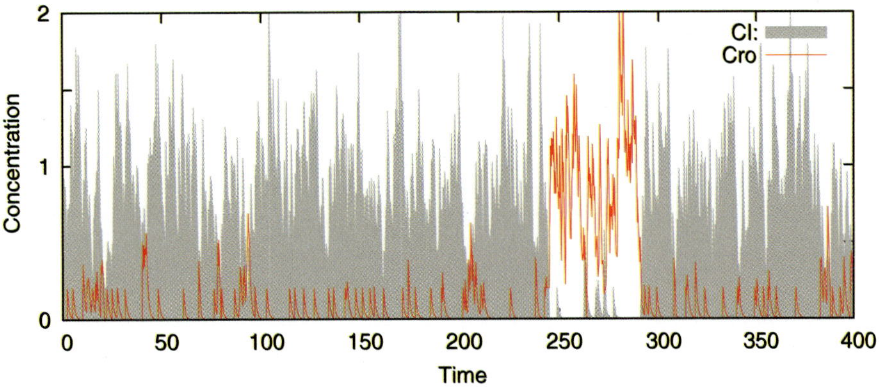

Figure 6.11 Simulation of the phage model from question 6.2.1 using Eq. (6.20) with proteins produced in bursts of size $\delta = 0.2$.

the probability that CI concentration drops below 10% of the average lysogenic level, a drop that, for example, will occur if there is no CI production events in about three consecutive cell generations.[2]

6.2.3 Questions

6.2.1 Consider the following model for a lambdoid like phage:

$$\frac{dCI}{dt} = \frac{0.1 + (CI/0.2)^2}{(1 + (CI/0.2)^2) \cdot (1 + (Cro/0.5)^2)} - CI \qquad (6.20)$$

$$\frac{dCro}{dt} = \frac{1}{(1 + (CI/0.2)^2)} - Cro$$

resembling PRM activated by CI and repressed by Cro, and PR repressed by CI (see Fig. 6.11). What does the product in the denominator of the "PRM term" correspond to in terms of operator design? Plot regions of (CI, Cro)
$\in [0, 1] \times [0, 1]$ where $dCI/dt > 0$, and $dCro/dt > 0$. Identify stable and unstable fixed points. Hint: identify null-clines by solving steady-state equations in terms of Cro as a function of CI.

6.2.2 Simulate the model in Question 6.2.1 using Gillespie algorithm with discretized production into units of CI and Cro of characteristic size $\delta = 0.1$, but with dilution being considered to occur in very small units, say 0.01. Discuss the timescale in the simulation, and the step size δ in terms of promoter activity. Hint: define four processes, two production events and two decay events, each happening with rates given by the respective terms in Eq. 6.20 divided by the size of the change ($\delta = 0.1$ for production and 0.01 for decay). If any variable becomes less than 0, then set it equal to zero, since concentrations cannot be negative.

6.2.3 Rewrite the equations in Question 6.2.1 in nM units, where we assume that CI binding has binding constant $K = 40\,\text{nM}$, and Cro has $100\,\text{nM}$, instead of 0.2 and 0.5, respectively. How many proteins per message does $\delta = 0.1$ then correspond to?

6.2.4 Rewrite the equations in Question 6.2.1 as a one-variable simulation, by assuming that Cro is a fast variable that instantly adapts to the steady-state value set the equation for Cro:

$$\frac{dCI}{dt} = \frac{(0.1 + (CI/0.2)^2) \cdot (1 + (CI/0.2)^2)}{(1 + (CI/0.2)^2)^2 + 4} - CI = \text{Prod} - CI \qquad (6.21)$$

Plot the production and decay terms separately, as well as their sum.

---

[2] If we assume that CI in individual cells is distributed as a Gaussian curve with CV = 0.3 [229], then the likelihood that CI becomes less than three standard deviations from the mean becomes 0.00003. This estimate for spontaneous lysis frequency in fact resembles that measured in RecA+ strains of *E. coli*. However, for RecA− strains, the lysis frequency is much much smaller [240].

6.3 The phage λ circuit beyond CI and Cro

Until now we have avoided discussing the developmental decision that λ makes within the first few minutes after injecting its DNA into the *E. coli* cell. The associated decision circuit is in fact substantially more complicated than the maintenance motif. Figure 6.12 follows the average expression of a few promoters after infecting the *E. coli* cell with a huge excess of phages. The plot shows two promoters, the promoter for establishment of lysogeny, PRE, and the promoter PR′ that transcribes late lytic genes.

The wild-type infection (wt) is characterized by an early burst of CII, measured by a burst of activity of the promoter PRE, which requires CII for its activation. In contrast, the lytic promoter PR′ remains at close to zero activity, indicating that only a small fraction of the cells lyse. The mutant phage CI− always chooses the lytic pathway, as

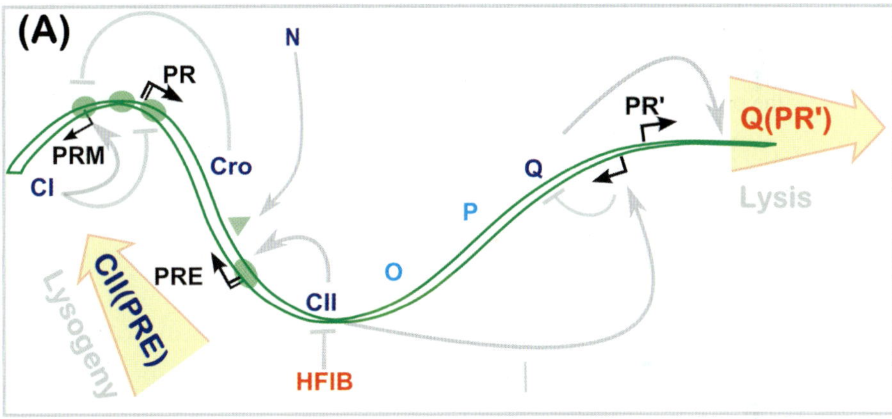

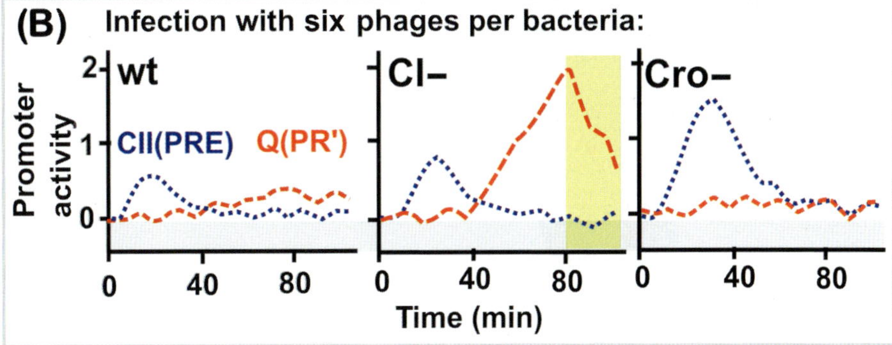

Figure 6.12 (A) Section of λ genome. (B) Activity of promoters in phage λ after infection with six phages per *E. coli* as recorded by Kobiler *et al.* [244]. The three plots examine wild-type, CI− and Cro− mutants. In CI− all phages choose the lytic pathway, and the decline in expression of Q(PR′) in the yellow area reflects declined expression due to death. In Cro− all phages choose the lysogenic pathway and Q(PR′) never becomes expressed. Notice that PRE is expressed the most in absence of Cro, indicating that Cro repression, of itself and CII, is substantial. Estimated absolute rates for activity of PR and PRE can be found in [245]. Copyright (2005) National Academy of Sciences USA.

signaled by large late activity of Q(PR′). The mutant Cro− always chooses lysogeny, and does this forcefully, as indicated by the especially large early activity of PRE.

Following [246, 247, 248], we will explore early infection dynamics in the regulatory network of phage λ and phages with the same functional properties. Instead of a detailed description of a given regulatory network, we will introduce an alternative approach based on sampling networks and parameters, in order to pinpoint essential features required to obtain a given functionality [247, 248, 249]. Essential for such sampling is to specify the functions that the system is supposed to perform. To understand this requirement for the λ phage we first describe its key ability in more detail: it counts.

Bacteriophage λ is in fact making its developmental decision as a function of the number of phages that infect the host simultaneously (see Fig. 6.13): Kourilsky found that a single λ always chooses lytic development, whereas two infecting genomes can choose lysogeny [150]. Thus the phage can count to two! This counting ability is remarkable, because of the replication of the phage genomes soon after infection. Thus after a time $t \sim \tau_{rep} \sim 5\,\mathrm{min}$, a single genome $n = 1$ has replicated to become $n = 2$, and therefore appear similar to $n = 2$ phages infecting at time $t = 0\,\mathrm{min}$. The only difference would be the proteins produced during this short time interval. This task is outlined in Fig. 6.14.

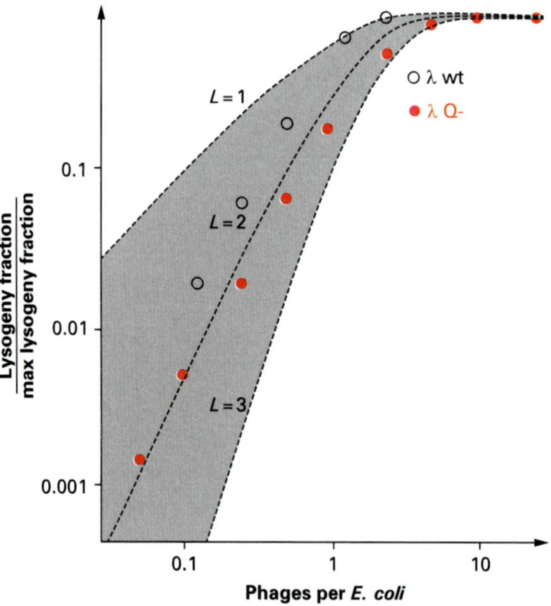

Figure 6.13 Kourilsky's data (red circles) for a replicating λ phage[150, 153, 250], showing the fraction of phage-infected bacteria that entered lysogeny as a function of the average phage input (API, defined as the ratio between total available phages and total numer of bacteria). The best fit is with slope = 2, indicating that one phage cannot enter lysogeny alone. Noticeably, the maximal lysogeny frequency reported by Kourilsky was a few percent. However Little *et al.* [166] obtained a maximum lysogenization frequency of 63% for high API.

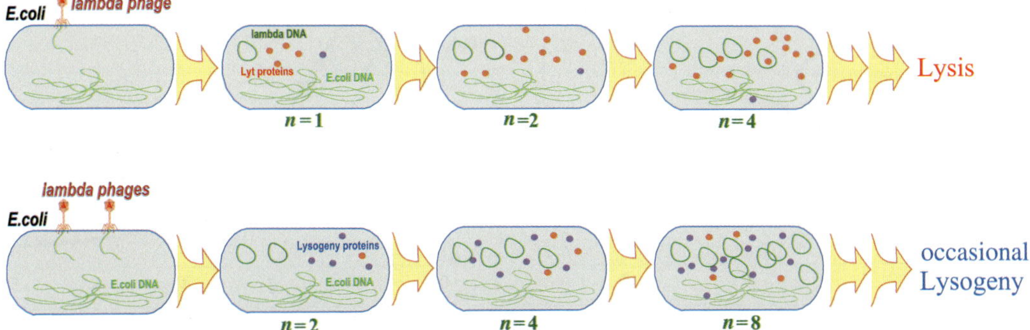

Figure 6.14 Functional task: find motifs that can count genomes and do so robustly so that they can remember the initial count even after resetting the genome count to $n = 1$. Notice that the above requirement refers to λ growth in rich medium. Grown in succinate, the phage may also chose lysogeny at MOI = n = 1 [150, 153, 250].

In the following we explore networks that are able to count the gene dosage of their own encoding, and do this within the relatively short time set by τ_{rep}. We sample two- and three-node networks of various topologies and logic, and at the same time sample parameters for repression, promoter strengths, Hill coefficients and decay times.

To illustrate the simulations, consider a case study with only two proteins, a CI-like protein and a Cro-like protein. A possible motif of this type could then be:

$$\frac{dCro}{dt} = n \cdot A_{\text{Cro}}(Cro, CI) - \frac{Cro}{\tau_{\text{Cro}}} \tag{6.22}$$

$$\frac{dCI}{dt} = n \cdot A_{\text{CI}}(Cro, CI) - \frac{CI}{\tau_{\text{CI}}} \tag{6.23}$$

where n is the present number of phage genomes inside the *E. coli* cell. The promoter activity $A_{\text{Cro}}(Cro, CI)$ is the production rate of Cro. This rate depends on the level of CI and/or Cro in a form that can be modeled using standard equations for activation and repression (much like Eqs. (3.18) and (3.19)). For instance, if Cro activates itself and is repressed by CI, then:

$$A_{\text{Cro}} = r \times \frac{Cro^{h_1}}{K_1^{h_1} + Cro^{h_1}} \times \frac{K_2^{h_2}}{K_2^{h_2} + CI^{h_2}} + \epsilon \tag{6.24}$$

where r is a rate constant and ϵ is a small "leak" that parameterizes the background activity of promoters that contribute to the production of Cro.[3]

The simulation of Eq. (6.21) is initiated with $CI = Cro = 0$ at time $t = 0$ and an initial value of $n = 1$, and $n = 2$, respectively. The gene copy number n will double at each

---

[3] Notice that the two regulatory inputs in Eq. (6.22) are multiplied. This reflects a logic gate where activity requires that Cro is present AND CI is absent. One may as well consider input functions that also reflect other logic gates, including, for example, an "or" gate where presence of Cro OR absence of CI secures a near to maximally active promoter.

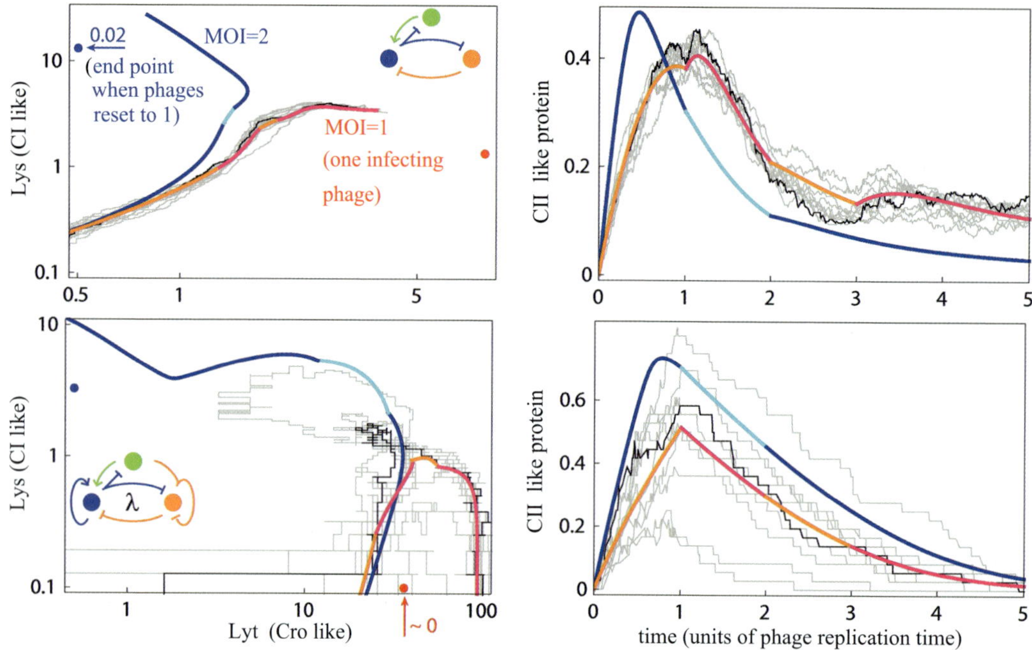

Figure 6.15 Examples of motifs that can count [248]. Left: Lys(CI)–Lyt(Cro) trajectories after infection. Right: Trajectories of CII with red-orange trajectories are for the multiplicity of infection MOI $= n = 1$ leading to lysis. Blue-cyan trajectories are for the $n = 2$ infections, leading to lysogeny. The color is set to change for each subsequent phage replication time. The lower set of panels simulate a λ-like circuit, with the additional assumption that CII is not produced at time $t > \tau_{rep} = 1$.

phage replication, an event occurring every $\tau_{rep} = 5$ minutes. In the simulation we let n grow to $n = 8$, and then first maintain phage copy number from further growth, then subsequently drop to $n = 1$ to test the ability of the phage to remember the initial count. The motif is simulated from an initial value $n = 1$ and also from $n = 2$, and a motif is classified as succesful if the end points of the two trajectories at time $t \gg \tau_{rep}$ are widely seperated.

Figure 6.15 shows examples of infection dynamics, starting with $n = MOI = 1$ (red-orange) and $n = 2$ (blue-cyan) infections. The timescale is measured in units of the phage replication time, τ_{rep}, believed to be about 3–5 min for λ infecting an *E. coli* in rich medium. One sees a parallel development of the two infection trajectories until about one phage replication time. After this point the trajectories diverge rapidly, reflecting a network that is able to count the initial gene dosage.

Sampling regulatory links with associated binding constants, protein lifetimes and rate constants, one finds that mutual repression between CI and Cro is essential for succesful counting. However, for some motifs it is easier to find working parameter sets than for other wiring diagrams. The two motifs that have the highest success rates are shown in Fig. 6.16. Remarkably, including noise in the dynamics, motifs without CII fail

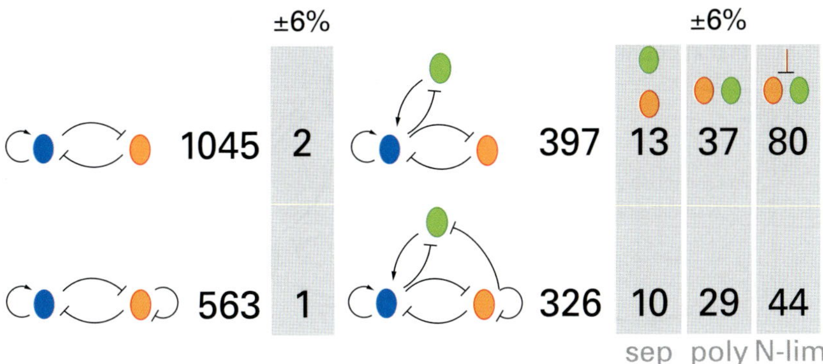

Figure 6.16 Robustness of network designs that function similarly to phage λ. The columns show the number of parameter sets (in ppm = parts per million) that work for the corresponding motif. The gray shading refers to stochastic simulations with spread/mean = 0.06. The "sep" column refers to simulations where Cro and CII are produced from independent promoters; "poly" refers to simulations where Cro and CII are produced in synchrony from the same polycistronic operon; "N-lim" includes the polycistronic design, and in addtion stops CII production after the first phage replication.

systematically.[4] The statistics of working motifs with noise is shown in the gray-shaded columns in Fig. 6.16.

Many of the three-node motifs in fact work robustly at quite high noise levels, a feature that is is closely coupled to the fact that succesful motifs typically split the decision and the maintenance tasks between a short-lived CII and a long-lived CI. This corroborates the more general observation of a wide class of decision motifs [251] that are found to be governed by division of tasks between a fast decision loop and a slow maintenance loop.

The robustness of a given motif is measured in terms of the number of working parameter sets. This robustness is increased when CII and Cro are produced from the same promoter (see "poly" column). Robustness is further increased if CII production is terminated after one phage replication time (see the last gray column), a regulation that may be associated with the short-lived anti-terminator N.[5]

The ability to count multiplicity of infection (MOI) reported by Kourilsky is remarkable, given the small number of molecules. Also, phage counting should be disturbed/influenced by variation in cell size at the moment of infection [252], as one phage in a small cell may act chemically as two phages in a big cell. Finally the lysis decision may be biased by the possibility that some phage genomes switch to lysis, even if the majority opt for lysogeny [253]. A possibility that would require substantial influence by

---

[4] Noise is simulated by a Gillespie algorithm, where protein concentrations are changed in discrete steps with a size that is calibrated to give noise CV = 6% for a steady state of proteins produced from unregulated promoters. This number is chosen to obtain at least one working two-node motif. At CV = 7% none of the two-node motifs work. In contrast, three-nodes with polycistronic Cro–CII messaging and a stop to CII protein production at time $t > 1$ can work for 30% noise.

[5] N is a protein that is expressed from PRE in early infection, but repressed later by both CI and Cro. The anti-terminator N allows RNAP initiated from PR to transcribe CII (as well as O, P and Q) (see Fig. 6.12).

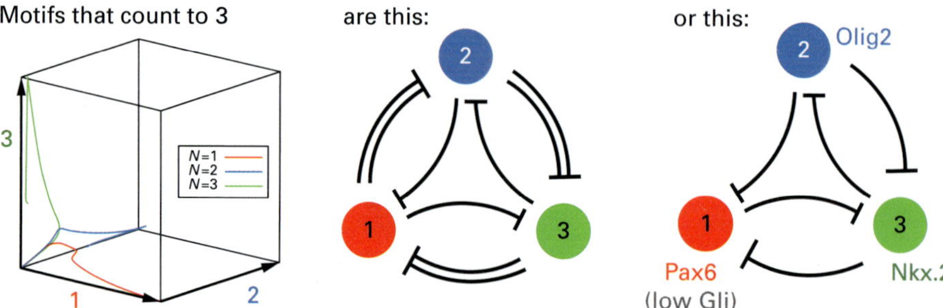

Figure 6.17 Genetic regulatory circuits that can count gene dosage [255], in one, two and three units and store the results in a persistently maintained high expression of one protein. Left: Low gene dosage of one (red) leads to domination of red protein, a gene dosage of two leads to domination of blue protein etc. Notice that simulation is continued for gene dosage reset to one, causing a sharp turn in $n = 2, 3$ trajectories. The middle and right panels show motifs that robustly count to three: a repressilator motif with breaks and a motif lacking repression from "1" to the "2" node, respectively [255]. The double arrows represent strong repression, maintained by high-affinity binding. The right-most motif is found in a developmental module that differentiates the front, middle and back of the spinal cord [256].

non-diffusable regulators, a role that may be associated to the *cis*-acting anti-terminator Q [254] located on the far right of Fig. 6.12A.

From a wider perspective, counting is associated with many developmental processes, including the outstanding problem of how vertebrates count to the exact number of vertebrate segments during their development (31 for humans). A more tractable problem is to construct networks that show more than binary sensitivity to a morphogen. Sampling possible three-node motifs, [255] found two classes that can perform such a task. These two network motifs are shown in Fig. 6.17. The right-most network is also found in a developmental module of vertebrates, which use it to differentiate between a low, a medium and a high dosage of the morphogen Gli [256].

6.4   Summary

- The discrete nature of single mRNA transcription can be captured by an event-driven simulation. For two processes happening with rates α and β, respectively, the next of these events events is selected according to the minimum of $-\ln(ran)/\alpha$ and $-\ln(ran)/\beta$, where *ran* is a random number $\in [0, 1]$.
- Bistability involves positive feedback and co-operativity between transcription factors.
- Molecular networks can remember states far beyond the lifetime of the proteins that regulate them.

7   *cis*-Acting gene regulation and epigenetics

7.1   Nucleosomes and their enzymes

Previous chapters explored gene regulation where the regulator was produced from one gene, and subsequently diffused across the cell to bind to suitable binding sites at operators for other genes. This type of *"trans"*-acting gene regulation is commonly observed in both prokaryotes and eukaryotes. However, there exist other and more local mechanisms, allowing more fine-grained regulation. Such so-called *"cis"*-acting gene regulation involves proteins that influence regulation locally on the genome (see Fig. 7.1).

One example of such local regulation is the anti-terminator protein Q in phage λ, which mainly influences the phage genome where it is expressed [257, 258, 259].[1] The known examples of *cis*-acting gene regulation in prokaryotes is associated with regulators with very short lifetimes that bind either DNA or RNA, and thus the locality may rely on competition between effective diffusion time and decay time [259]. We will see that eukaryotes may instead use positive feedback as a powerful localization strategy.

Mechanisms for localized regulation of gene activity in eukaryotes include DNA methylation of CG sites [261, 262] and gene silencing by nucleosome modifications [263, 264, 265].[2] Interestingly, the nucleosome mediated gene silencing is so strong that it in itself can be used to sustain a genome in one of two bistable systems across many cell generations. A convincing experimental proof of such locally manitained epigenetics is found in mutants of *Saccharomyces cerevisiae*, where [266] demonstrated that two copies of the same genomic region within the same cell can be maintained independently in different states. The experiment is outlined in Fig. 7.2. Similarly Thon and Friis [264] demonstrated that mutants of *Schizosaccharomyces pombe* can locally sustain an active genomic region on one chromosome, while the genomic locus on the duplicate copy of this chromosome is silenced. Accordingly, it has been demonstrated that epigenetics can be sustained within a local (*cis*-acting) region in the genome.

---

[1] Other examples of *cis*-acting genes is the DNA nicking protein A in phage ϕX174 and the regulatory protein A in phage P2 [260]

[2] Eukaryotes are larger and perhaps more complicated than prokaryotes, with membranes seperating the cell nucleus, the cytoplasm and the endoplasmatic reticulum. Eukaryotes originated 2 billion years ago as a succesful merger between archea and bacteria. The volume of a eukaryotic cell is 100-to 1000-fold larger than the bacterial volume.

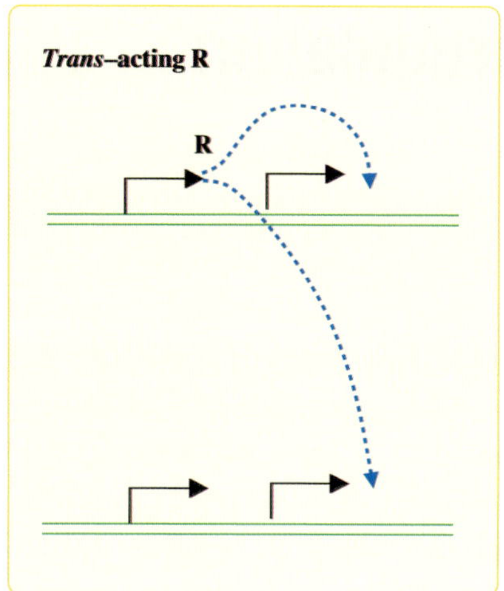

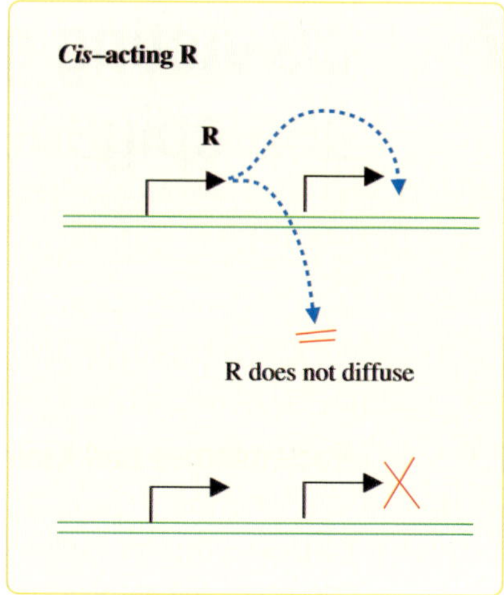

Figure 7.1 Left: Most genes are regulated by diffusible regulators, which implies that proteins produced from one gene regulate a nearby promoter and a distant promoter equally well. The figure illustrates such a setup, where the gene producing the regulator R is removed from the lower gene locus. Right: If the regulator R is *cis*-acting it only regulates a promoter positioned close to where R is produced. Such *cis*-regulation is reported for the anti-terminator protein Q for phage λ, which under normal expression only acts on the phage genome where it is produced.

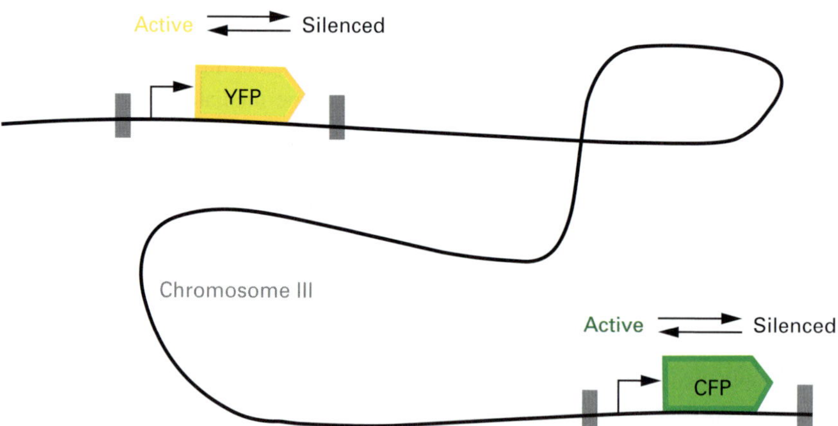

Figure 7.2 *cis*-Regulation as demonstrated by ref [266]: the mating type of *S. cerevisiae* is copied, and a YFP, or a GFP is inserted at each locus. For a certain mutant of this yeast, one observes that each yeast locus is bistable, with states that each are inherited for 5–80 generations. This bistability cannot be associated with any diffusible transcription factor because the state of one locus is found to be independent of the state of the other [266].

Regulatory systems with a local component are needed for example in the olfactory system, where hundreds of nearly identical genes each express their particular receptor protein [267, 268, 269, 270], such that one and only one gene is transcribed in each cell. Thus, in some way, these membrane-bound receptor proteins conspire to express only one of these genes in each cell, and silence all the other genes. To regulate this, the cell needs a combination of a repressor that acts on all genes, and a local positive feedback acting independently around each gene. The local positive feedback may involve nucleosomes [268, 271, 272], which on the scale of individual genes can maintain themselves in a persistently active or persistently inactive state. The end of this chapter will discuss olfactoric differentiation.

Another important *cis*-acting system is associated with polycomb silencing of *Hox* genes, which plays a central role in the development of mammals [273]. Polycomb proteins interact with chromatin, modify nucleosomes and thereby play a part in stem-cell differentiation during the early embryonic state [274, 275, 276].[3] The polycomb complexes in fact work together with the *trans*-acting transcription factors OCT4, SOX2 and NANOG that govern early stem-cell differentiation. Thus, when considering *cis*-acting gene regulation, one should keep in mind that they represent a sub-section of a larger regulatory system, that incorporates *trans*-acting diffusible regulators.

We here focus on nucleosomes as the main substrate for *cis*-acting regulation. Nucleosomes are protein complexes that package eukaryotic DNA, with a density of about one nucleosome per 200 bp (see Fig. 7.3). The core nucleosome is composed of two half-nucleosomes, each consisting of four core histones (H2A, H2B, H3 and H4), around which DNA is wrapped about two times. A nucleosome is 6×11 nm and binds to its 142 bp DNA segment with about 27 ± 1 $k_B T$ free energy [279, 280, 281].[4] The nucleosomes also interact with each other; *in vitro* the stacking energy is measured at $14\,k_B T$ [283].

Nucleosomes may carry a number of alternative chemical modifications (e.g. acetylation and methylation) at different amino-acid positions. The modifications allow recruitment of proteins/enzymes that bind to the nucleosomes in a modification-sensitive manner ("readers"), and modify other nucleosomes in turn ("writers"). The "writer" enzymes add or remove modifications on nucleosomes that change the binding energy and recruitment ability of the modified nucleosome. The "writer" enzymes incude histone acetyltransferases (HATs) [284, 285], histone methylases (HMTs) [286, 287, 288, 289, 290, 291, 292], histone deacetylases (HDACs) [293, 294, 295] and histone demethylases (HDMs) [296, 297, 298]. See Fig. 7.3 for examples and actions of read–write enzymes in two yeast model organisms.

---

[3] Polycombs organize into two key complexes, PRC1 (polycomb repressive complex 1), and PRC2 (polycomb repressive complex 2). PRC2 is used in the maintenance of repressive chromatin states, on which PRC2 binds, a binding that in turn allows activation of the methyltransferase activity of PRC2 on nearby nucleosomes, facilitating a positive feedback circuit for a sustained repressive chromatin state [277]. This repression is supported by methylated nucleosomes recruiting PRC1, which in turn facilitate chromatin compaction. In contrast, a group of proteins named thritorax favors active marks on nucleosomes [278].

[4] The total binding energy comes from a DNA–protein binding energy of $85\,k_B T$ that is reduced due to the cost of a bending energy of $58\,k_B T$ [281]. The resulting $27\,k_B T$ binding energy implies that nucleosomes are already fully bound to DNA at a concentration of 0.5 M. Nucleosomes can be actively removed from DNA by the abundant RSC [282].

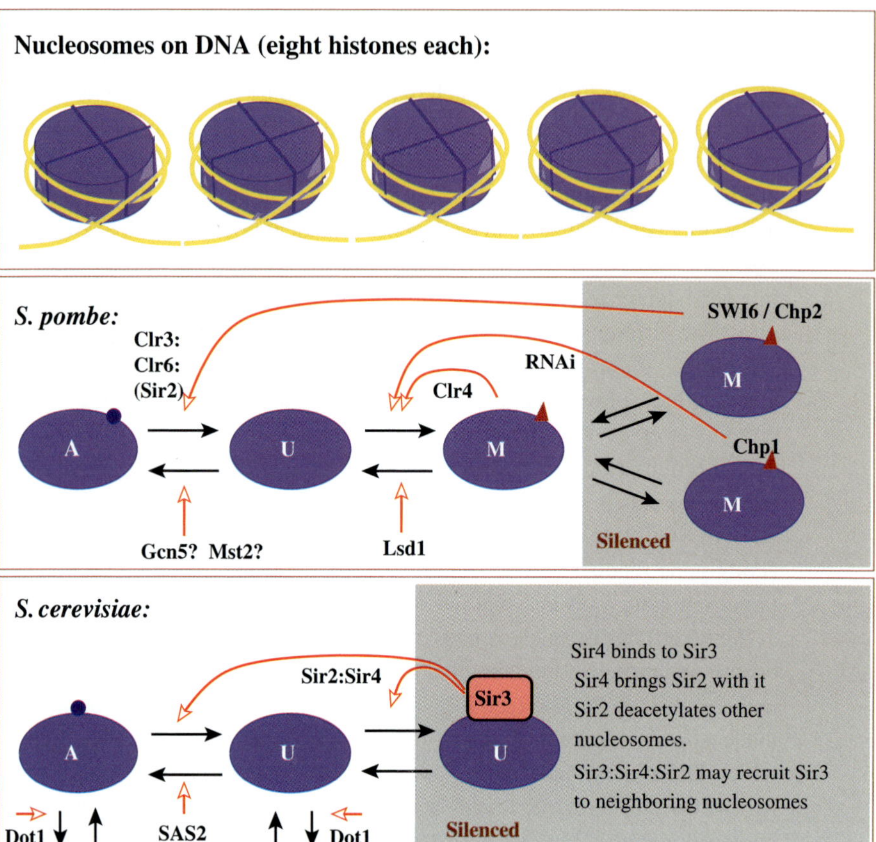

Figure 7.3 Top panel: A histone octamer on DNA forming core nucleosomes that each consist of four pairs of histones H2A, H2B, H3 and H4. The mass of these proteins is similar to the mass of the DNA involved. Each histone has residues that can be modified, for example by methylation (on H3K9) or by acetylation. Middle panel: Methylated nucleosomes in *S. pombe* recruit certain enzymes ("readers"), which in turn modify other neighboring nucleosomes ("write") [286, 287, 288, 289, 290, 291, 292, 293, 294, 295, 296, 297, 298, 299, 300, 301, 302, 303]. This allows positive feedback and epigenetic memory. The opposing demethylation [304, 305] and acetylation [284, 285] may not be recruited. DNA covered by acetylated nucleosomes can be transcribed, whereas DNA covered by (H3K9me) methylated nucleosomes can be silenced [306]. Lower panel: Nucleosomes in *Saccharomyces* are also involved in positive feedback [307, 308], involving enzymes other than *S. pombe*.

The read–write enzymes create the possibility of positive feedback and bistability [263, 265, 277, 306, 309, 310, 311]. The idea is that modified nucleosomes recruit enzymes that cause other nearby nucleosomes to become similarly modified. Indeed, several histone-modifying enzymes involved in epigenetic regulation are known to recognize and create the same modification [277, 312, 313, 314]. The positive feedback is illustrated by "red" recruitment links in Fig. 7.3, forming self-reinforcement dynamics around nucleosomes that are in the "silenced state."

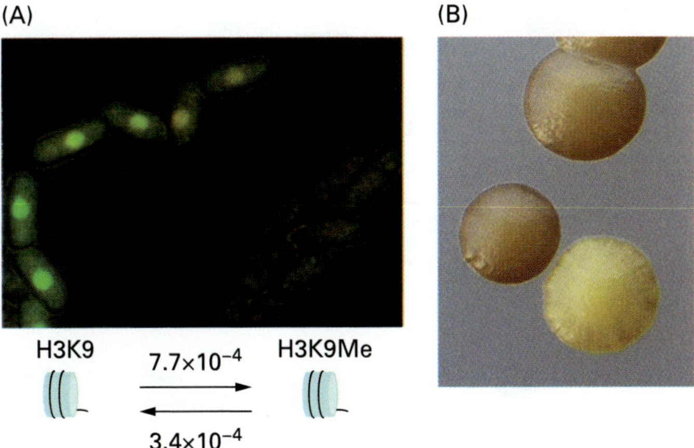

Figure 7.4 Epigenetic states in mating-type mutant of *S. pombe* [264, 265]. (A) The two states of *Pombe*, with cells having transcription of GFP from mating locus, and cells without any transcription. Lower part shows the measured switching rates, and, in addition, illustrates that the trancsription activity is systematically repressed by methylated nucleosomes (that furthermore bind SWI6) [315]. (B) Epigenetics at the colony level, emphasizing that the stability of each state is so large that a whole colony typically will be in one or the other. Figures courtesy of G. Thon.

7.2 Nucleosome-mediated epigenetics

A remarkable example of chromosomal epigenetic memory is provided by certain mutants of the silenced mating-type region of the fission yeast *Schizosaccharomyces pombe* [264, 265] (see Fig. 7.4). This mutant has a 12 kb DNA region where the expression is bistable, switching occasionally between states where the *ura4* gene in the region is expressed (active) or silenced. Each state is stable and heritable, with stochastic transitions occurring on a timescale of 2000 cell generations.

To understand how this works, we will introduce a model of the *S. pombe* mating-type system with its ~ 60 nucleosomes. We assume [306]: (1) only three relevant kinds of nucleosomes are considered, unmodified (U), methylated (M) and acetylated (A); (2) nucleosomes are actively interconverted by read–write enzymes, forming the positive feedback in the system (for simplicity we include four symmetrical positive feedback loops in the model (Fig. 7.5) and quantify all reactions by one parameter α); (3) nucleosomes can also be interconverted in a recruitment-independent manner, for example due to the activity of modifying enzymes that are not part of the mating-type region (for simplicity, we consider equally strong direct (non-recruited) interconversions); (4) rates of interconversion reactions are independent of position, thereby disregarding effects of silencer elements, which play an important role in the wild-type mating system.

7.2.1 Model of a system with L nucleosomes

The stochastic model is executed in steps, where at each step a random nucleosome $n_1 \in [1, L]$ is selected:

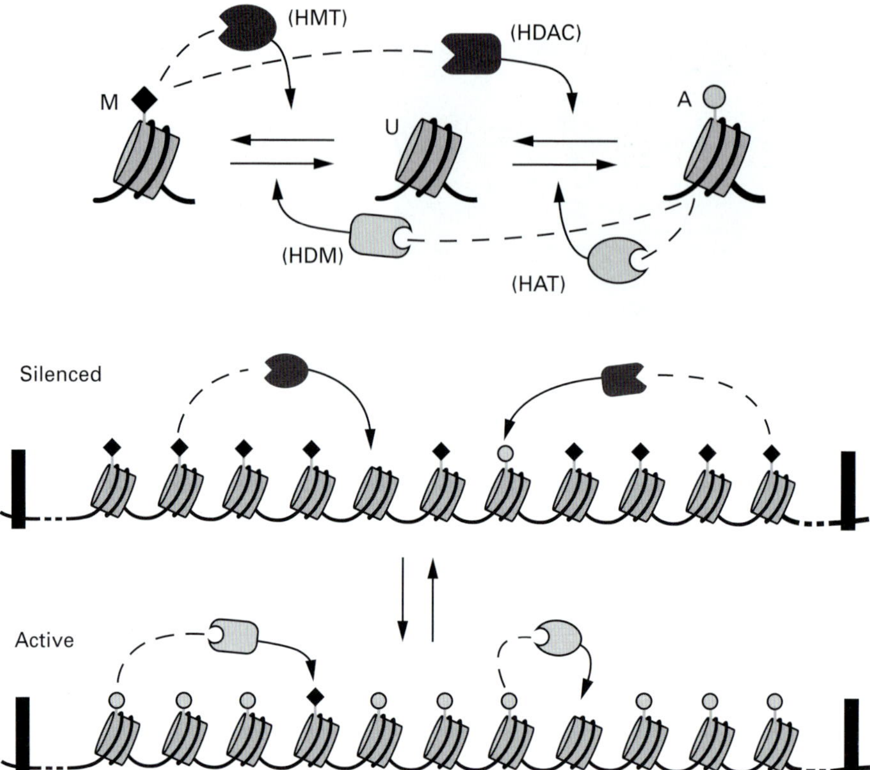

Figure 7.5 Each site represents a nucleosome that can be methylated (M), unmodified (U) or acetylated (A). Transitions between the three states are in part non-recruited (controlled by direct transitions $1 - \alpha$), and in part recruited through other nucleosomes in either the M or the A state. The corresponding mating-type switch in a mutant of *S. pombe* consists of $L = 60$ nucleosomes that switches between a silenced (methylation-dominated) state and an active (acetylation-dominated) state with a lifetime of each state of about 2000 cell generations [264].

- A recruited conversion is attempted with probability α. A second random nucleosome $n_2 \in [1, L]$ is selected ($n_2 \neq n_1$). If n_2 is in either the M or the A state, nucleosome n_1 is changed one step toward this state. For example, when nucleosome n_2 is an M, if n_1 is an A, then it is changed to a U and if n_1 is a U it is changed to an M. If nucleosomes n_1 and n_2 are in the same state, or if n_2 is a U, then no changes are performed.
- A noisy change is attempted with probability $1 - \alpha$. A nucleosome is selected and converted to one of its neighbor states along the A–U–M axis.

We found it convenient to define a feedback-to-direct ratio, $F = \alpha/(1 - \alpha)$, to emphasize the importance of the relative strength of the positive feedback. The time, t, in the model is measured as average attempted nucleosome conversions per nucleosome: one timestep is $L = 60$ attempted conversions of randomly chosen nucleosomes.

Figure 7.6 shows the dynamics of the system at three different F values. At low feedback F, the fraction of M nuclesomes in the system, m, fluctuates around 0.25, as

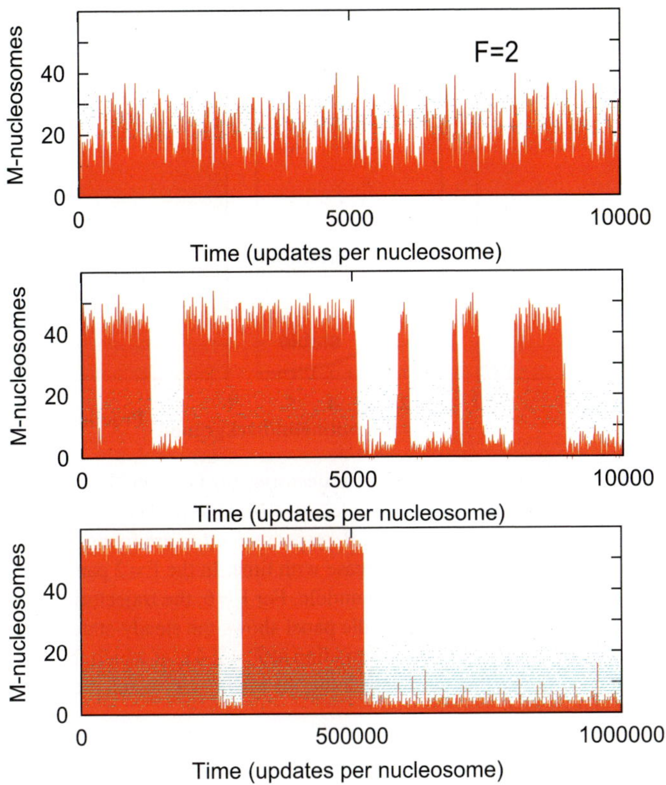

Figure 7.6 Time development of the standard model for a system consisting of $L = 60$ nucleosomes. The three plots illustrate the onset of bistability when $F = \alpha/(1 - \alpha)$ increases from $F = 2$ (top) to $F = 4$ (middle panel). Further observe that the lifetime of each epigenetic state increases dramatically when changing from $F = 4$ to $F = 6$) (bottom panel). The small dots show occupation of the intermediate U state.

do the number of A nucleosomes. As F increases, the system becomes more bistable, tending to exist in either a high M state or a low M state. The average lifetime of the high M state depends strongly on F. In particular, a sufficiently high F can stabilize the epigenetic state to an arbitrarily high degree.

The critical F can be calculated analytically, using the deterministic description of the algorithm:

$$\frac{da}{dt} = f = \alpha[a \cdot (1 - a - m) - a \cdot m] + (1 - \alpha)\left(\frac{1 - a - m}{2} - a\right) \tag{7.1}$$

$$\frac{dm}{dt} = g = \alpha[m \cdot (1 - a - m) - a \cdot m] + (1 - \alpha)\left(\frac{1 - a - m}{2} - m\right) \tag{7.2}$$

Here, m is the fraction of nucleosomes in the M state, a the fraction in state A and $1 - a - m$ the fraction in state U. Positive signs of f and g are shown by shaded areas in Fig. 7.7. The condition for steady state is that *both* of these functions are zero. The steady-state solutions are shown in Fig. 7.7. Stability of the fixed points can be

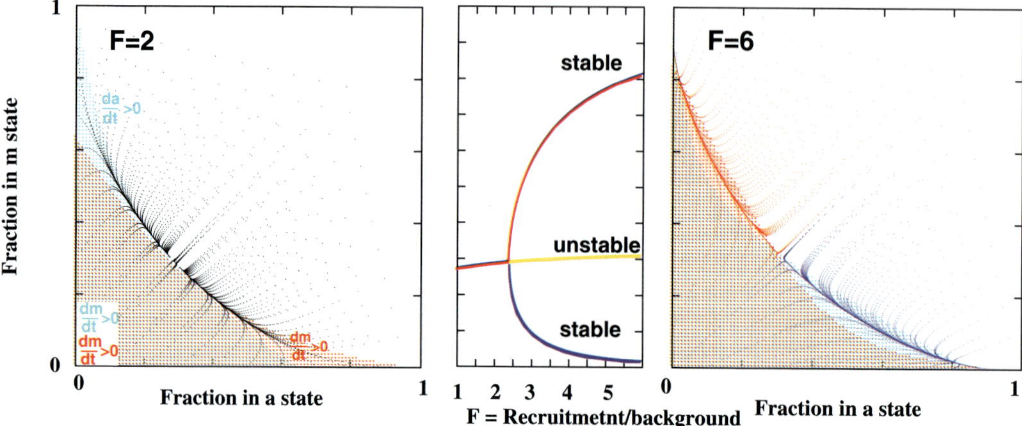

Figure 7.7 Left and right panels: Deterministic trajectories for $F = 2$ and $F = 6$ as a function of $a = A/L$ and $m = M/L$, from starting at points $(a, m) \in [0, 1] \times [0, 1]$. The shaded regions mark areas where $f = da/dt > 0$, and $g = dm/dt > 0$, respectively. f, g are found in Eqs. (7.1) and (7.2). In non-shaded areas both a and m decrease with time. In the $F = 2$ panel, the trajectories converge around the stable fixed point in the middle. For $F = 6$, the trajectories diverge and separate into two metastable states. The middle panel shows the steady-state solution to Eqs. (7.1) and (7.2). Notice the pitchfork bifurcation at $F = 2.41$, at which the mixed state becomes unstable. Also notice that the stable solutions for $F \to 6$ match the fixed points of trajectories on the right panel.

determined from the sign of flow in each region. This identifies the middle intersection at $F > 2.41$ as unstable. The stability of a fixed point can also be determined by linear stability analysis, as discussed in Question 7.2.3 at the end of this subsection, pinpointing the critical value $F = F_c = 2.41$ (see Fig. 7.7).

7.2.2   Result 1: Epigenetics need co-operativity

Bistability requires not only positive feedback but also non-linearity in the feedback loop. One way in which this can occur is through direct co-operativity (high Hill coefficient), if two or more nucleosomes are needed for each recruitment event. This will be demonstrated later in the simplified two-state model.

In the presented model, each recruitment reaction requires only one other nucleosome to be in an A or M state in order to cause conversion. Thus the rate at which any U nucleosome converts to an M nucleosome increases only linearly with m. However, the two-step reaction model (Fig. 7.5) is implicitly co-operative: a transition from, for example, an A to an M nucleosome, requires two consecutive recruitments by M nucleosomes (one for de-acetylation and one for methylation) and therefore has a rate proportional to m^2. If we remove this two-step positive feedback by eliminating either the recruited de-modification reactions or the recruited modification reactions, we eliminate bistable behavior (middle panels in Fig. 7.8).

Bistability could, however, be restored to single-feedback systems by adding explicit co-operativity into the recruitment reactions, see Fig. 7.8. Consider, for example, the model in the middle panels of the figure, where the red histogram refers to the case

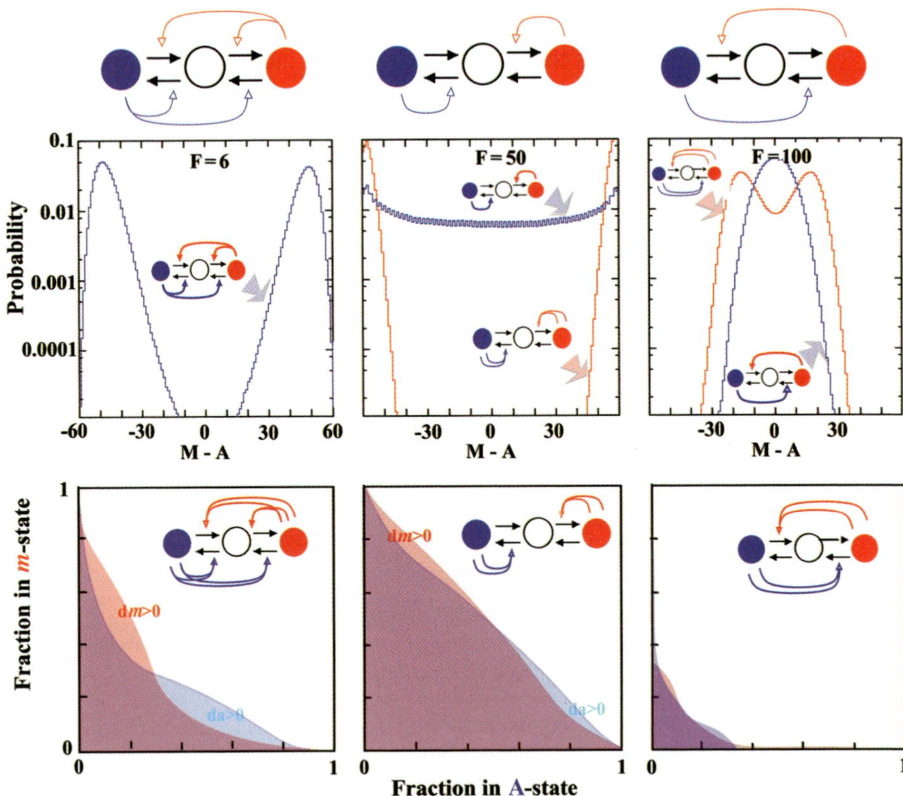

Figure 7.8 Variation of recruitment model where some recruitment links are removed. Simple models without consecutive recruitment do not work. However, bistability can be obtained if recruitment demands two random nucleosomes to be in the same state before the indicated conversion can take place. Lower panels show the region where $dm/dt > 0$ and $da/dt > 0$ for the corresponding models.

where a recruitment reaction consists of first selecting two nucleosomes; if both are in the A state, a third nucleosome is selected and if this is in the U state it is converted to the A state. If the third nucleosome is in the M state, then no recruitment is done. In the case where both of the first nucleosomes are in the M state, recruitment is attempted in the opposite direction. Figure 7.8 analyzes this and other variants of co-operative recruitments, emphasizing that two-step sequential linear recruitment can indeed be replaced by one-step co-operative recruitment.

7.2.3 Result 2: Epigenetics need non-local interactions along the DNA

Nucleosome modifications are envisioned to spread along the DNA. This mechanism is supported by studies of telomeric silencing in *S. cerevisiae* [316]. In contrast, our standard model assumes that, in the recruitment reaction, any nucleosome can act on any other nucleosome in the region (perhaps by DNA looping).

We tested the neighbor-only contact mechanism by drawing the recruiting nucleosome, n_2, randomly from the neighbors of the nucleosome to be converted, n_1. We found that this constraint eliminated bistability in our model (Fig. 7.9). Effectively, the local model concentrates the stability to be nearly independent of system size and linearly dependent of F.

The neighbor-only system spends a large fraction of its time in states in which m ranges from 0.05 to 0.95; there is no clear threshold between silenced and active states. When the neighbor-only system is in the intermediate region of the probability distribution, boundaries between large patches of M or A nucleosomes wander along the DNA as a random walk (see Fig. 7.10). Defining M as the number of nucleosomes in

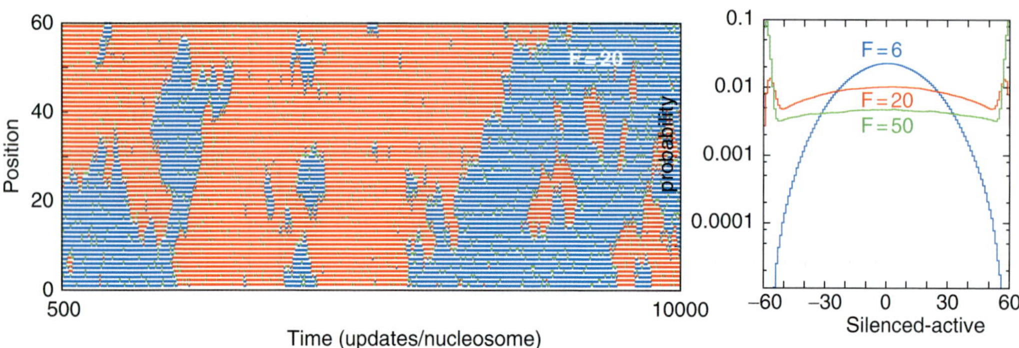

Figure 7.9 A local version of our standard model: here the recruiting enzyme can only modify a nearby nucleosome. The left panel shows space-time development with $F = 20$, whereas the right panel shows histograms for different F values. One never obtains bistability, in the sense that the stabilities in all cases are equal to the probability that one nucleosome "switches." Thus there is no stability gained by even a very large system size.

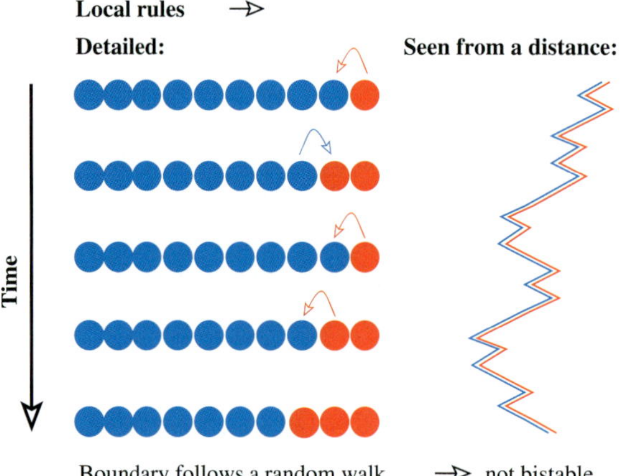

Figure 7.10 Local recruitment cannot give bistability. Assume a balanced situation, where A and M are equally strong: the interface between the M part of the system and the A part is insensitive to the majority state and therefore follows a random walk. As a consequence, the stability will be proportional to the stability of the single nucleosome (The chance that one is converted is $\propto L$, and that a single random conversion grows to dominate will be $1/L$. Together these predict a stability that is independent of system size L.)

the M state, all M values become equally likely when there is only one transition boundary. In contrast, the non-local model system is strongly pushed away from intermediate states and spends the vast majority of its time in a low M or a high M configuration.

The difficulty of obtaining clear two-state behavior in the neighbor-only model reflects transition dynamics that are similar to those found in the one-dimensional Ising model [317] or the helix–coil transition in polymer physics [318]. Although nucleosome-mediated epigenetics is clearly a result of non-equlibrium dynamics, the requirements for bistability resemble classical arguments for phase transitions in one-dimensional versus higher-dimensional systems.

7.2.4 Robust bistability across DNA replication if the generation time is > 10

A critical feature of an epigenetic system is inheritance. We will therefore investigate the ability of both the high-M and the low-M states to be maintained through DNA replication. Upon DNA replication the parental nucleosomes are believed to be partitioned randomly between the two daughter strands and new nucleosomes are inserted or assembled to fill the gaps.

Cell division is included by supplementing our standard model with cell divisions at certain fixed time intervals. The generation time is measured in units of the number of attempted nucleosome updates per nucleosome. At these fixed times each nucleosome is replaced by a U nucleosome with probability $1/2$.

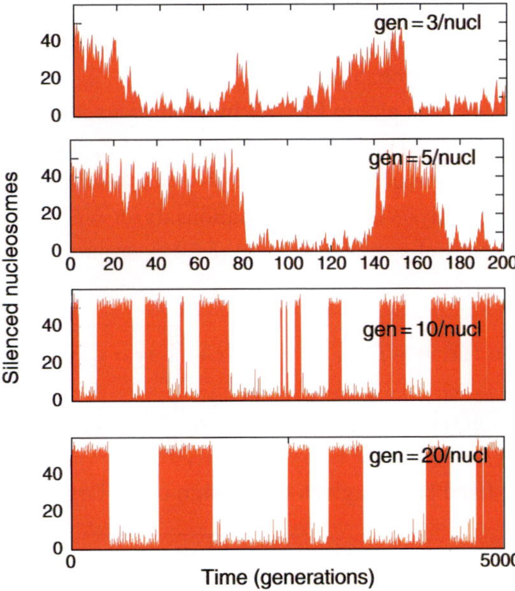

Figure 7.11 Inheritance of the epigenetic states across cell divisions, investigated for a recruitment to non-recruitment ratio of $F = 6$. A cell division is simulated by replacing about half the nucleosomes with unmodified ones. Generation time g is measured by number of attempts to update each nucleosome. For cell generation times < 10 nucleosome updates per cell generation, the stability deceases sharply.

Figure 7.11 shows the dynamics of the system as a function of cell generation time, for $F = 6$. The figure shows that high stabilities can be achieved, despite the destabilizing effect of frequent cell divisions. The observed $\sim 5 \times 10^{-4}$ transitions per cell generation can be achieved with feedback-to-non-recruitment ratios of $F \sim 6 - 8$ and about 10 updates per nucleosome per generation. Thus epigenetics require about one nucleosome modification once every 10 minutes in *Pombe* (which has a 120 minute generation time). When there is more than this number of modification attempts per nucleosome per cell division, then the number of switches per generation is independent of cell generation. The model then predicts a stability of epigenetic states that is robust to cell generation time.

7.2.5   Questions

7.2.1   Simulate the A–U–M model from Fig. 7.5 without cell divisions, but with the modification that recruitments $U \to A$ and $U \to M$ are local, whereas the other recruitments are global. Use an $L = 60$ nucleosome system, ignore cell divisions and select a feedback strength that gives reasonable bistability. Examine behavior in a space-time plot of nucleosome states as a function of position and time measured in updates per nucleosome.

7.2.2   Simulate the A–U–M model above, but with the modification that recruitments $A \to U$ and $M \to U$ are local, whereas the other recruitments are global. Use the same overall approach as above.

7.2.3   Repeat the above simulations, but now where global recruitments are acting across a distance x with probability $1/x$.

7.2.4   Reproduce the bifurcation plot in Fig. 7.7 by plotting fixed points for the model defined by Eqs. (7.1) and (7.2) as a function of F. Show analytically that bistability is only possible when $F > \dfrac{1}{\sqrt{2}-1} = 2.4142$.

7.2.5   The linear stability analysis of a set of equations $dx/dt = f$ and $dy/dt = g$ consider the dynamics of a small deviation $(\delta x, \delta y)$, around a fixed point (x_0, y_0):

$$\frac{d(\delta x)}{dt} = f(x, y) - f(x_0, y_0) = \frac{df}{dx}\delta x + \frac{df}{dy}\delta y = f_x \delta x + f_y \delta y \tag{7.3}$$

$$\frac{d(\delta y)}{dt} = g(x, y) - g(x_0, y_0) = \frac{dg}{dx}\delta x + \frac{dg}{dy}\delta y = g_x \delta x + g_y \delta y \tag{7.4}$$

The stability of a fixed point is given by the sign of the largest eigenvalue:

$$\lambda = \frac{f_x + g_y}{2} + \sqrt{\frac{(f_x + g_y)^2}{4} - f_x g_y + f_y g_x} \tag{7.5}$$

Insert Eqs. (7.1) and (7.2) into this equation and plot the largest eigenvalue as function of F for the $m = a$ fixed-point solution.

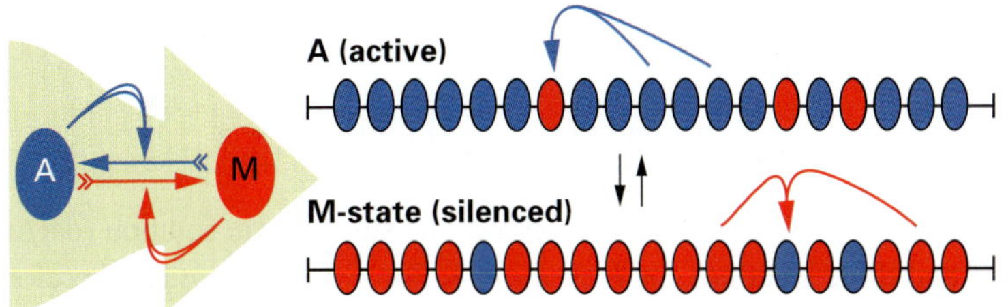

Figure 7.12 Two-state model, with two arrows indicating that two M are needed to convert one A into an M. Each site represents a nucleosome that can be methylated (M) or acetylated (A). Transitions between these two states are in part non-recruited, and in part auto-regulated by recruitment of histone-modifying enzymes: at each update of a nucleosome i it is set to either A or M randomly, with probability $1 - \alpha$. Two other nucleosomes are chosen with probability α, and if these two are in equal states then the state of nucleosome i conforms to this state.

7.3 A regulated two-state model

7.3.1 The two-state model

In order to analyze more complex situations, this section introduces a simple two-state version of the three-state model [319]. These two states denote the modified (M) and anti-modified (A) states. Instead of the previously used linear recruitment scheme, the two-state simplification requires direct co-operativity, in the sense that two nucleosomes with the same modification are needed for active conversions. For example, two Ms are needed to convert one A→M.

The two-state model is described in Fig. 7.12. The system can be mathematically characterized by the fraction of methylated states $m = M/L$, where L is number of nucleosomes in the considered system. Given a certain value of m, the acetylated fraction is then $1 - m$. If recruitment is attempted, then there is probability m of picking a methylated site, and a probability $(1 - m)^2$ that two subsequent random sites are acetylated. Thus the probability that $m \to m - 1/L$ due to recruitment is $\alpha m (1 - m)^2$. If a direct (non-recruited) event is attempted, then $m \to m - 1/L$ with probability m. Thus the probability that $m \to m - 1/L$ due to direct transition is $(1 - \alpha)m$.

Adding up all recruitment and direct transitions per unit time, the change in m is given by a Langevin equation (with a noise term denoted ξ that takes into account that events are discrete and in random order):

$$\frac{dm}{dt} = \alpha(m^2(1 - m) - m(1 - m)^2) - (1 - \alpha)m + (1 - \alpha)(1 - m) + \xi(t) \qquad (7.6)$$

Here the average noise $\langle \xi \rangle = 0$ and variance of the noise $\langle \xi\xi \rangle_t \propto 1/L$.[5] The above equation can be rewritten:

---

[5] In fact $\langle \xi(t)\xi(t') \rangle_t = 0$ when t differs from t', whereas $\langle \xi(t)\xi(t) \rangle_t = \int_{t=0}^{t=T} dt \xi(t)\xi(t)/T \propto 1/L$ when average over long time T. This last expression states that although ξ can be both positive and negative (so the average $= 0$), the square of the noise is finite. Further, this noise gets smaller as the system L gets larger. The reduction in noise is due to the L random recruitment choices at each timestep.

$$\frac{dm}{dt} = \alpha m(1-m)(2m-1) + (1-\alpha)(1-2m) + \xi$$

$$= \alpha\left(m(1-m) - \frac{1}{F}\right)(2m-1) + \xi \tag{7.7}$$

with $1/F = (1-\alpha)/\alpha$. This equation has one steady-state solution ($dm/dt = 0$) when $F < 4$ and three solutions for $F > F_c = 4$. For $F > F_c$ there are therefore two stable solutions: One at low m, the other at high m, separated by a barrier at the unstable state with $m = 1/2$.

7.3.1.1　Mathematical argumentation for co-operativity

The two-state model allows a simple argument for the need for co-operativity. If one only required one methylated site to make the recruitment, Eq. 7.7 would be replaced by:

$$\frac{dm}{dt} = \alpha(m(1-m) - m(1-m)) + (1-\alpha)(1-2m) + \xi$$

$$= (1-\alpha)(1-2m) + \xi \tag{7.8}$$

which would have only one solution, namely $m = 1/2$. Co-operativity is explicitly needed to drive the system away from the intermediate state.

7.3.1.2　Epigenetic landscapes

Modifications in one or the other direction may well be asymmetric. This asymmetry of activity of read–write enzymes could be under biological control, for example through the presence of localized silencers, or globally by changing the concentration or activity of one of the modifying enzymes.

In terms of the two-state model, asymmetry may be introduced by modifying the transition rates:

$$\frac{dm}{dt} = 2(1-\eta)\alpha(1-m)m^2 + (1-\alpha)(1-m),$$

$$- 2\eta\alpha m(1-m)^2 - (1-\alpha)m + \text{noise}. \tag{7.9}$$

Here, a new parameter η, defined in the range $0 < \eta < 1$, sets the relative strength of modification versus anti-modification. The symmetric case corresponds to $\eta = 1/2$.

In order to illustrate the asymmetric model we now introduce the reader to the concept of epigenetic landscapes, suggested by Waddington [320]. In this, the state of a cell is supposed to be confined by potentials that contain multiple valleys each of which correspond to a particular cell type. Strictly speaking, the dynamic system that defines the state of a cell is not describable by a potential. However, for systems in steady state one may define an effective potential from long-term sampling of the steady-state distribution P of states the system visits:

$$V = -\log(P) \tag{7.10}$$

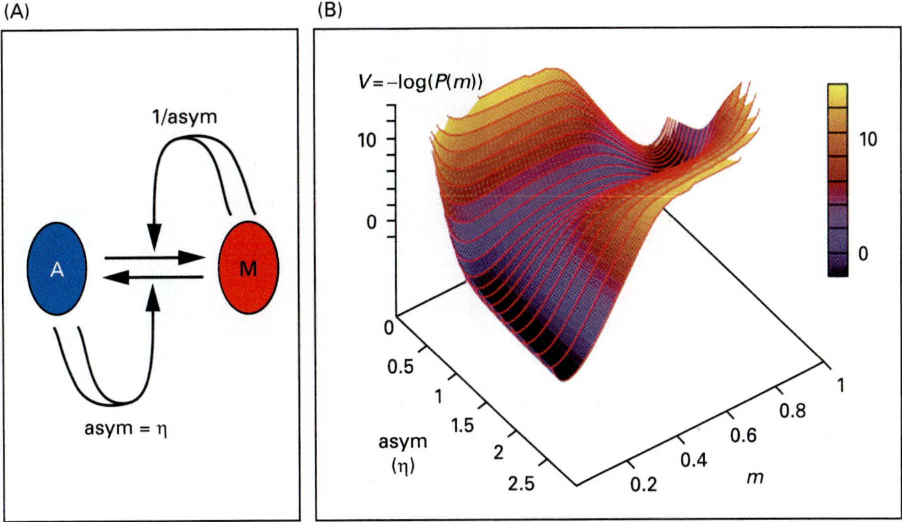

Figure 7.13 Epigenetic landscape obtained from the two-state model with asymmetric transition rate (the landscape is identified by the log of probabilities, $-\ln(P(m))$). The asymmetry is included via the parameter η, which is under biological control, and larger η indicates that the anti-modified state is more favorable. The total recruitment-to-direct-transition ratio is fixed at $F = 20$. The figure illustrates a dramatic tilting of the epigenetic landscape by a moderate change in nucleosome modification rates.

Therefore, very likely states will have low values of the "potential" V, reflecting the tendency that the system prefers such states.

Figure 7.13 shows how the potential landscape $V(m) = -\ln P(m)$ changes as a function of η, where $P(m)$ is the steady-state distribution. For simplicity, cell division is not taken into account here. For small η, the valley corresponding to the M state is much deeper than that for A state. However, the landscape changes dramatically as η increases past the critical value $\eta = 1/2$, at which point the system is symmetrically bistable. Increasing η above $1/2$, develops an increasing domination by the A state. Importantly, changing η slightly around the transition point gives an ultra-sensitive switch from an M-dominated state to an A-dominated state.

7.3.2 Indirect regulation through a system of nucleosomes

The properties of nucleosomes with their read–write enzymes may be used in *trans*-acting transcription regulation [321]. Figure 7.14 shows a thought experiment on how such a *trans*-acting nucleosome-modifying transcription factor may indirectly regulate the activity of a nearby gene [322]: by tilting the balance between an M-dominated state and an A-dominated state, the majority of the nucleosomes may shift to the active state, and subsequently allow replacement of some of the nucleosomes with transcriptional activators and RNAP.

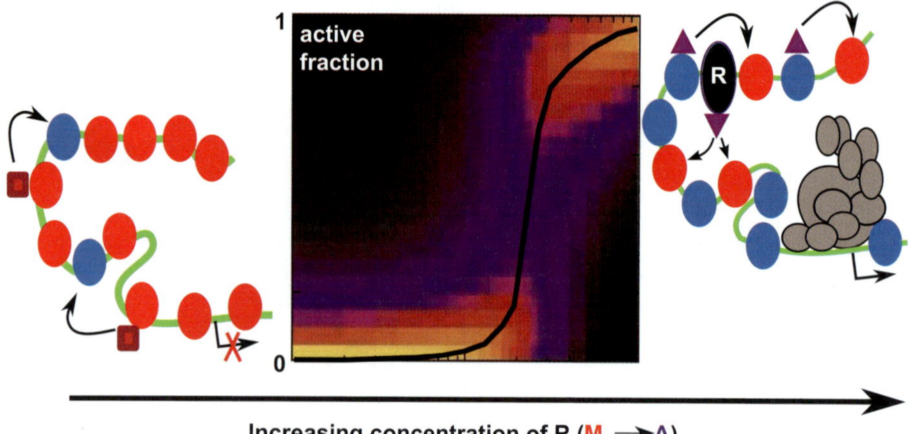

Figure 7.14 Models of indirect regulation by transcription factors. Indirectly acting transcriptional regulator R (black oval) which affects promoter activity by binding to DNA and recruiting a histone-modifying enzyme (squares) that modifies local nucleosomes (circles). When modified nucleosomes themselves recruit enzymes that foster the same modification on nearby nucleosomes (orange arrows), in a positive feedback loop, then even a regulator that binds non-co-operatively can produce an ultra-sensitive response.

For genes embedded in a system of nucleosomes, indirect gene regulation could, in principle, be conducted without any direct contact between the regulator and the RNAP at the promoter. Thereby, histone-modifying transcription factors can influence promoters without crowding the relatively small volume around the promoter. Further, it will allow a linearly acting regulator to provide an ultra-sensitive response. Possible examples of this mechanism for ultra-sensitivity may be found in [323, 324].

Indirect regulation through modifying a whole system of nucleosomes has not been proven, and at present remains a proposal for how upsteam operators and distant enhancers may modify the state of nucleosomes around the promoter. Noticeably, the activity of a promoter can be boosted by enhancer elements that are as much as 150 000 base pairs upstream of the promoter [325]. It is believed that gene activation and removal of the nucleosome at the promotor is tightly coupled [325], and removal should be easier if the nucleosome is in the active state. Gene activation is also often associated with transcription factors binding to nucleosome-depleted regions closely upstream of the promoters. Thereby the interplay between nucleosome states and competing transcription factor binding may in itself add additional dynamic feedbacks to gene regulation.

7.3.3   Toolkit for nucleosome-mediated epigenetics

Central to the local feedback is the assumption of a well-defined system of a number of L nucleosomes that can be described in itself. In reality, any stretch of nucleosomes is part of the larger cellular system, and will interact with it. In the above models, the only way to include these interactions is through non-recruited or spontaneous conversions,

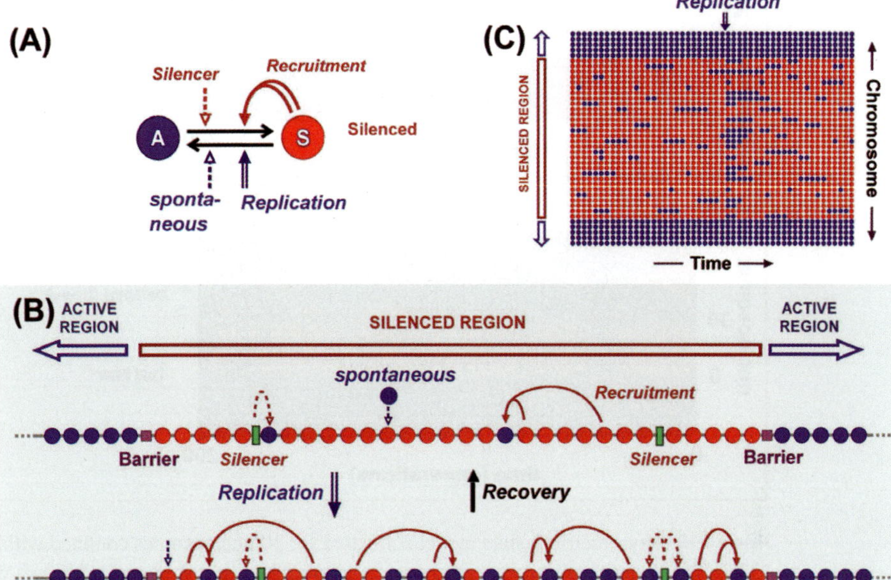

Figure 7.15 Asymmetric two-state model that in its simplest form recapitulates what is known from the Sir2:3:4 system in *S. cerevisiae*. Recruitments here only act from the unmodified nucleosomes, which bind Sir3:Sir4 co-operatively, and in turn recruit Sir2 that de-acetylates A nucleosomes. In the above model these complications are included by a recruitment action where two nucleosomes are selected randomly in $[1, L]$, and if both are silenced, a neighbor of one of them is set to the silenced state. The figure further illustrates a number of tools that the cell may use to confine activity to a region in the chromosome.

assigned a strength set by $1/F$. A more complete model should address the boundary between the system and its surroundings.

The nucleosome-mediated epigenetic systems in *S. cerevisiae* and *S. pombe* are systems that may be silenced, although they are surrounded by a large region of nucleosomes in the active state. This raises the question of how to localize a bistable system, a problem that is most easily investigated in terms of the two-state model. Figure 7.15A outlines a simple stochastic model [326], with silenced nucleosomes acting co-operatively to silence "active" nucleosomes. The nucleosome changes in the opposite direction due to cell divisions (which is assumed to convert each nucleosome to the active state with probability 1/2), as well as *trans*-acting enzymes. That this type of asymmetric model indeed works is demonstrated in Fig. 7.16. Importantly, if there are only recruitments from the silenced state, then the active nucleosomes throughout the surrounding genome will not disturb or destroy a local epigenetic system.

A more complete model for a confined bistable region could consider [326]:

*   The co-operative recruitment should be both local and global. For example one selects two nucleosomes to recruit, and if they both are silenced, one of their neighbors is set to the silenced state.

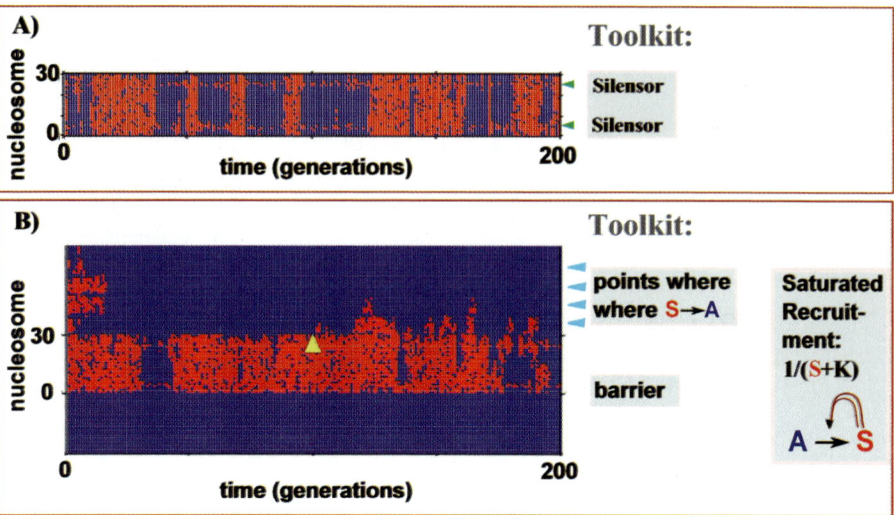

Figure 7.16 Asymmetric 2-state model simulated for 30 nucleosomes confined within two barriers [326]. The system parameters, are chosen to give a bistable system, provided two weak silencers within the 30 nucleosome region (at position 5 and 25). (A) Isolated system, emphasizing the role of silencers. (B) 30 nucleosome region embedded in larger active region, highlighting barrers and the possibility to confine activity when recruitment from silenced nucleosomes is limited by read-enzymes, and when outside region disfavors nucleosome in silenced state due to, for example, actively transcribed genes.

- The chance of interaction decreases with distance x along the genome. The frequency of recombinations is $\propto 1/x^{0.75}$ for distances up to about 10 kb (corresponding to $x \sim 50$ nucleosomes) [327]. For larger distances the decrease is consistent with the fractal globule scaling ($\propto 1/x$) [193, 194].
- There exist silencer elements [328, 329]. These are binding sites for transcription-like factors, which persistently spread the silenced state to neighboring nucleosomes. The silencer is an ongoing source for silencing within the considered region, a source that is absent outside the region.
- There exist barriers that prevent the spreading of the silenced state along the chromosome. The fact that spreading can be stopped challenges the possibility of recruitment at a distance. One way that a barrier can act is if just one of the necessary recruitments is strictly limited to the nearest neighboring nucleosomes, as, for example, suggested in the left-hand panel of Fig. 7.17.
- The maintenance of the silenced state may require persistant binding of a "reader" protein in limited supply. When such a limit is approached, the recruitment strength decreases and the number of silenced nucleosomes becomes self-limiting.
- Active promoters may suppress spreading of the silenced state [330], possibly by recruiting alternative modifications. Thereby some regions of the chromosome may be inherently more prone to silencing than others.

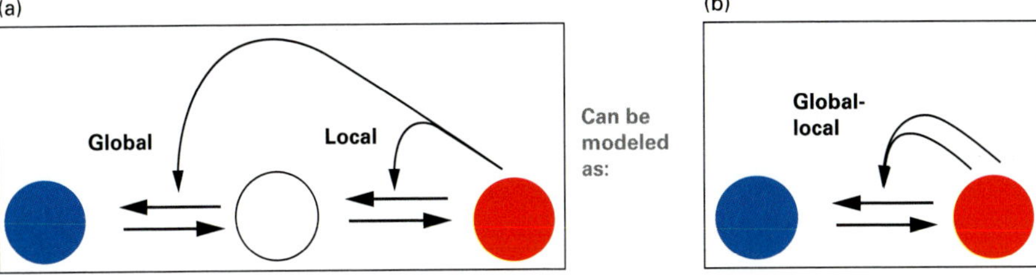

Figure 7.17 Simplifying the three-state model into a simple two-state model with co-operative recruitment. Notice that both models can exhibit bistability, when allowing one recruitment to be restricted to the nearest neighbors (local), and the other to interact with nucleosomes at distance x away with probability $1/x^{0.75}$ (global). In the two-state model, the global-local recruitment is simulated by selecting two nucleosomes, and if in the same state, then a random neighbor of one of them is converted.

Figure 7.16 shows the spatio-temporal dynamics of an extended system that includes these basic elements [326]. For subsequent model building we emphasize that although real nucleosome recruitment presumably involves sequences of state-to-state conversions, these processes can be modeled in terms of a simple two-state model with co-operative recruiment, see Fig. 7.17.

7.3.4 Questions

7.3.1 Simulate the two-state model for different F values for an $L = 30$ system.

7.3.2 Discuss the two-state model in terms of movements in an epigenetic landscape of the form:

$$V \sim - \int \langle \mathrm{d}m/\mathrm{d}t \rangle \, \mathrm{d}m$$

$$\propto \int \left(m(1 - m) - \frac{1}{F} \right)(1 - 2m)\,\mathrm{d}m \tag{7.11}$$

$$= m(1 - m) \cdot \left(\frac{1}{2}m(1 - m) - \frac{1}{F} \right)$$

Plot the potential, and discuss how residence time in one of the "potential wells" depends on the system size L.

7.3.3 Implement and simulate gene regulation in a two-state version of the nucleosome model where the co-operative recruitment only acts in one direction. Thus the co-operative recruitment acts by converting S to A, when two random nucleosomes are in the A state, and the effect of the transcription factor is to convert the A state to S with some adjustable rate R.

7.3.4 Consider a two-state nucleosome model with co-operative recruitment from state S to state A and with passive conversion of nucleosome in the opposite direction. Assume direct transitions β per active recruitment attempt, and that directed drift (toward "y") has strength *drift* per active recruitment attempt. Derive equations for the model and plot regions in x, y where $dx/dt > 0$ and $dy/dt > 0$. Verify that there are parameters where the model is bistable, i.e. where there are two stable fixed points separated by an unstable one.

7.4   Coupled epigenetics in olfactoric differentiation

The olfactory system presents a case for a coupling between *trans*-acting regulation and local positive feedback. Each olfactory neuron expresses only one allele out of a large and highly homologous gene family, comprising almost 1400 olfactory neuronal receptor genes (OR) in mice [267, 268]. In this choice, each olfactory neuron represents a developmental system with multiple stable states. Mechanisms have been put forth to describe the underlying nature of this seemingly stochastic multistability [269], including the possible involvement of chromatin remodeling in regulation of OR gene selection [268]. Figure 7.18 illustrates conceptually how selection of one of many can be accomplished by coupling a local positive with globally negative feedback.

Figure 7.19 from [333] illustrates a model of OR choice using an extended two-state model for the *cis*-acting positive feedback. In this model, each OR could be in one of two general states; either activated or silenced. To mimic the mutual exclusion of expression amongst OR genes they are assumed to be coupled through a hypothesized X-factor, governed by the sum of all active OR genes, see Figure 7.19A. This X provides global negative feedback between genes, an assumption supported by [334, 335, 336, 337].

The negative feedback should favor the silenced state of all nuclesomes with a strength that increases with the sum of the activity of all genes. The negative feedback may, for example, be mediated by a factor that binds to silent nucleosomes and prevents

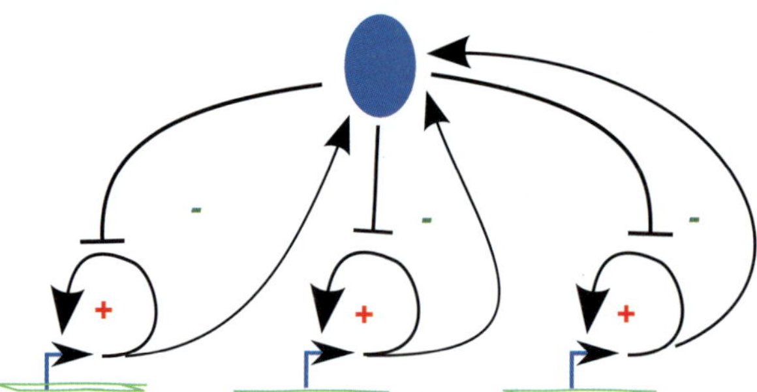

Figure 7.18 Local positive feedback and global negative feedback, visualized conceptually in analogy with Turing patterns [5, 331, 332]. For simplification we have not shown that the genes in fact are different, as, in respect to the negative feedback, all are assumed to behave similarly.

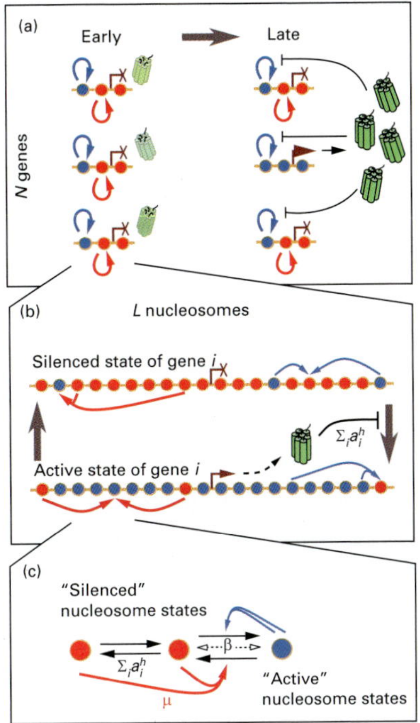

Figure 7.19 Local positive feedback and global negative feedback. (A) N genes are each covered by nucleosomes (red indicates silent). The nucleosomes on each gene form local positive feedback. The active gene expresses a receptor protein that indirectly favors the silencing of all nucleosomes. (B) Each gene is covered by L nucleosomes that can each be "active" or "silent." The nucleosomes with associated "read–write" enzymes form positive feedbacks. The green gene product indirectly pushes all genes towards the silent state. (C) Nucleosome states catalyze modification of each other co-operatively, with "silent" recruitment reduced by the activity of all genes, $\sum a_i^h$. (Figure from [333].)

their conversion to active ones. This will repress the local positive feedback towards active nucleosomes (reducing μ in Fig. 7.19C), and maintain all other genes covered by nucleosomes in the silenced state, in agreement with observation [272].

A new olfactoric neuronal stem cell is initiated by assigning all nucleosomes in all OR genes to the silent state. This is consistent with the observations of ref. [272]. Within the simulations we identify the state of the individual gene, i, as active, when the active fraction of nucleosomes, a_i, exceeds a threshold that can be either modeled as a Hill function (activity of gene $i \propto a_i^h$ with $h > 1$) or as a sharp threshold where activity = 1 when more than 2/3 of the nucleosomes are in the A state. As each gene is modeled as a 50-nucleosome system, the sharp threshold would set gene number 5 in Fig. 7.20 as active when the number of A nucleosomes exceeds 33. This happens at time ~ 60.

Stochastic fluctuation combined with an internal local bias toward the active state tends to move each gene into a dominant active state. When this state is reached for one gene, the activated gene increases the global negative feedback and thereby prevents a second gene becoming active. Noticeably, the activated gene also attempts to inhibit its

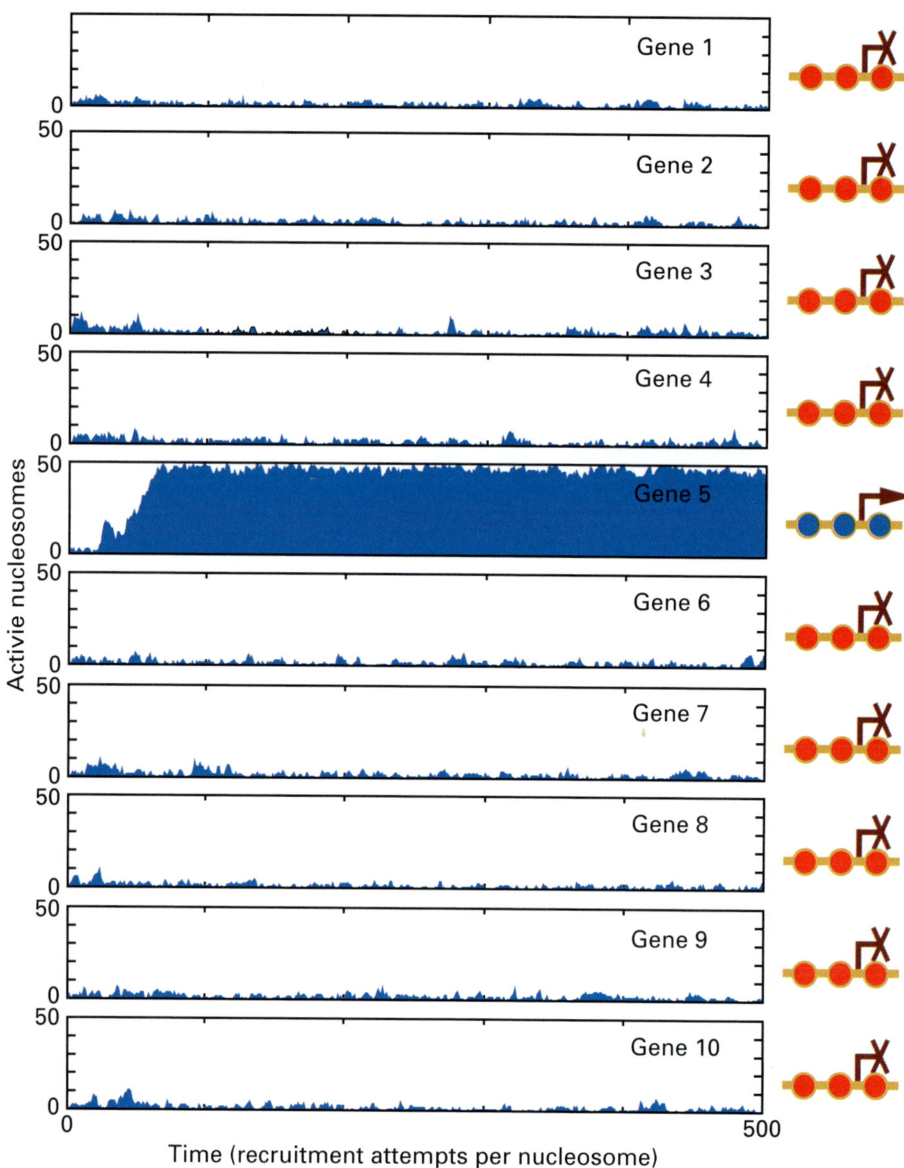

Figure 7.20 Simulation of $N = 10$ genes, each covered by $L = 50$ nucleosomes. The simulation shows that one gene is turned on quite early, while all other genes remain silenced. Sketches in the right-hand panel illustrate the OR gene state at the final time of the simulation. Crossed promoters indicate silent genes. Other parameters of the simulation are $\mu = 0.50$, overall repression factor $r = 1$, Hill coefficient of repression $h = 2$ and direct conversion $\beta = 0.03$. Figure from [333].

own activity, but the model allows for a broad range of parameters where this inhibition is too weak to push it back over the barrier in its epigenetic landscape. This can also be seen from Fig. 7.21, which shows the "epigenetic landscape" of two of the many genes in OR differentiation. The landscape is coupled, in the sense that the epigenetic

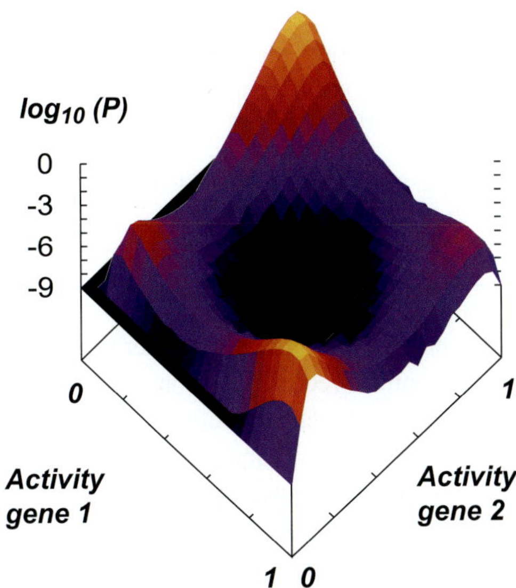

Figure 7.21 The fraction of active nucleosome for the two most active genes of a N = 100 gene system. The vertical axis is log$_{10}$ of probability for the states for the two genes on the horizontal axes. Each gene is covered by $L = 50$ nucleosomes, and exposed to bias $\mu = 0.50$, a threshold activity with $h = 2$ and an external/direct conversion rate $\beta = 0.03$. One can see that the most likely state has one active gene and the rest are silenced, but also that there is a metastable state with two simultaneously active genes. The valleys define barriers between preferred states. Figure from [333].

landscape of "gene 1" strongly depends on the activity of "gene 2": that is, if "gene 2" has activity ~ 0 then "gene 1" prefers to remain at full activity ~ 1, which in turn constrains "gene 2" to remain inactive. As an overall result, the coupled landscapes reinforce the selected dominant gene by decreasing the probability of local activation of other genes. An analytic version of the model is considered in the questions.

7.4.1   Questions

7.4.1   Consider the model for olfactoric differentiation from Fig. 7.19, with nucleosomes in the silenced state (red in panel C), being partially protected from conversion to the a state (blue) by a protein P, produced by all active genes. Argue for negative feedback through P that reduces the rate of recruitment from S to A:

$$\text{Rate}(S \rightarrow A) \rightarrow \text{Rate}(S \rightarrow A) \cdot \frac{1}{1 + r \cdot P} \qquad (7.12)$$

where P is the total activity of all olfactory genes.

7.4.2   Assume that each gene has an activity proportional to the fraction of active nucleosomes of that gene to an exponent h, and use this to find an expression for the amount of a protein P that is produced due to the activity of all genes. Combine this expression with the results from Question 7.4.1 to argue for the following equation for differentiation between the $i = 1, 2, \ldots N$ genes from Fig. 7.19:

$$\frac{da_i}{dt} = \frac{1}{1 + r \cdot \sum_j a_j^h} \cdot a_i^2 \cdot (1 - a_i) - \mu \cdot a_i \cdot (1 - a_i)^2$$
$$+ \frac{1}{1 + r \cdot \sum_j a_j^h} \cdot \beta \cdot (1 - a_i) - \beta \cdot a_i \qquad (7.13)$$

Here $\mu < 1$ describes an inherent bias, giving a weaker basic recruitment from the silenced state. In contrast, recruitment from active nucleosomes is reduced due to influence from other genes. β represents passive conversions. Analyze the model for $h = 4$, $r = 3$, $\mu = 1$ and $\beta = 0.03$.

7.4.3 Simulate the two-gene version of the equations from Question 7.4.2, using a Gillespie algorithm on the four different terms in the equations (making a total of eight rates for the two genes). Use an update where each of the state fractions is changed in steps of size 0.1.

7.5 Other models for *cis*-mediated gene regulation

7.5.1 Combining read–write enzymes with protein binding

As outlined in Fig. 7.3, the epigenetic switch in *S. cerevisiae* differs from that seen in *Pombe*. In fact, most main players are different,[6] and the feedback is between an active acetylated "state" and a silenced state consisting of an unmodified nucleosome binding Sir3. The feedback in this system involves the nucleosome-bound Sir3, which recruits a Sir4:Sir2 complex, that in turn: (1) de-acetylates other nucleosomes, (2) prevents dissociation of Sir3 from nearby nucleosomes. The feedback that reinforces dominance of nucleosomes with Sir3 combines: *(1) a read–write recruitment step, (2) a protein–protein binding that silences the nucleosomes by covering them with a Sir4:Sir2:Sir4 complex* Fig. 7.22 outlines and simulates a simplified version of this model.

7.5.2 Epigenetics with extended modification space

Consider now a wider region of nucleosome modifications with an associated variety of feedback circuits [338, 339]. In the minimal extension consider a system where each nucleosome has two modifications that each can either be on (1) or off (0). There will then be four nucleosome types: 00, 10, 01 and 11. The "read–write" enzymes that bind to

---

[6] In fact Sir2 in the two very distantly related yeast organisms is the same, but the role of Sir2 seems less pronounced in the *Pombe* system.

Saccharomyces:

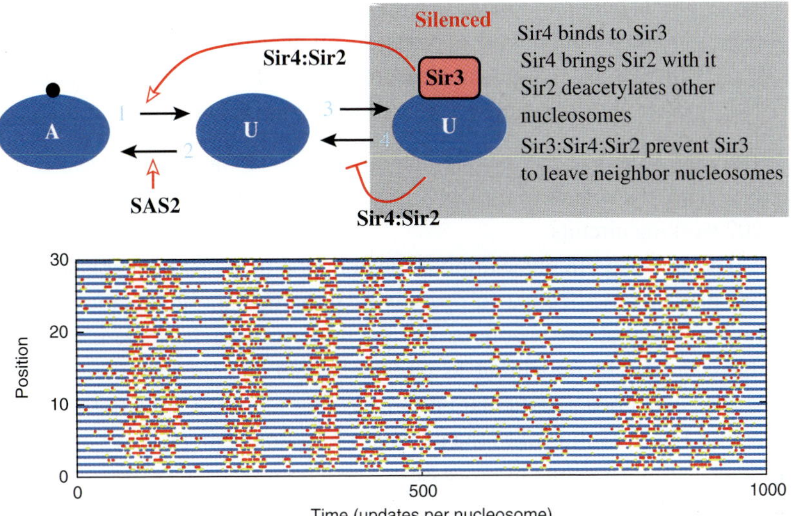

At each time step: (1) Select nucleosome, in US then select another. If A then $\to$ U
(2) select nucleosome and if U, change to acetylated (A)
(3) select nucleosome and if U, change to US = U:Sir3
(4) Select nucleosome and if US, count number of neighbors in
state US = n. Change the US to U with probability 0.40^n
(5) select nucleosome with probability 0.01. If this is A then $\to$ U

Figure 7.22 Model for epigenetics in *S. cerevisiae*, taking into account that Sir3 tends to spread linearly along the DNA, whereas de-acetylation can be more non-local. Silenced nucleosomes (US) are marked red, active nucleosomes are blue. Acetylation and loss of Sir3 is primarly assumed to be due to non-recruited processes The simulations in the lower panel were done without cell division. (Model by I. Dodd, D. Moazed and K. Sneppen.)

and modify these nucleosomes should be capable of distinguishing these four combinations. Further, there are eight specific modification or de-modification reactions capable of causing interconversions between the four types.

Following [338] we examine possible read–write combinations that allow an extended system of such nucleosomes to be bistable, switching rarely between an epigenetic state E0 = 00 and an alternate epigenetic state E3 = 11. The considerations are limited to motifs where each conversion can only occur from one of the four nucleosome states, or alternatively be exposed to low-level direct transitions. This results in $(4 + 1)^8$ different circuits. Each circuit was tested by iteration of recruitment reactions, non-recruitment reactions and DNA replication for a system of $N = 30$ nucleosomes. At each step:

(1)   *Recruitment:* a nucleosome i and nucleosome j are selected randomly. If j is a type that can catalyze modification of i (determined by the circuit), then i is changed accordingly (if j can modify i in two ways, one is randomly chosen). If j is in no such state, then nothing happens.

(2)   *Direct conversions:* with probability β, a random nucleosome is changed to one of its neighbor types.

(3)   *Replication:* at DNA replication, all nucleosomes have a probability 0.5 of being replaced by a nucleosome of the 01 type. The replication step happens after $N \cdot r$ iterations of steps 1 and 2 (one generation). We used $r = 50$.

Testing for the stability of both the 00 and the 11 states for at least 250 generations at $\beta = 0.01$, the scanning for the nearly 400 000 different circuits allowed us to identify 202 working circuits.

The working circuits exhibit some common patterns that expand the ways in which one may obtain effective co-operativity. In particular, 190 circuits contain the two-step recruitment that was a core part of the three-state nucleosome model proposed first in this chapter. The remaining 12 circuits did not contain such two-step recruitment, but instead contained a particular push–pull–drag sequence of recruitments. The right-hand part of Fig. 7.23 illustrates the two main ways to obtain effective co-operativity through sequences of recruitment processes:

•   The upper panel emphasizes two-step sequential recruitment, supplemented by some counteracting recruitment of the intermediate state.
•   The lower panel contains the push–pull–drag sequence.

Interestingly, the push–pull–drag motif may play a major role in another epigenetic system, namely in maintenance of DNA methylation of CpG islands [340, 341, 342, 343, 344, 345, 346]. DNA methylation has the ability to support different cell states provided there is some sort of co-operativity [347]. The recruiting enzymes for DNA methylation are DNMT1 and DNMT3A/B whereas TeT1 and TeT2 "guard" the non-methylated CpG islands [348, 349]. The maintenance of DNA-methylation is challenged by cell division, where all fully methylated sites are divided and thereby become methylated on one

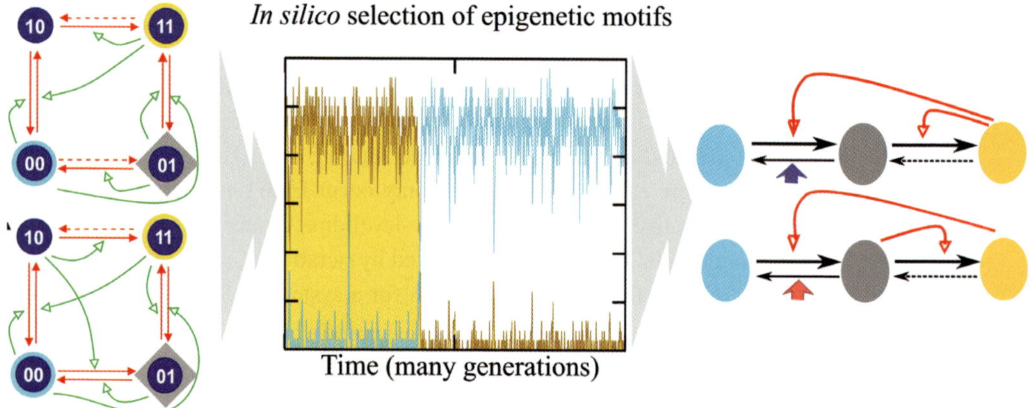

Figure 7.23 Two recruitment motifs that support epigenetics. The upper one is central to the standard model.

strand (hemi-methylated),[7] and $\sim 1/2$ of hemi-methylated sites become un-methylated. For a consistent model consult [764]. Noticeably, DNA methylation and nucleosome silencing are expected to be coupled to each other [350, 351, 352].

7.6 Summary

- Epigenetics = bistability that is robust to cell divisions.
- Epigenetics and bistability need positive feedback.
- Nucleosome-mediated gene regulation can facilitate bistable gene expression if the dynamics of the silenced nucleosome fraction m fulfill:

$$\frac{dm}{dt} \propto m^h \cdot (1 - m) - (1 - m)^h \cdot m \qquad (7.14)$$

with a value $h > 1$, implying co-operativity.
- Nucleosome-mediated epigenetics need "read–write" enzymes that can act across distances of tens of nucleosomes along the DNA.
- Limits to spreading of nucleosome modifications require some strictly local interactions.

---

[7] Hemi-methylated DNA is methylated on one strand and non-methylated on the other.

8 Feedback circuits

8.1 Small regulatory subnetworks

The interior of the cell is crowded with proteins, nucleic acids and metabolites; molecules with widely different properties and which each interact specifically with a few of the other ones. These interactions define a network of molecular species connected by links that represent interactions and flow of information. The links are often directed, and may differ widely in the time they need to transmit a "message." For example, enzymatic conversion of small metabolites is usually fast, while synthesis of biopolymers is a much slower process. Importantly, most biological processes are regulated through feedback circuits, which involves both activating and inhibiting links.

Feedback loops (FLs) are formed when regulatory links are combined into a closed cyclic chain. Thus, the influence of each regulator in this cycle eventually loops back onto itself. Depending on the types of links in the cycle, this effective self-interaction can be positive or negative, and this strongly determines the behavior of the FL.

Feedback circuits and small regulatory networks (RNs) have been extensively studied, both experimentally and theoretically [25, 116, 142, 175, 238, 360, 361, 362, 363, 364, 365, 366, 367, 368, 369, 370, 371, 372, 373, 374, 375, 376, 377, 378]. Figure 8.1 shows RNs with various combinations of FLs. Panels A and B show case studies where a small molecule is "harvested" from the exterior of the *E. coli* cell, and then subsequently utilized within the cell. In both cases, the uptake system is regulated through an interplay of transcription factors and small regulatory RNA (Spot42 in panel A, and RhyB in panel B [120]). Figure 8.1C is the SOS network in *E. coli*, a network where the central regulator LexA can be degraded by an activated RecA protein at a very high rate [115]. The network also contains two competing protein complexes that set a threshold for the onset of the error-prone polymerase [354]. Finally, panel D shows an example of regulation in a eukaryotic system that incorporates traffic between cellular compartments [118, 379]. The dynamics of the networks are shown in Fig. 8.2, presenting various aspects of response dynamics to change in nutrients (A,B) or stress (C,D).

Instead of the detailed modeling explored for the λ phage, this chapter advocates simplifying networks to a core regulatory logic [380, 381].[1] One example of how a

---

[1] This strategy has also been advocated under the term "nework motifs" [382, 383], with emphasis on signal propagation across feed-forward RNs.

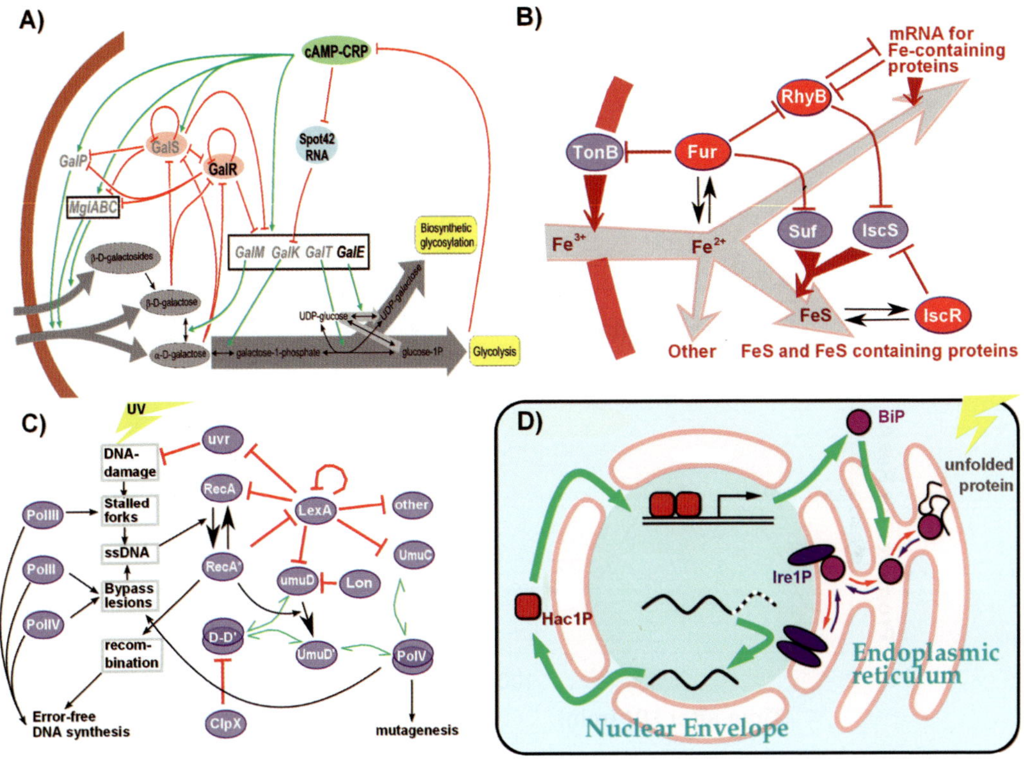

Figure 8.1 Regulatory networks: (A) Galactose enters the *E. coli* cell from the left and is converted to glucose on the right [353, 354]. (B) Fe uptake network in *E. coli* [355, 356]. (C) SOS system in *E. coli*, which is responsible for repair of single-stranded DNA breaks [115, 357, 358, 359]. (D) Unfolded protein response in *S. cerevisiae* [118], a network which in mammals is associated with insulin production and diabetes.

complicated RN could be simplified to FLs is shown in Fig. 8.3. The figure illustrates how regulation of transport across cellular compartments, together with protein–protein binding may be simplified to one repressive link that maintains the overall negative feedback.

This chapter describes the functional behaviors of some isolated FLs, as well as examples of two entangled FLs with applications in excitable media and metabolism. As an overall guideline we advocate the sign of FLs, and how FLs of different signs are combined, as the prime determinant of the function of an RN.

8.2   Negative feedback

An FL is negative if it has an odd number of repression links. The logic of negative feedback makes it ideal for stabilizing systems and minimizing fluctuations [367]. Thus, negative feedback is often associated with maintenance of homeostasis. A common negative feedback is found in metabolism regulation, where a small molecule binds

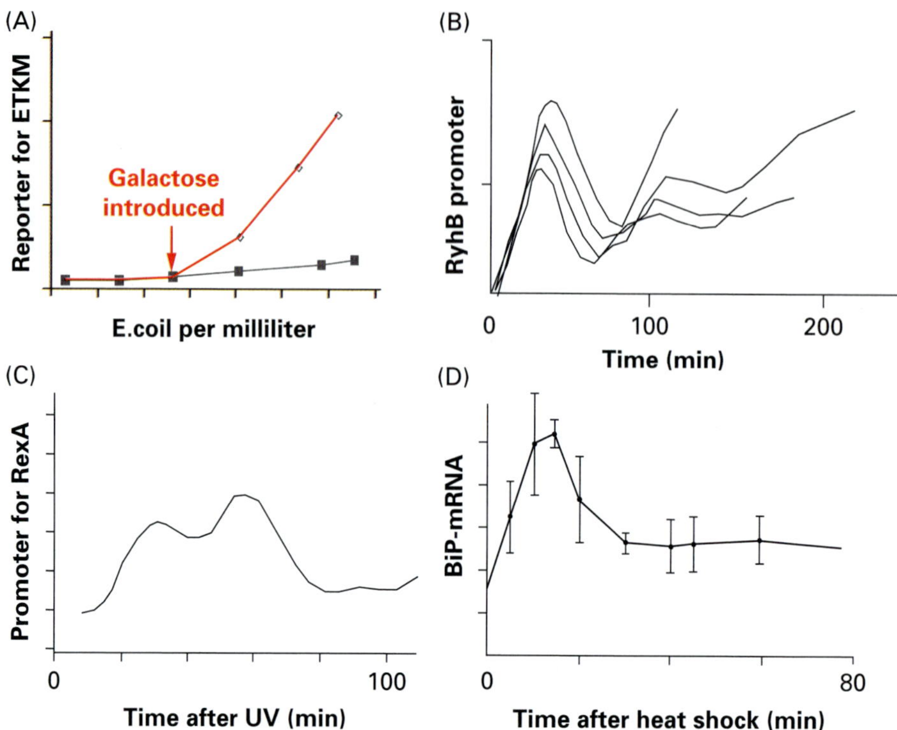

Figure 8.2 Examples of the dynamic responses for the networks from Fig. 8.1: (A) Response of the accumulation of reporter for enzymes ETK and M on introduction of galactose (red curve). The black curve is the accumulated count without galactose (data courtesy of S. Semsey [353, 354]). The x-axis measures bacterial density and can be translated into a measure of time where the exponentially growing bacteria have a doubling time of one hour. (B) The response of the promoter for RhyB on starvation of Fe at all times > 0 [356]. The different curves denote different individual bacteria. (C) The response of the RecA promoter to UV radiation, measured by [358]. (D) The unfolded protein response [118], quantified by the amount of mRNA for the chaperone BiP after a sudden increase in temperature at $t = 0$. In all cases the y-axis is in arbitrary reporter units, and is therefore un-numbered.

to and inhibits an enzyme that catalyzes one of the earlier steps in its metabolic pathway [360]. A similar effect is mediated by ribo-switches, where a metabolite binds to and inactivates the mRNA of the necessary enzyme [387].

8.2.1   Homeostasis

Common motifs for negative feedback involving gene regulation are outlined in Fig. 8.4. Many regulators repress the transcription of their own gene, as illustrated in Fig. 8.4A. Such direct self-repression allows the regulator to stabilize other genes against global variations in protein–DNA binding, for example caused by changing salt concentration [388]. Negative auto-regulation of a regulator R can be readily modeled using a

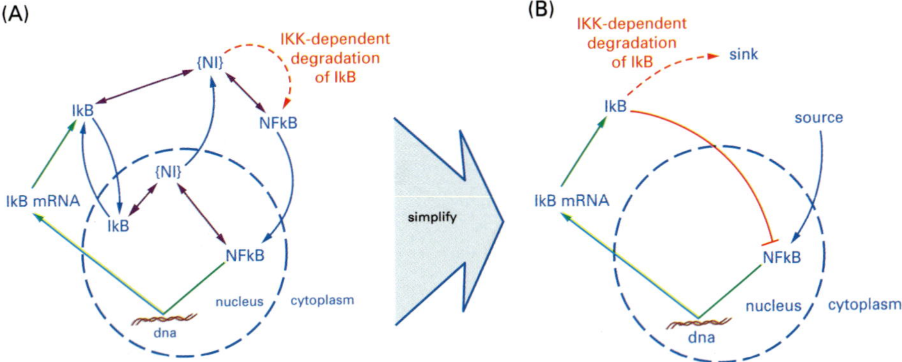

Figure 8.3 The NF-κB network in mammalian cells [384] that is involved in sensing stress and excreting cytokines, a type of signaling molecule that acts between cells [386]. The stress response of this network is associated with movement of the transcription factor NF-κB in and out of the cell nucleus, see Fig. 8.6C.

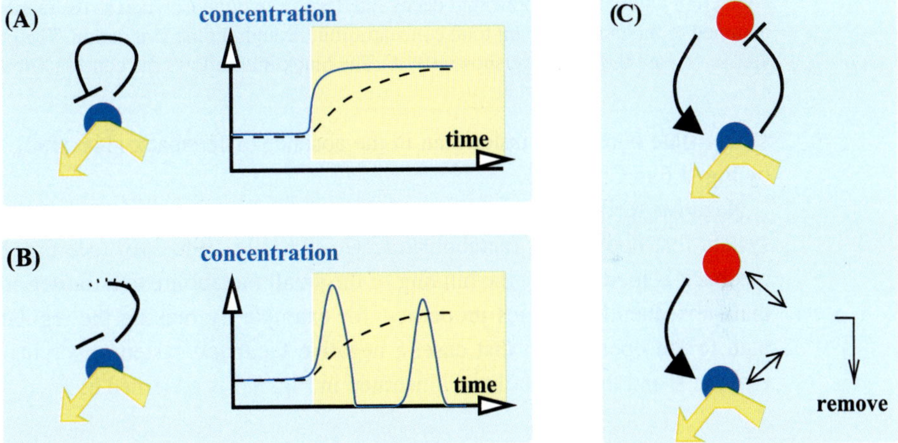

Figure 8.4 Negative FLs together with their response dynamics. (A) Negative feedback may stabilize the output to a near constant level, and allows for a fast transient increase in production in response to stress or perturbations. (B) If the negative feedback is delayed, the protein concentrations may oscillate in time. (C) Two types of indirect negative feedback, emphasizing, respectively, feedback through transcriptional repression, and the common feedback acting through protein–protein binding to a transcriptional activator.

straightforward modification of Eq. (3.20) [389]:

$$\frac{dR}{dt} = \text{leak} + \frac{\text{capacity}}{1 + (R/K)^h} - \frac{R}{\tau} \tag{8.1}$$

In a steady state, this results in an R that grows slower than linearly with increased capacity. Therefore the steady-state R is buffered against changes in cellular factors that affect capacity, a feature that makes a negative FL good for homeostasis. In addition, the

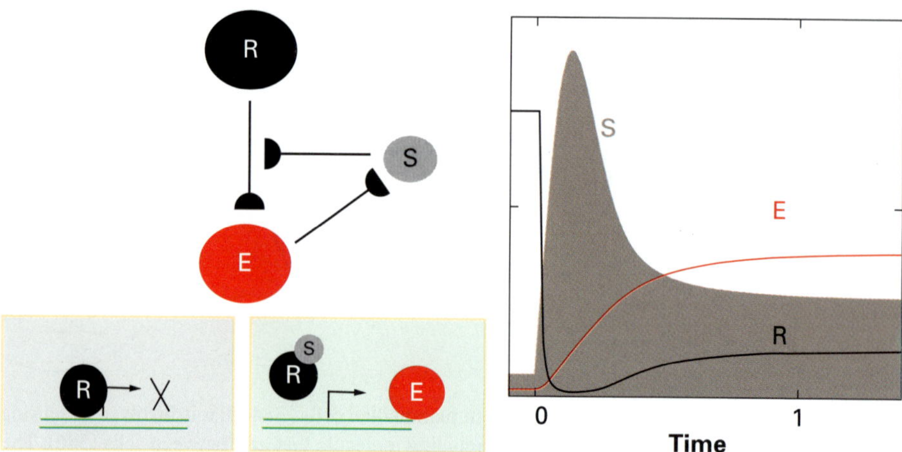

Figure 8.5 Negative feedback in regulation of a metabolite "*s*" that is degraded/converted to another metabolite by an enzyme E [392]. The right-hand panel shows the response to a change in the source terms for s, from K_s to $100\,K_s$, at time > 0, where $\gamma = 100$ and $R = 10$. Time is counted in units of the exponential decay rate for the enzyme E, whereas the regulator is assumed to maintain constant total concentration throughout the simulation. The gray shaded area is "*s*" and the red curve shows $10 \cdot E$. The black line follows the concentration of free R.

steady state is reached faster than in the absence of feedback [116, 390], as illustrated by Fig. 3.6 in Chapter 3.

Negative feedback systems are often seen between a metabolite and the regulatory system that controls its metabolism [392, 393, 394, 395, 396] (see Fig. 8.1A,B). The feedback is mediated by the binding of the small metabolite to a transcription regulator that subsequently changes properties, for example by making the regulator unable to bind to the operator. In that case, a negative feedback system involving regulator R, enzyme E and the metabolite s illustrated in Fig. 8.5 is governed by:

$$\frac{dE}{dt} = \frac{1}{1 + R_{\text{free}}^2} - E \tag{8.2}$$

$$\frac{ds}{dt} = \text{source} - \gamma \cdot E \cdot s \tag{8.3}$$

Here R_{free} is the concentration of the regulator that is not bound by the small metabolite s ($R_{\text{free}} \sim R/(1 + s)$). Further, the above equation assumes that R_{free} binds co-operatively to the operator, mimicking the typical dimerization of many regulators.[2] This concentration, in turn, decreases with s, as discussed in Question 8.2.1 below. As a consequence, an increase in the source of s causes a decrease in R, which in turn secures an increase in E. Therefore, the overall feedback is negative: an initial increase in s is counteracted by an increased removal/conversion of s to something that does not influence R.

---

[2] R_{free} is expressed in units of its binding constant to the operator. Also, the equation is expressed in time units, where E is diluted by rate 1, whereas E is given in units of its maximum possible value.

The right-hand panel in Fig. 8.5 shows typical response dynamics to a sudden increase in the source term for the metabolite s.

8.2.2 Negative feedback in stress response

Cells often need to respond to environmental stresses, like osmotic stress, heat shock, DNA damage, presence of toxic substances or mechanical damage. For each of these, the cell has a repertoire of proteins that can mitigate the stress. Accordingly, when stress is experienced, the cell has to communicate the need for stress-response proteins to the transcription/translation factors that can activate production of the required proteins – this is a situation suitable for negative feedback regulation.

In fact, a simple model for a canonical stress-response system could be mathematically equivalent to the one used for the metabolic response, Eq. (8.3). Then E should be read as the enzyme that counteracts the stress, R its regulator and s the stress product that needs to be removed. The coupling between the stress product and the central regulation should then occur through, for example:

$$R_{\text{free}} = \frac{R}{1 + (s/K_s)} \tag{8.4}$$

This direct coupling[3] from s to the regulator R of its enzyme closes the FL. In real life the situation is often more complicated. For example, in heat shock in *E. coli*, the coupling between the actual stress of unfolded proteins U and the regulator σ^{32} is indirect: the signal from s (= U) to R (= σ^{32}) goes through the need for E (= DnaK) and the protease HflB [397, 398], thereby mimicking the last motif in Fig. 8.4C.

A stress response usually needs to be initiated as quickly as possible to minimize the damage the stress can cause. Therefore, one usually finds that negative FLs in stress-response systems involve protein–protein interactions, where R regulates the production of a central enzymatic protein that, in turn, catalyzes the proteolytic inactivation of R [114] [397].[4]

8.2.3 Oscillations

Negative feedback can cause oscillations. For example, oscillations are reported for the developmental gene Hes1 in mammalian embryos, the DNA-stress-sensitive p53-Mdm2 system in mammalian cells and perhaps also in the cytokine-regulating NF-κB system in mammalian cells. Figure 8.6 shows the observed dynamics for all these systems. The systems are all part of a larger RN, with many components that interact in various ways,

---

[3] Other functional ways to couple s and R may be envisioned, for example $dR/dt = 1 - s \cdot R$, which, however, also would imply an active $R \propto 1/s$ for large *s*.

[4] An example of this is found in the SOS system in Fig. 8.1C, where RecA degrades its repressor LexA [357, 359]. Similarly, in mammalian systems, the protein mdm2 binds to its activator p53 and mediates its degradation when the DNA is damaged. The above-mentioned heat shock in *E. coli* [25, 399] uses this feedback strategy, where again a target of transcription regulation catalyzes the degradation of its activator, σ^{32}.

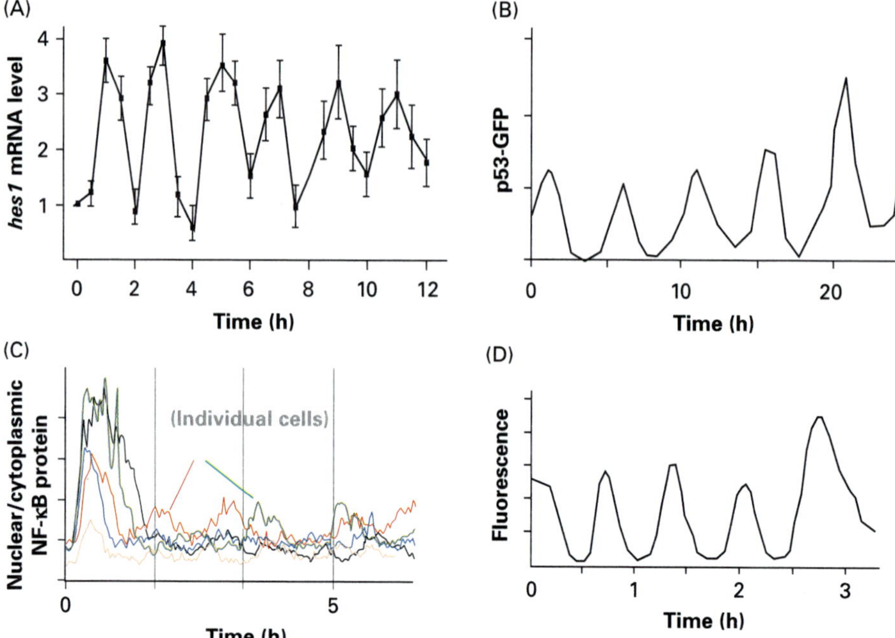

Figure 8.6 Experimentally observed oscillations in concentrations of GFP-labeled transcription factors in different RNs. (A) Hes1 expression, taken from [372], (B) p53 oscillations after DNA damage from [400, 401]. (C) NF-κB protein, which is reshuffled between nucleus and cytoplasm, from [386] (see also [385]). Oscillations are seen on a single-cell level in [402]. (D) Oscillations in a synthetic two-protein circuit in *E. coli*, with LacI repressing araC, while AraC activates the lacI (and the fluorescent protein) [403]. Figure reprinted with permission from AAAS .

such as transcriptional activation/repression, translation and post-translation regulation, protein–protein interaction, targeted protein degradation and active nuclear-cytoplasmic translocation.

Although negative feedback tends to drive a system to steady state, it only does so if the feedback is fast. A large time delay may produce oscillations. Consider a person who walks along a straight line to a given point, marked on the ground. If able to take instantaneous decisions, this person will approach the mark and then stop. This is a stationary solution to the walk kinetics. If, on the other hand, it takes some time to realize that the mark has been reached, the person will not stop at the mark, but cross it. When eventually the information that the mark has been crossed is processed, the person will turn back and walk in the opposite direction. The mark will again be reached and overshot, and so on. Depending on the size of delay, the resulting kinetics will be damped or amplified oscillations. In cellular systems the time associated with several subsequent processes could produce time delays, and thus cause oscillations.

In terms of equations, a damped motion would be:

$$\mathrm{d}x/\mathrm{d}t = -x(t) \qquad\qquad (8.5)$$

with exponential damping. The corresponding minimal model with time delay takes the form:

$$\mathrm{d}x/\mathrm{d}t = -x(t - \tau) \tag{8.6}$$

an equation that for delay τ larger than a certain critical size leads to both oscillations and exponential growth. Using negative feedback in gene regulation, oscillations can similarly arise from the equation [377]:

$$\frac{\mathrm{d}R}{\mathrm{d}t} = \text{leak} + \frac{\text{capacity}}{1 + (R(t - \tau_d)/K)^h} - \frac{R}{\tau} \tag{8.7}$$

The system approaches homeostasis and shows no oscillations when the time delay, τ_d, is small compared to the lifetime τ of the regulator. However, the system exhibits oscillations when the delay is larger than a critical amount (see questions).

Of course, assuming an explicit time delay does not really shed light on the mechanism producing the time delay. There are several possibilities that can be used to produce oscillations in a negative feedback loop:

(1) A process that takes a finite minimum time,
(2) Many intermediate steps, like the repressilator (see questions),
(3) A sharp response by some of the variables,
(4) Saturated degradation, included by replacing R/τ with $R/(\tau + R/\kappa)$ in Eq. (8.1). Saturated degradation is suggested to be the source of delay in the negative feedback systems of p53, Nf-κB, and Wnt [362, 384, 404].

When a regulatory system oscillates, the period is set by the slowest rate in the system, which is usually the decay time of a protein.

Time-delayed differential equations have been used to model oscillations in the developmental regulator Hes1 [377], which inhibits its own transcription [372, 378], as well as oscillations in the p53 stress response system [376, 404].

8.2.4 Questions

8.2.1 If total concentration of a transcriptional regulator R is partitioned between the part that is free, R_{free}, and the part that is bound to a metabolite, then argue for conditions where $R_{\text{free}} = R/(1 + (s/K_s)^h)$, and identify the meaning of h and K_s.

8.2.2 Simulate Eqs. (8.2) and (8.3) with $\gamma = 100$, $K_s = 100$ $h = 2$, $R = 10$ and "source" changing from 10 000 to 1 000 000 per time unit (parameters for which $s \gg R$). Hint: start by simulating the equations until steady state is reached, then change "source" and follow the time development of E and s. Repeat the simulation for $h = 1$, and $\gamma = 1000$.

8.2.3 Simulate the repressilator: $\mathrm{d}A/\mathrm{d}t = \epsilon + 1/(1 + (C/0.2)^h) - A$, $\mathrm{d}B/\mathrm{d}t = \epsilon + 1/(1 + (A/0.2)^h) - B$, $\mathrm{d}A/\mathrm{d}t = \epsilon + 1/(1 + (B/0.2)^h) - C$ with $h = 3$, $\epsilon = 0.01$ and using simple integration with $\mathrm{d}t = 0.005$.

8.2.4   Simulate the above repressilator using event-based simulation (Gillespie algorithm), updating A, B, C in units of $u = 1/N$ with $N = 100$, according to production and decay events for each protein type, i.e. for A according to $A \rightarrow A + u$ with rate $r = (\epsilon + 1/(1 + (C/0.2)^h)) \times N$, and $A \rightarrow A - u$ with rate $r = A \times N$. Study the obtained oscillations for other N values.

8.2.5   Simulate the idealized Goodwin model [361, 365]:
$\mathrm{d}m/\mathrm{d}t = 1/(1 + c^9) - k_\mathrm{m} \cdot x$, $\mathrm{d}p/\mathrm{d}t = m - k_\mathrm{m} \cdot p$ and $\mathrm{d}r/\mathrm{d}t = p - k_\mathrm{r} \cdot r$ using $k_\mathrm{m} = k_\mathrm{p} = k_\mathrm{c} = 0.5$ a degradation rates of mRNA, protein and the regulator, respectively. The regulator r represses production of m, closing the negative feedback loop.

8.2.6   Simulate the simple time-delay equation:

$$\frac{\mathrm{d}x}{\mathrm{d}t} = -x(t - \tau) \tag{8.5}$$

and convince yourself that it either gives exponentially damped, or exponentially amplified oscillations.

8.2.7   Simulate the time-delay equation:

$$\frac{\mathrm{d}p}{\mathrm{d}t} = \frac{1}{1 + (p(t - \tau)/0.1)^2} - p(t) \tag{8.8}$$

for $\tau = 1$, $\tau = 2$ and $\tau = 5$.

8.2.8   Simulate the time-delay equation:

$$\frac{\mathrm{d}p}{\mathrm{d}t} = \frac{1}{1 + (p_\mathrm{old}/0.1)^2} - p(t) \tag{8.9}$$

where p_old at time t is the proteins produced from time $t' \in [t - 3 \cdot \tau, t - \tau]$ weighted by $w \propto \exp(-(t - t')/\tau)$, i.e. production takes at least τ to contribute, but after this time the delivery time is smeared out with a characteristic time τ. The time-delay equation should be simulated for $\tau = 1, 2$ and 5.

8.2.9   Maintenance of homeostasis is essential for health, and degenerative processes associated with diseases should be be counteracted by cellular repair. An example may be Parkinson's disease [30] where growth of fibrils F may be counteracted by cellular (protease) proteins P:

$$\frac{\mathrm{d}F}{\mathrm{d}t} = \frac{m}{1 + P} - P \cdot F \tag{8.10}$$

$$\frac{\mathrm{d}C}{\mathrm{d}t} = P \cdot F - v \cdot C \tag{8.11}$$

$$\frac{\mathrm{d}P}{\mathrm{d}t} = 1 - P - P \cdot F + v \cdot C \tag{8.12}$$

Here m parameterizes the influx of new potential fibrils, an influx that is partially repressed by P. $1/v$ is the lifetime of the complex between mature fibrils and the cellular repair machinery P. Draw the implied network, and identify positive and a negative feedback. Simulate the equations for $v = 1$ and $m = 10$, and $m = 25$, respectively.

8.3 Positive feedback

In contrast to negative feedback, positive feedback often amplifies perturbations. Positive feedback can be caused by self-activation [112, 408], or a double negative feedback, as in phage λ. A positive FL often shows an ultra-sensitive response where the steady state of the protein concentration shifts sharply and non-linearly from a low level to a high level, as factors affecting one of the proteins are varied [408]. Positive FL may also allow the system to persist in one of two very distinct states for extended periods of time [409] (see Fig. 6.4). The behavior of such bistable systems is characterized by long periods of time where the system is in one state, interrupted by rapid transitions between these states [175, 405]. The states may easily be sufficiently stable to form the basis for epigenetics and persistent cell differentiation [143, 144, 145, 146].

Figure 8.7a reiterates the common transcriptional positive FL consisting of two mutually repressing Rs. Extending the formalism of Eq. (3.20), the equations for this λ-like motif are simply:

$$\begin{aligned}
\frac{dR}{dt} &= \text{leak} + \frac{\text{capacity}}{1 + L^{h_L}} - \frac{R}{\tau_R} \\
\frac{dL}{dt} &= \text{leak} + \frac{\text{capacity}}{1 + R^{h_R}} - \frac{L}{\tau_L}
\end{aligned} \tag{8.13}$$

If the Hill coefficients are greater than one, the system may be bistable, with one state where R is large and another where R is small.

Conceptually, it is important to remember that some sort of co-operative interactions are essential for bistability. Co-operativity may be in the form of homodimers or tetramers, as in the phage λ switch (Fig. 8.7A), it may be hidden in the ultrasensitivity of the free regulator when it binds strongly to another protein, as in phage TP901-1 [126] (Fig. 8.7B), or it may rely on sequences of auto-catalytic reactions, as found in nucleosome-mediated epigenetics (Fig. 8.7C). The importance of co-operativity relies in the more than linear increase of effect with the dose. This means that the winner of the two "battling" states has more than a proportional advantage for being dominant. If there was no co-operativity, production and degradation in Eq. (8.13) can only balance each other for one value of the concentration, which then becomes the only sustainable state.

Bistability is, in fact, a pre-requisite for epigenetic "memory" – the ability of genetically identical cells to maintain different distinct phenotypes from one generation to the next. Positive feedback is therefore at the core of stem-cell differentiation [411, 415, 416], an RN that shares remarkable similarities with the RN for phage λ

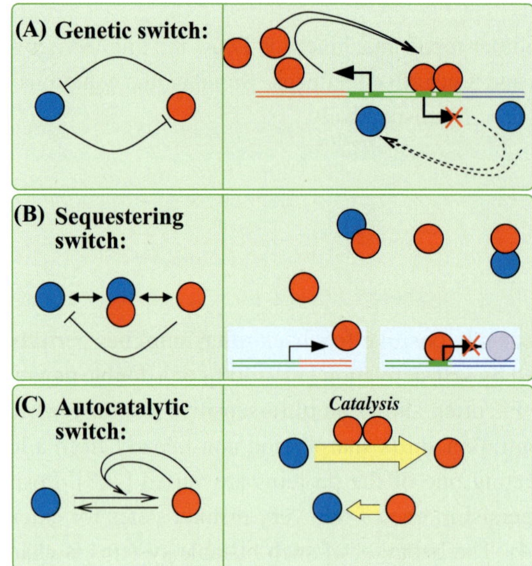

Figure 8.7 Bistable FLs: (A) Mutual transcriptional repression [405], also known as a "genetic switch," seen in phage λ [112]. (B) A positive FL that uses strong protein–protein binding so that only the abundant protein species are available in the free form [406]. Here the "blue" promoter is stronger and the heterodimer does not act as a transcriptional repressor. This motif is found in the phage TP901-1 [126, 407]. (C) A switch consisting of catalyzed modifications, where, for example, the proteins in the "red" state can direct a "blue" protein to change to red. The directed processes compete with background random conversions. This mechanism is found in the nucleosome-mediated epigenetics from Chapter 7.

(see Fig. 8.8).[5] Bistable switches are also involved in the development of sharp spatial boundaries between tissues in all types of animals, see, for example [418, 419, 420].

8.3.1   Questions

8.3.1   Model the sequestering switch between two proteins A and B produced according to $dA/dt = P_A/(1 + B_f^2) - A$, $dB/dt = P_B - B$, and where A and B form a complex with binding constant $K = 0.01 = A_f B_f/[AB]$. Here free concentrations are $A_f = A - [AB]$ and $B_f = B - [AB]$ binding to each other with $K_{AB} = 0.01$, whereas basal production rates are, respectively, $P_A = 10$ and $P_B = 5$. Show that the system is bistable, by starting the simulation from two different initial conditions (one may also perform 100 simulations, starting at points where both A and B are between 0 and 10).

---

[5] The genetic regulation of OCT:Nanog may, in addition, be modulated by the protein–protein complexes β-catenin:Oct4, Oct4:Nanog and β-catenin:Nanog, protein complexes that may well modulate degradation rates. *In vitro* stem cells seem to be captured in ongoing quasi- oscillatoric/chaotic dynamics. Such behavior can also be obtained from a moderate extension of the network in Fig. 8.8, provided that delayed negative feedbacks play a strong role [417].

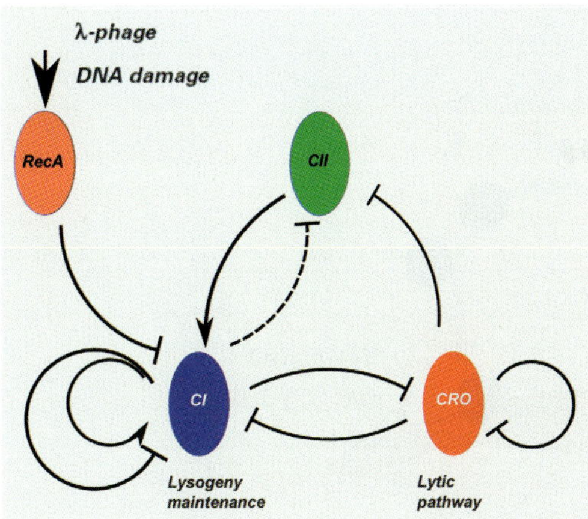

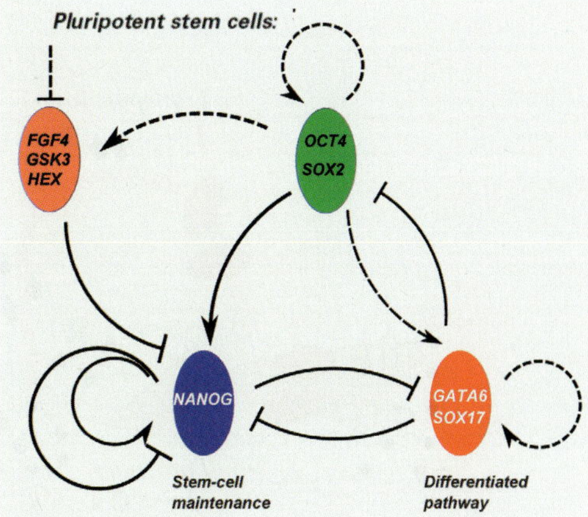

Figure 8.8 Contemporary model of stem-cell differentiation (right) compared to the phage λ decision (dashed regulation highlights differences). The bistable switch between GATA6/SOX17 and Nanog is modulated by the heterodimer Oct4:Sox2 [410], modeled by [411]. GATA6 denotes proteins in the differentiated pathway, whereas non-differentiated stem cells have high Nanog. Remarkably, the similarities suggest the stem-cell-maintaining gene Nanog replacing CI, GATA6 positioned as Cro, Oct4:Sox2 taking a network position that partly resembles CII (with the caveat that Oct4 and Sox2 each favor their own type of downward differentiation [412]). Evidence for positive and negative auto-regulation of Nanog is provided by [413, 414]. Noticeably, the pluripotent stem cell only stay as such for a few generations, and thus the stability of this state may be of little practical concern.

8.4 Combining feedback in small-molecule regulation

8.4.1 Regulated fluxes

Apart from Eqs. (8.2), (8.3) and (8.4), our RNs have missed small molecules and their metabolism. Substantial fractions of life are devoted to conversion of food molecules into raw materials for anabolic processes. Metabolic networks may be viewed as a mixture of intertwined "production lines" with flows of matter being conserved at each reaction node. RNs control these reaction nodes through binding between the small metabolites and enzymes [360] or the regulators of enzymes. An FL is formed for example when a small molecule binds to a regulator that represses an enzyme, which, in turn, affects the metabolism of the small molecule. In *E. coli*, about 50% of all transcription factors bind small molecules, thus forming a "zoo" of feedback circuits at the interface between the RN and the metabolic network [421, 422].

A common "feedback motif" associated with metabolism is shown in Figure 8.9: a transcriptional regulator (R) senses the intracellular concentration of a particular small molecule (s) and regulates transcription of the transport proteins (T) using one FL, facilitating the influx of s. With a second loop, R controls transcription of the enzyme (E) responsible for the metabolic conversion of s. Figures 8.1A,B show two examples of such networks, involving, respectively, a sugar-uptake system and the Fe-uptake system in *E. coli*.

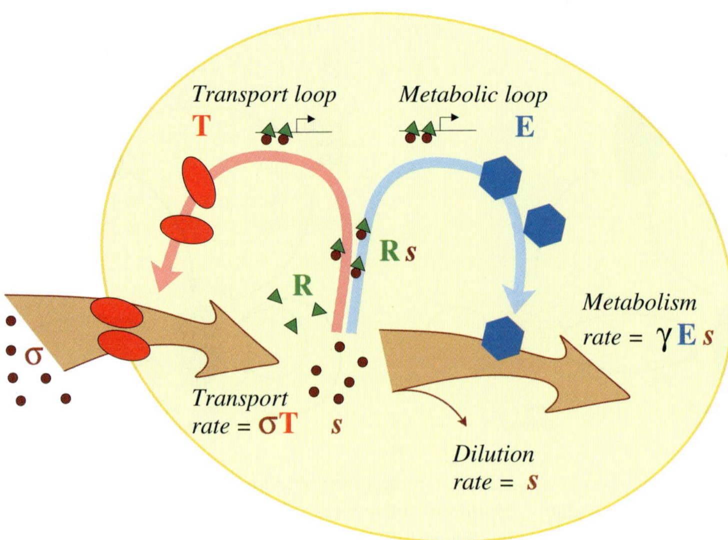

Figure 8.9 Entangled FL motifs for regulation of uptake and metabolism of a small molecule with extracellular concentration σ and an intracellular concentration s. It is imported into the cell by transport proteins, T, where it is consumed in reactions catalyzed by enzyme, E. In addition, it may be diluted by cell growth. A regulator, R, binds to the small molecule, thereby sensing its concentration. The regulator in turn affects production of T and E, thereby closing two FLs. Several logical possibilities exist for all these interactions, resulting in different combinations of the two FLs.

The dynamics of the intracellular concentration s of the small molecule is approximately given by:

$$\frac{\mathrm{d}s}{\mathrm{d}t} = \sigma \cdot T - \gamma \cdot E \cdot s - s. \tag{8.14}$$

where time is counted in time units of the overall dilution of s due to cell growth. Rates are simplified from their more correct Michaelis–Menton form, implying that $\gamma \cdot E \cdot s$ is an approximation of $k_{\mathrm{cat}}^{\gamma} \cdot E \cdot s / (K_M^{\gamma} + s)$.[6] σ is the extracellular concentration of the small molecule, and γ is the metabolic rate in units of the dilution rate due to cell growth.

γ is a central parameter because it controls the ratio between the flux of small molecules and the size of the intracellular pool: in steady state, the intracellular pool will be:

$$s = \frac{\sigma \cdot T}{\gamma \cdot E + 1} \tag{8.15}$$

with a flux $= \sigma \cdot T$ of small molecules coming into the cell. Typically, small metabolites are processed very fast, and at any one time only a tiny fraction is freely available in the

---

[6] k_{cat} is the maximum rate at which the enzyme E can act on s, a rate that for real enzymes varies hugely, between $0.1\,\mathrm{s}^{-1}$ and up to $\sim 10^6\,\mathrm{s}^{-1}$. Examples are lysozyme, with $k_{\mathrm{cat}} = 0.5\,\mathrm{s}^{-1}$, and carbonic anhydrase, with $k_{\mathrm{cat}} = 600\,000\,\mathrm{s}^{-1}$ [423]. Equation 8.14 makes the simplifying assumption that the enzymes E and T are functioning in a linear, un-saturated regime.

	Free molecules	Consumed in one cell generation	$\gamma \cdot E$
Number of amino acids	6 000 000	1 000 000 000	200
Number of ATP	2 000 000	15 000 000 000	8.000
Fe	10 000	1 200 000	120
Glucose (if main food source)	$\sim 200\,000$	500 000 000	2.500
Galactose (if main food source)	$\sim 20\,000$	500 000 000	25.000

Table 8.1 Active versus total amount of different metabolites in one *E. coli* cell. In addition, *E. coli* has 12 tRNAs per ribosome, which are recycled roughly every second, giving a turn-around factor for tRNA of $\gamma \cdot E \sim 2000$ per cell generation. The data is from the cyper cell database (CCDB) [63, 64].

cell. In other words, $\gamma \cdot E$ is typically very large, see Table 8.1. This implies that any change in the uptake or usage of a small molecule occurs at a timescale of $\sim 1/\gamma E$, set by the fast enzymatic conversion of metabolites into other metabolites. This conversion rate is estimated for some metabolites in Table 8.1 by use of Eq. (2.2) from Chapter 2.

For iron in *E. coli*, for example, when growing in rich medium, a cell contains $10\,000-20\,000$ free or loosely bound Fe^{2+} ions, whereas the total content of Fe^{++} is 10^6, corresponding to a "flux" of 10^6 per cell generation [424]. Therefore $\gamma_{Fe} \sim 100$. That is, most of the iron in an *E. coli* cell is integrated into proteins, which effectively removes them from the available pool. For nutritional molecules like glucose, galactose and lactose, the estimate for γ is even higher [424] (see Table 8.1).

8.4.2   "Consumer" and "fashion" in metabolic regulation

To model the feedback from the intracellular small molecule s to the transport and metabolic systems, one must consider equations for changes in T and E. As explained above, the feedback typically uses a transcriptional regulator that senses the small-molecule concentration by forming a complex with it.

Assuming that s is always much larger than the regulator concentration, R, we obtain the concentrations of free (R_{free}) and bound regulator (Rs):

$$R_{\text{free}} = R^{\text{tot}} \cdot \frac{1}{1+s}; \quad \text{and} \quad \{Rs\} = R^{\text{tot}} \cdot \frac{s}{1+s} \tag{8.16}$$

Here R^{tot} is the total amount of regulator and s is counted in units of the dissociation constant of the $\{Rs\}$ complex. For a more detailed expression see Eq. (8.18) in the question section.

The derivation of the above equation is similar to the derivation of the fraction of Rs bound to an operator site, see Eq. (3.5) in Chapter 3. After all, the two situations are analogous to an abundant molecule (here s) that bind to a much less abundant molecule (here R).

The concentration of the active form of R, denoted R^*, is either R_{free} or $\{Rs\}$, depending on the details of the regulation. If R^* is a repressor, the dynamics of T and E are given by:

$$\frac{dT}{dt} = \text{leak} + \frac{1}{1 + (R^*/K_T)^h} - T$$

$$\frac{dE}{dt} = \text{leak} + \frac{1}{1 + (R^*/K_E)^h} - E \tag{8.17}$$

where again "leak" represents a small basal activity,

Each FL in Fig. 8.9 can be characterized by a sign that denotes whether the loop implements positive (+) or negative (−) feedback on s. Note that a positive metabolism FL means that an increase in s leads to a decrease in the metabolic rate, hence a self reinforced increase in s itself. In describing the logic of the entangled loop motifs, we will use the notation of two signs, e.g. (+ −) meaning that the transport loop has a positive feedback and the metabolism loop a negative one.

There are four possible logical structures [425] with architectures that each define a particular function illustrated in Fig. 8.10:

- (− −) Maintains homeostasis of intracellular small-molecule concentration, i.e., exhibits relatively small variation in s with σ. Useful for essential molecules that are harmful at high intracellular concentrations. It is a prominent part of the iron

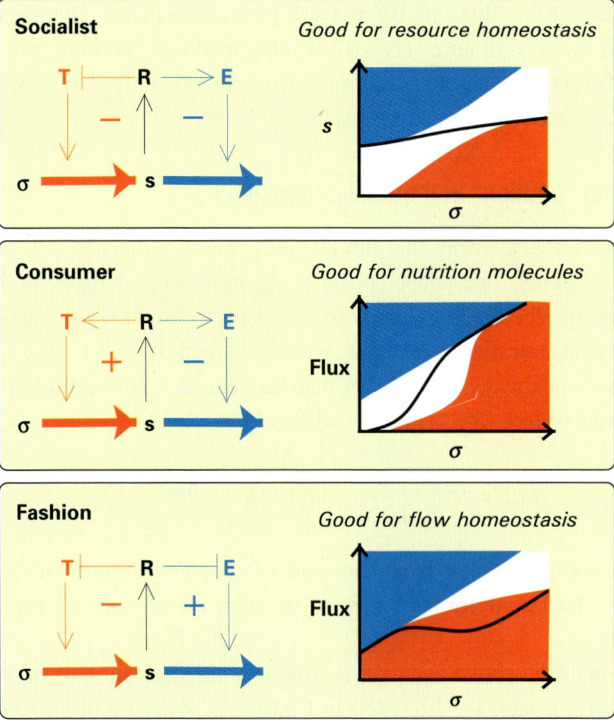

Figure 8.10 Steady-state values of s or flux ($\sigma T = \gamma Es + s$) as a function of σ, for each of the three used for feedback systems. The black curve corresponds to standard parameters assuming two loops of equal strength. The boundary of the red (blue) region is the behavior when E (T) is fixed at unity, i.e. when only the transport (metabolism) feedback is active.

uptake system in Fig. 8.1B, involving in particular the well-studied sRNA RhyB [120].

- (+ −) Differentiates discretely between low and high consumption states such that consumption is initiated when a food molecule is sufficiently abundant. At low external σ the positive feedback will maintain a low internal s, whereas an increase in σ above a certain threshold will turn on transport and thereby secure large consumption. Found in many sugar-uptake systems, including the galactose system in Fig. 8.1A and the lactose-uptake system [426]. When abandoning the enzyme loop (red curve), the remaining system shows sign of bistability [426].

- (− +) When external σ gets "too" large, the internal s becomes so large that the enzyme is repressed, and thereby the conversion of s is stopped. This motif is represented in the Suf-regulated part of the Fe^{++}- uptake system in *E. coli* (see Fig. 8.1B).

- (+ +) Causes bistability in both the flux and the intracellular small-molecule level, to an extent that makes it useless (therefore not shown in the figure).

The functional behavior of each two-loop motif has led us to name them, respectively, the "socialist" (− −), the "consumer" (+ −), the "fashion" (− +) and the "collector" motif (+ +) [425].

In addition to steady-state behavior, each motif exhibits a dynamic response to sudden changes in resources. The response time is primarily limited by the relatively long time it takes to stabilize to new levels of E and T. As E and T are usually abundant "workhorse proteins", they are not actively degraded and decay times are set by dilution during one cell generation. For some particularly valuable metabolites, the relaxation time may be modulated by regulating the decay times of E. For example, in the Fe^{++}-uptake system in *E. coli*, the role of E is taken by mRNA for proteins that incorporate Fe as part of their structure, which in turn is under fast regulation by the small RNA called RyhB [355].

8.4.3   Integrating motifs into larger networks

Real regulatory networks naturally contain more than one or two feedback loops. This can be illustrated by inspecting the iron-uptake system or the galactose system, both outlined in Fig. 8.1.

The iron-utilization network in Fig. 8.1B, involves fast decay of key mRNA in Fe utilization [355]. To maintain constant internal Fe^{++}, the RN includes several negative feedbacks, with a core consisting of the (− −) motif, with the last negative feedback being associated with degradation of the mRNA that encodes for proteins that need iron. But the network also involves additional positive feedback on iron utilization acting through Suf (see Fig. 8.1B). The purpose of this mixing of the "socialist" and the "fashion" motifs (Fig. 8.11B), is to prioritize during iron depletion: when iron is scarce the negative feedback (socialist) decreases its utilization, whereas the positive feedback (fashion) tends to strengthen the depletion by using the remaining iron.

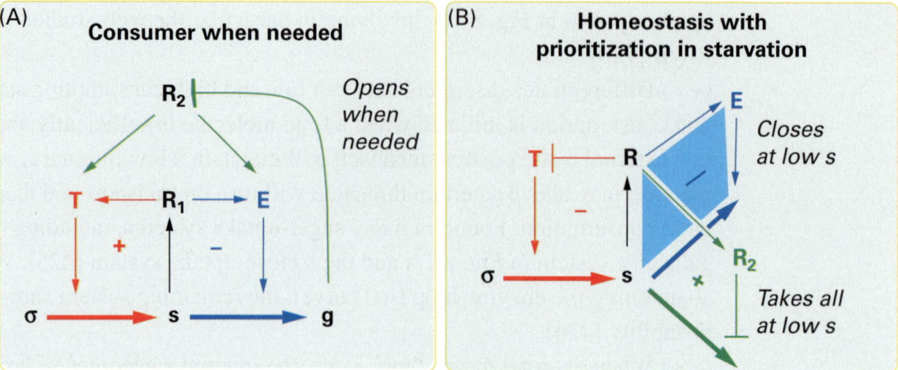

Figure 8.11 Different ways to combine multiple feedback loops. The left-hand panel shows a subset of a sugar-uptake system (galactose in *E. coli*). The red FL regulates the inport of sugar, whereas the blue FL is its metabolism. The green FLs use a second R to sense carbon starvation through the level of glucose, and down-regulates uptake when food is not limiting for growth. This overall negative feedback is a regulatory example of feedback inhibition [360], which is a common motif acting between the end product and the beginning of a metabolic pathway. The right panel resembles the Fe-uptake system in *E. coli*, with the red FL regulating transport, whereas the blue and green FLs redirect the flux of Fe.

Another example is the galactose-uptake system from Fig. 8.1A, which at its core is a "consumer" motif, but supplemented with an overall negative FL which increases E and T production during carbon starvation. This design is highlighted in Fig. 8.11A. In contrast to the iron three-loop motif of Fig. 8.11B, the sugar uptake uses a second regulator (CRP taking the role of R_2 in Fig. 8.11B) to sense the level of another small molecule (cAMP) to measure overall starvation. Both GalR/S and CRP regulate the same set of genes, and the logic of signal integration for production of E and T [353] therefore also comes into play [354]. The functional motifs above are also seen in amino-acid metabolism [427], where the transport loop may sometimes be replaced by regulated fluxes from other types of amino acids [428].

8.4.4   Questions

8.4.1   Simulate the "consumer motif" where the small molecule activates the regulator. The concentration of the complex $[Rs]$ is given by:

$$K^h = \frac{(R^{\text{tot}} - [Rs]) \cdot (s - h[Rs])^h}{[Rs]} = \frac{(R^{\text{tot}} - [Rs])s^h}{[Rs]} \Rightarrow [Rs] = \frac{R^{\text{tot}} \cdot (s/K)}{1 + (s/K)}$$

for $h = 1$, where we use $R = R^{\text{tot}}$ as the total R, and assume R to be much smaller than s. K_s sets the binding strength of the {Rs} complex.[7] Assume

---

[7] If multiple s bind to R, the concentration of free R (not sequestered by s) is $R_{\text{free}} = R^{\text{tot}} - h_s[Rs_{h_s}] = R^{\text{tot}}(1 + (1 - h_s)(s/K)^{h_s})/(1 + (s/K)^{h_s})$.

that Rs is the concentration of the active form of R. Consider the consumer motif:

$$\frac{dE}{dt} = \epsilon + \frac{[Rs]^2}{[Rs]^2 + K_E^2} - E \ ; \quad \frac{dT}{dt} = \epsilon + \frac{[Rs]^2}{[Rs]^2 + K_T^2} - T \tag{8.18}$$

with $R = 1$, $\gamma = 1000$, $\epsilon = 0.001$, $K = 10$ and $K_E = K_T = 0.1$. Simulate the system until steady state is reached for a range of values of the source σ between 1 and 1 000 000 (i.e. set $\sigma = 2^i$, $i = 0, 1, 2 \ldots 20$). For each value of E and T, s is determined from the flux Eq. (8.14).

8.4.2 Simulate the "consumer motif" above, but using R_{free} as a repressor instead of Rs as an activator, and compare the steady-state plot in the two cases. Use $R = 1$, $\gamma = 1000$, $\epsilon = 0.001$, $K = 10$ and $K_T = 0.1$, but $K_E = 0.8$. Simulate the motifs for various values of the source σ between 1 and 1 000 000 (i.e. set $\sigma = 2^i$, $i = 0, 1, 2 \ldots 20$).

8.4.3 Simulate the "consumer motif" above if Rs regulates itself as activator and repressor, and compare the steady-state plot in the two cases. Use $R = 1$, $\gamma = 1000$, $\epsilon = 0.001$, $K = 10$ and $K_T = 0.1$, but $K_E = 0.8$. Set $K_R = 0.1$ and use the same equation for R as for T. Simulate the motifs for various values of the source σ between 1 and 1 000 000 (i.e. set $\sigma = 2^i$, $i = 0, 1, 2 \ldots 20$).

8.4.4 Simulate the "fashion motif" for steady-state values of both internal s and the flux $= \sigma \cdot T$ through the system. Use $R = 1$, $\gamma = 1000$, $\epsilon = 0.001$, $K = 10$ and $K_T = K_E = 0.1$ and let Rs act as a repressor. Simulate the motifs for various values of the source σ between 50 and 1 000 000 (i.e. set $\sigma = 2^i$, $i = 0, 1, 2 \ldots 20$).

8.4.5 The fashion motif is so named so one may view s as a product, T as producer, and E as consumer. The positive feedback around E reflects the tendency of consumers to want what is scarce. Adding the additional positive feedback of R being activated by itself (Rs) may then reflect a tendency of consumers to communicate and enhance their common fashions. Simulate the "fashion motif" with Rs activating production included. Plot state values of both interval s and the flux $= \sigma \cdot T$ through the system. Use $R = 1$, $\gamma = 1000$, $\epsilon = 0.001$, $K = 10$, $K_T = K_E = K_R = 0.1$ and let Rs act as a repressor, except for itself. Simulate the motifs for various values of the source σ between 50 and 1 000 000 (i.e. set $\sigma = 2^i$, $i = 0, 1, 2 \ldots 20$).

8.5 Combined feedback loops and spatial organization

8.5.1 Feedback across space

FLs may extend across space, combining the local action of some molecules with the spatially extended action of molecules that are designed to diffuse more freely. By combining production with secretion, or cellular growth with mutual inhibition of growth, one may obtain a large variety of both frozen and dynamic patterns [5, 429, 430,

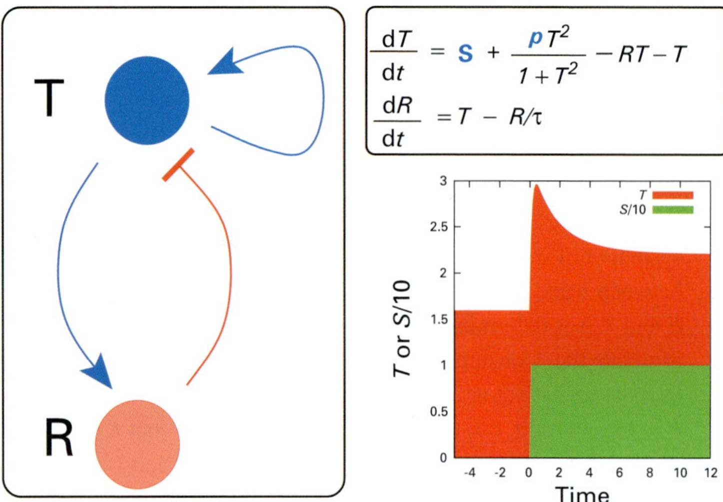

Figure 8.12 A simple motif with many properties, including its ability to generate a transient response to a change in external stimuli, which can be used to facilitate wave propagation. In the example shown we use $\tau = 5$, $p = 20$ and change S from 0 to 10 at time $t = 0$.

431, 432], including patterns as advanced as segmentation and somite formation in vertebrates [433, 434, 435]. At the population level one may also understand the topology of growing bacterial cultures [436] or growing yeast colonies [437].

We will see that the dynamics of a combination of positive and negative feedback depends crucially on whether the components defining the positive feedback have larger or smaller spatial extents than that of the negative feedback. We turn to model systems that are larger than bacteria, with the primary focus on an idealized model [438] of inflammation spreading across mammalian tissue.

8.5.2   Trigger waves from spatially extended positive feedback

8.5.2.1   Excitable medium, information spreading and inflammation

Propagating waves are seen in many biological systems, presumably because these waves facilitate transfer of information over large distances [41, 439].[8] These excellent reviews contain a number of references to classical examples of excitable media in chemical [440] and living systems, ranging from the Hodgkin and Huxley action potentials along neurons [26] and electrical excitation of the cardiac muscle, to reaction-diffusion waves on the skin of angelfish [429]. Recent observations [432] have found such trigger waves of the cytokine Cdk1 that move across the 1.2 mm frog egg with propagation speeds of about 1 µm s^{-1}.

Waves can be generated by small regulatory circuits, like the one shown in Fig. 8.12, provided that they are coupled to each other in space. This can most easily be done

---

[8] The classical FitzHugh–Nagumo model [26, 41] deals with switching of states in spatially coupled double well potentials $du/dt = u \cdot (u - a) \cdot (1 - u) - v + d^2u/dx^2$ supplemented with slow relaxation $dv/dt = \epsilon \cdot u + \delta d^2v/dx^2$ with $\epsilon, \delta \ll 1$ is a simple model for an excitable medium.

through a diffusible substance. Here we will discuss one such potential system for generating waves in living tissue using the NF-κB system, as proposed by [438]. The key idea is that this system may allow the cells in tissue to form an excitable medium, thereby priming it for propagation of waves of cytokines.

An excitable medium comprises locally excitable regions – in our case it is a cell with tissue – which all have the ability to be induced (excited) and inhibited. Such systems are characterized by the "excitation threshold," so that subthreshold stimuli are rapidly damped, and the system persists in a resting state. Stimuli that exceed the threshold trigger a sharp local response and the system transits into the excited state. Shortly after, the region becomes insensitive to further perturbation and is said to be in a refractory period, after which it slowly relaxes back to the resting state where it is again sensitive to perturbations.

Figure 8.13 outlines a minimal example of an excitable medium, presenting a combination of an excitable protein T that subsequently can be repressed by a slower-responding regulator R. T is excited by an external stimulus $S = 1$ at one point in a

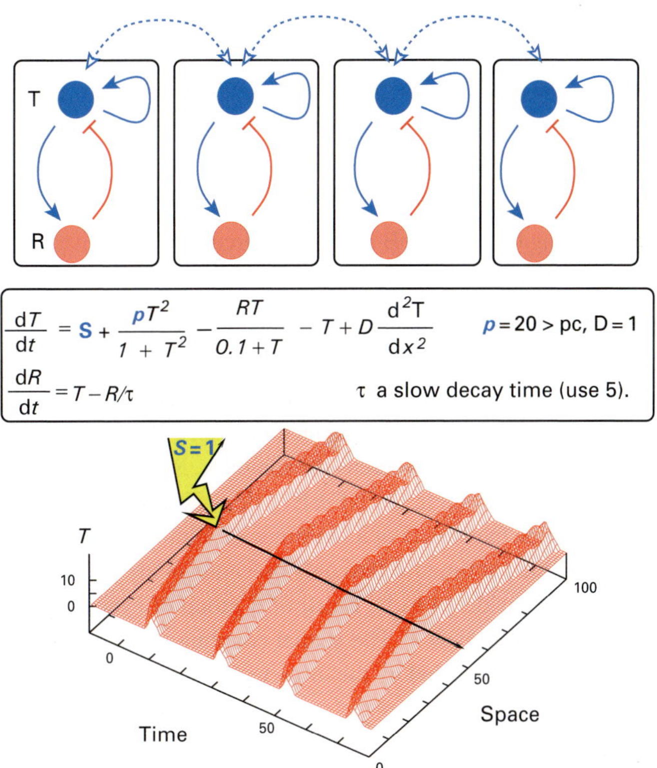

Figure 8.13 A minimal model of an RN-based excitable medium, consisting of a diffusible substance T, which catalyzes its own production, supplemented by delayed inhibition through a slowly degrading R. The system is exposed to an external stimulus $S = 1$ at cell number $x = 0$ for all times larger than $t = 0$, and as a consequence, waves of T are repeatedly emitted from this point. $S = 0$ everywhere else. p parameterizes the strength of the positive feedback.

one-dimensional model of tissue, causing a wave to propagate from this point outwards. The propagation of the wave requires positive feedback that can sustain two separate states. The strength of the positive feedback is parameterized by p.

Following [438], we reformulate the simple excitable medium into the language of a more realistic NF-κB system, known to play a key role in inflammatory response in cells. The NF-κB system is known to activate extracellular diffusing cytokines [441], which in turn are known to activate the NF-κB system [442]. This "activation spreading" feedback suggests that signals of inflammation may use spatially coupled NF-κB as the "excitable medium," that facilitates waves and distant signaling across a living tissue.

Inflammation is the initial reaction of the mammalian host to pathogens and damaged cells. In response to inflammatory stimuli, neutrophils chemotax up the gradients of cytokines and chemokines and, once localized to the affected region of tissue, they eliminate the offending agents by phagocytosis. This process is highly dependent on the successful recruitment of neutrophils from the bloodstream through a chemotactic signal. If the cytokines are not amplified continuously, but transiently, the tissue cells will avoid sustained exposure to toxic cytokines while the signal – the chemo-attractant gradient – can still penetrate far in the tissue.[9]

Consider a simplified model of one-dimensional tissue, i.e. a row of tissue cells that each activate NF-κB in response to an extracellular cytokine stimulus (T), as shown in Fig. 8.14. Further, in each cell, active NF-κB (N) up-regulates T synthesis, which is then exported out and diffuses into the extracellular space. The key features of the model are:

(1)   Slow NF-κB negative feedback,
(2)   Fast cytokine positive feedback,
(3)   Spatial coupling of the NF-κB regulatory circuits by diffusive cytokine in spatially distributed cells.

In Fig. 8.14, the FLs for NF-κB regulation [375], are reduced to a single negative feedback [384]. The main purpose of this feedback, $N \rightarrow R \dashv N$, is to reproduce the transient dynamics of NF-κB in response to TNF stimulation, see Fig. 8.6C. A transient spike occurs about 30 minutes after induction [375, 386]. In the NF-κB system, N represents the activity of NF-κB, whereas the inhibitor, R, represents a more complex agglomerate of the other players in the NF-κB system. Thus R should be thought of as a regulator that combines the effects of a multitude of inhibitors in one variable [438].

The stimulus S is added at time $t = 0$ to only one node in the middle of the simulated system. Each node is considered to be a cell or a group of synchronized cells and is assigned a spatial extension $\delta x = 1$. The spatial coupling is done through diffusion of TNF, denoted as T. This is modeled by the discrete diffusion equation:

$$T_{t+dt,x} = T_{t,x} + D \cdot (T_{t,x+1} - 2T_{t,x} + T_{t,x-1}) \cdot dt \qquad (8.19)$$

---

[9] Notice that the signaling wave does indeed give a gradient far away, but also that on the backside of the wave it naturally gives the opposite gradient. Thus a cell that senses the signal should be de-sensitized when first walking up the gradient on one side of the wave. When the cell then reaches the backside of the wave, it should be in a state where it is does not sense gradients at all. For the model see Question 8.5.4.

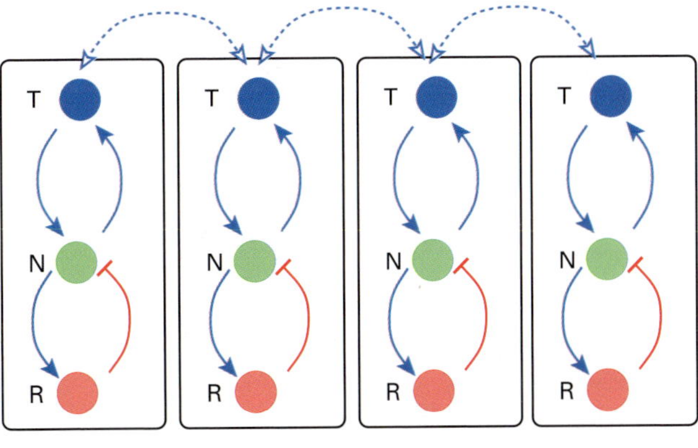

$$\frac{dT}{dt} = S + \frac{pN^2}{1+N^2} - T + D\frac{d^2T}{dx^2} \qquad p = 50 > pc,\ D = 1$$

$$\frac{dN}{dt} = \frac{T}{1+T}(1-N) - R \qquad (1-N)\ \text{term not essential for waves}$$

$$\frac{dR}{dt} = N - R/\tau \qquad \tau\ \text{a slow decay time (use 5).}$$

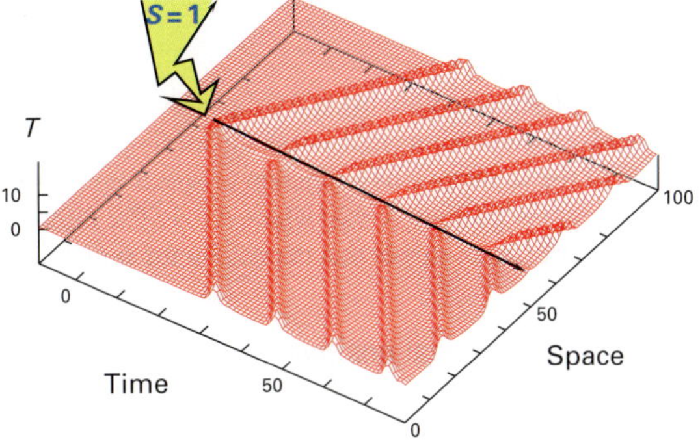

Figure 8.14 Model of cell tissue where each cell contains an NF-κB circuit. N is the fraction of active NF-κB, T = TNF is the cytokine concentration normalized to the "activation threshold" and R is a negative regulator of N. The system is exposed to an external stimuli triggered by inflammation $S = 1$ at cell number $x = 0$ for all times larger than $t = 0$, and, as a consequence, waves of T are repeatedly emitted from this point.

where t refers to time and x to space, simulated in units of a node diameter $\delta x = 1$. We use the diffusion constant $D = 1$ in the figures.[10]

In terms of simple cellular automata modeling (see Chapter 11), an excitable medium is like a forest, primed to propagate a fire. When the fire hits, it will spread as a wave leaving behind a system that needs time to regenerate before a new fire can spread. In the language of Fig. 8.13, T is then a fire that feeds on itself (and the absence of R), while causing bare ground R. This in turn represses back-propagation of the fire, before it is slowly converted into forest again.

[10] The TNF protein diffusing between cells in reality would be characterized by D around $\sim 20\,\mu\mathrm{m\,s^{-1}}$ whereas the mammalian cell size would be about $15\,\mu$m.

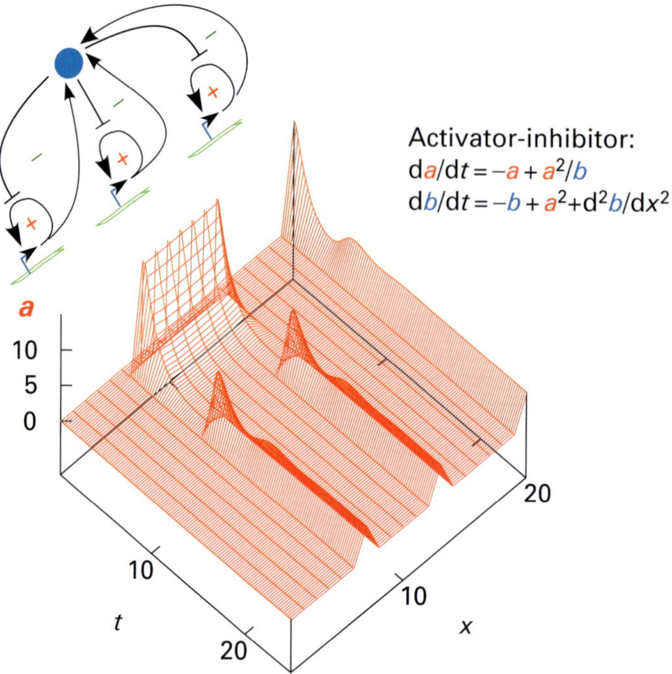

Figure 8.15 A minimal model of segmentation using local positive feedback combined with diffusing inhibitor. The network illustration emphasizes the local–global feedback combination. The figure shows the level of the activator a as a function of space at different times since the start, where $a = b = 0.01$, except at position X = 20 where $a = 1$. The diffusion constant $D = 1$. The model results in a pattern of frozen stripes that depend on the initial distribution of a and b.

8.5.3 Developmental boundaries from local positive feedback

Interestingly, the combinations of positive and negative feedback used above can be switched around to generate frozen patterns. That is, if one uses local positive feedback combined with a diffusing inhibitor, one may naturally generate patches of alternating states, as originally suggested by Allan Turing [331] and subsequently explored as models for developmental systems [332, 443, 444, 445]. A simple model for such an activator–inhibitor system was suggested by Meinhardt in his beautiful book about patterns on sea shells [5]:

$$\frac{da}{dt} = \frac{a^2}{b} - a$$

$$\frac{db}{dt} = a^2 - b + D \cdot \frac{d^2b}{dx^2} \tag{8.20}$$

where a is an activator, b an inhibitor and for simplicity we have removed saturation constants.[11] Figure 8.15 shows the resulting pattern with space discretized in units of cells of size 1 and a diffusion constant $D = 1$. In contrast to excitable media where positive feedback is acting between cells, negative feedback acting between cells enforces a

---

[11] Thus terms like $a^2/(a^2 + K_a^2)$ are replaced by a^2, and $1/(b + K_b)$ is replaced by $1/b$.

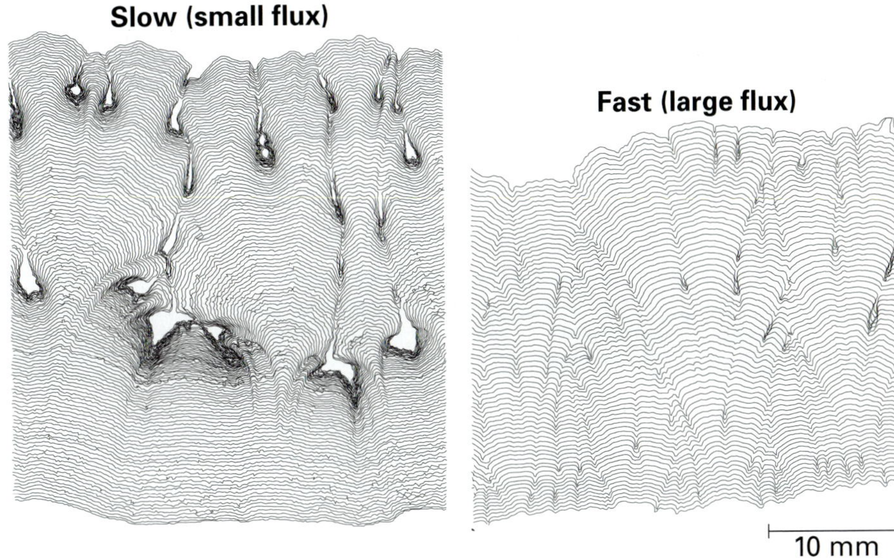

Figure 8.16 Experimentally measured growth of a yeast colony [437], where each front corresponds to the edge of a colony at one hour intervals. For the fast-growth case, the colony is growing on a 1 mm agar layer that separates it from a replenishing liquid. For slow growth the growing yeast is separated from 2.6 mm from this continous food supply and waste product removal [437].

frozen final pattern that may relate to stages of development [443, 444, 445]. Importantly, pattern formation by activator–inhibitor systems also requires that the dynamics of the inhibitor is not to slow (Question 8.5.7).

In real systems, inhibition between neighboring cells may take place on a scale that is set directly by communication across cell walls, as exemplified in a recent study on the Notch–Delta system [446]. Here the transmembrane Notch and transmembrane Delta may interact and effectively annihilate the Delta, and the Notch on the neighboring cell, thereby forming mutually exclusive states on neighboring cells.

8.5.4 Competition and order in disorder

Growth is often inhomogeneous, as it inherently contains instabilities associated with relative positioning in space. Figure 8.16 shows an example of such instabilities, where a growing yeast colony develops self-amplifying spatial inhomogenities that show a characteristic pattern of deep and slow-growing valleys, contrasting with the fast-moving front of the colony.

The instability associated with growth and its mutual inhibition is, in its simplest form, expressed by the Kuramoto–Sivashinsky equation [447, 448, 449, 450, 451, 452], often associated with propagating flame fronts. The central step in the derivation of this equation is outlined in Fig. 8.17. This figure illustrates an instability associated with mutual inhibition between segments of a growing front: the front ahead has better access to resources, and at the same time it may excrete diffusible waste products that

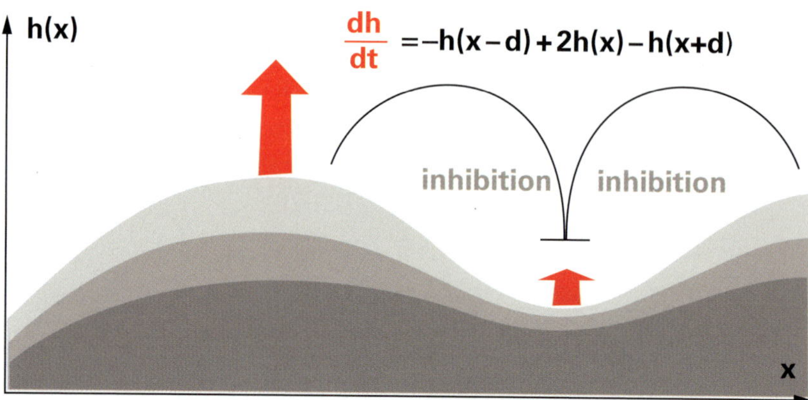

Figure 8.17 Growth with an instability caused by an "unfair" advantage to the part of the growing colony that is ahead of the rest. The front at some horizontal co-ordinate x will grow at some speed (here denoted v) plus some correction associated with whether the neighborhood is ahead or behind $h(x)$.

inhibit growth of the front. This mutual inhibition results in a self-amplifying inequality of the front profile $h(x)$, where x is the position along the front $x \in [0, L]$:

$$\frac{\mathrm{d}h}{\mathrm{d}t} = -h(x-1) + 2h(x) - h(x+1) + \ldots = -\frac{\mathrm{d}^2 h}{\mathrm{d}x^2} \ldots .. \tag{8.21}$$

Here $h(x) - h(x-1)$ and $h(x) - h(x+1)$ represent inhibition by the neighbors to position x.

Equation (8.21) is very unstable, in fact acting opposite to normal diffusion. Any modulation of h, like $h = \epsilon \cdot sin(qx)$ is amplified exponentially:

$$\frac{\mathrm{d}\epsilon}{\mathrm{d}t} = q^2 \cdot \epsilon \Rightarrow \epsilon \propto \exp(q^2 \cdot t) \tag{8.22}$$

This instability expresses the exponential advantage of being ahead: individuals that are at higher h positions are at an advantage and thereby grow faster, which in turn amplifies the initial difference. To stabilize this tendency of the "winner takes all," Kuramoto [447] and Sivashinsky [448] added two stabilizing terms:

$$\frac{\mathrm{d}h}{\mathrm{d}t} = -\frac{\mathrm{d}^2 h}{\mathrm{d}x^2} - \frac{\mathrm{d}^4 h}{\mathrm{d}x^4} + \left(\frac{\mathrm{d}h}{\mathrm{d}x}\right)^2 \tag{8.23}$$

which includes negative feedback on small scale, $-\mathrm{d}^4 h/\mathrm{d}x^4$,[12] and couples it, by $(\mathrm{d}h/\mathrm{d}x)^2$, to the larger- scale positive feedback. Intuitively, the fourth-order term speeds up the front movement in deep grooves, whereas the term $(\mathrm{d}h/\mathrm{d}x)^2$ closes the deep grooves by advancing the near vertical segments of the front.

---

[12] Inserting $h(x,t) = \epsilon(t) \cdot sin(qx)$ into $\frac{\mathrm{d}h}{\mathrm{d}t} = -\frac{\mathrm{d}^2 h}{\mathrm{d}x^2} - \frac{\mathrm{d}^4 h}{\mathrm{d}x^4}$ one obtains:

$$\frac{\mathrm{d}\epsilon}{\mathrm{d}t} = (q^2 - q^4) \cdot \epsilon \Rightarrow \epsilon \propto \exp(q^2(1-q^2) \cdot t)$$

which implies that ϵ only grows exponentially when $q < 1$, which means that the wavelength $= 2\pi/q > 2\pi$. Thus, the fourth-order term stabilizes all shorter wavelengths and thus tends to smooth small-scale differences.

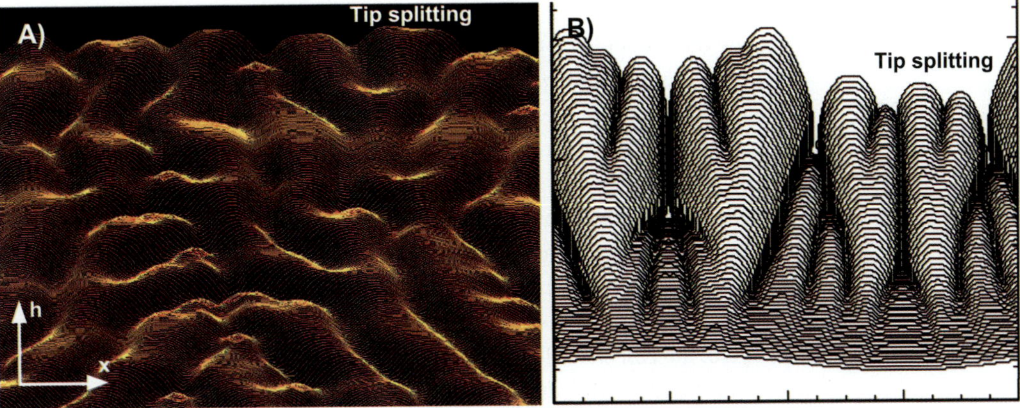

Figure 8.18 Left panel shows subsequent snapshots of a front governed by the Kuramoto-Sivashinsky equation with its characteristic tip splitting dynamics. On larger scales the Kuramoto equation exhibits chaos and randomness[452]. Right panel illustrates a sequence of a growing front of many cells, with the rule that every cell is a source of a diffusible molecule that inhibits the growth of all other cells [437]. Allowing such growth with overhangs, still exhibit the tip splitting of the Kuramoto equation, as well as the deep valleys from the experiment of [437].

The overall Equation (8.23) generates ongoing chaotic growth with characteristic "tip-splitting," where the organisms or parts of the system that are ahead amplify and spread laterally. The dynamics are illustrated in Fig. 8.18. When compared to our actual data, shown in Fig. 8.16, one can indeed see qualitatively similar dynamic behavior with the caveat that the real yeast colony allows overhangs and tends to leave the deep grooves behind. The right panel of Fig. 8.18 illustrates a growth model that allow for overhangs.

The Kuramoto–Sivashinsky equation is one of the simplest equations that generates chaos and noise from dynamics that are formally deterministic [450, 451, 452]. Thus large-scale stochastic behavior can emerge without use of the randomness of individual molecular events.

8.5.5 Questions

8.5.1 Simulate the model in Fig. 8.13, that is, update the following equations by using a timestep $dt = 0.001$, for $x = 1, 2, 3 \ldots 100$:

$$\Delta_T(x) = S(x) + \frac{p \cdot T(x)^2}{1 + T(x)^2} - \frac{R(x) \cdot T(x)}{0.1 + T(x)} - T(x)$$
$$+ D \cdot (T(x-1) + T(x+1) - 2T(x))$$
$$\Delta_R(x) = T(x) - R(x)/\tau \tag{8.24}$$
$$T(x) = T(x) + \Delta_T(x) \cdot dt \quad \text{and} \quad R(x) = R(x) + \Delta_R(x) \cdot dt$$

where one should remember to update all Δ_T and Δ_R before finally updating T and R. The inflammation is introduced through a persistent $S = 1$ at position $x = 50$, whereas $S = 0$ otherwise. Use parameters $p = 10$, $D = 1$, $\tau = 5$. Explore the behavior for other values of S.

8.5.2   Simulate the model from Question 8.5.1 for various values for the strength of the positive feedback p.

8.5.3   Simulate the model in the previous questions for parameters $p = 50$ and $S = 1$ at $x = 0$, for different values of the degradation time of R, $\tau = 5$ to $\tau = 20$.

8.5.4   We have not described how a cell can chemotax through a series of waves. The main idea is that it should be more sensitive to the positive gradient than the negative. This is obtained by a chemotaxing cell that is sensitive at low concentrations, and insensitive when the concentration is higher. To propagate, the cell should remember extreme positive changes for some time. Make a model for the random walk of a chemotaxing cell, assuming this behavior.

8.5.5   Explore timescales for remembering the large positive changes in T in previous questions.

8.5.6   Simulate the model in Fig. 8.15 and Eq. (8.20) on a one-dimensional line $(1, 2, 3, \ldots, 20)$ with $D = 1$. First use initial conditions $a = b = 0.01$, except $a(20) = 1$, then use initial conditions $a = b = 0.1$, except $a(20) = 1$.

8.5.7   Repeat the previous question for initial conditions $a = b = 0.01$, except $a(20) = 1$, but with half, and then a third of production/decay rates for the dynamics of b.

8.5.8   Estimate the wavelength for the fastest-growing wave of the linear part of the KS equation.

8.6 Summary

- Feedback allows the cell to adjust the repertoire of functional proteins to current needs.
- Negative feedback can cause oscillations if signal propagation around the feedback loop is slow: $\mathrm{d}x/\mathrm{d}t = -x(t - \tau)$ oscillates when τ is large.
- Positive feedback can come from strong self-activation, mutual repression between proteins or by auto-catalytic processes, cases that all allow decision-making, bistability and epigenetics.
- Regulation of metabolism involves two widely different timescales:

$$\frac{\mathrm{d}s}{\mathrm{d}t} = \sigma T - (\gamma \cdot E)s - s \qquad (8.25)$$

  with $\gamma E \gg 1$, an inequality that can be derived from the fact that the intracellular pool of small molecules is smaller than the total amount consumed during a cell generation. T and E change on a timescale of order 1 (cell generation).
- Combinations of positive and negative feedback can be used to make spatial patterns, as well as sending signals across space.

9 Networks

9.1 Convergent computation

The examples of small networks in previous chapters typically consisted of two to three proteins, which together sensed and governed one particular function in a living system. In this and the next chapter we will zoom out and view larger-scale network properties, with the guiding perspective that these networks, in the end, form some functional coherent whole. Figure 9.1 illustrates this main philosophy by showing the regulatory network of two evolutionarily unrelated temperate phages, the enterobacteria phage 186 and the λ phage. Apart from having different numbers of proteins in the outer parts of the networks, the two networks look remarkably similar. Thus, the functional requirement over-rides the randomness of evolution: convergent evolution has directed both phages to the same "computational" design.

The visual similarity of the two networks shown in Fig. 9.1 can be formalized using a network alignment algorithm [157], a comparison that is based entirely on the wiring diagram and the sign of the regularity links. These signs are indicated by different colored links, with green links referring to activation of the production of one protein due to the activity of another, and red links to an inhibition of the downstream protein.

Importantly, in order to align the networks, one needs to take into account indirect regulation, where one identifies a repression of a repression with an activating link. Thus, for example, the RecA degradation of CI in the λ network, in phage 186 is replaced by a RecA that degrades LexA, which in turn represses Tum, which finally binds to the CI in 186 and prevents its function as a repressor. The important lesson is that:

$$(-1) \cdot (-1) = +1 \tag{9.1}$$

for total action of regulatory links in a series.[1] Notice that we obtained a similar result earlier, in, for example, the consumer motifs. There the the small metabolite s can be to bind and allow the Rs complex to act as a repressor, or equivalently, s could bind to an activator R and convert it to the form Rs where it cannot act as activator. In both cases

---

[1] Formally for a series of repressions a ⊣ b ⊣ c one writes $db/dt = k/(k+a) - b$ and $dc/dt = 1/(1+b) - c$, which for steady-state $b = k/(k+a)$ inserted in $dc/dt = 1/(1 + k/(a+k)) - c$ gives $dc/dt = (k+a)/(1+k+a) - c \sim a/(1+a) - c$ when a binds strongly to the operator for b ($k \ll 1$). Thus in effect a acts as an activator for c, with a time delay that will depend on the timescale for b to reach its steady-state value. If a is not a regulator but a protease, then $db/dt = 1 - \beta \cdot a \cdot b - b \rightarrow b = 1/(1 + \beta \cdot a)$ and again one obtains an activation curve for c with a.

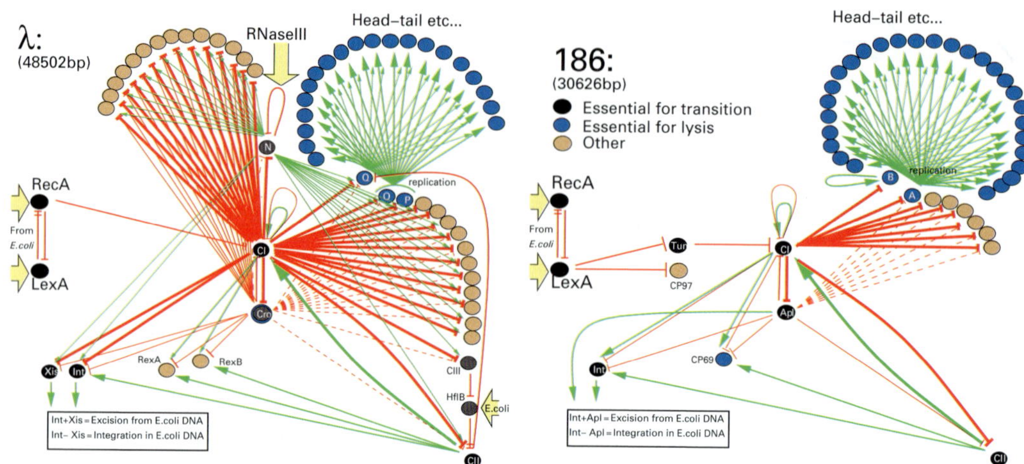

Figure 9.1 Genetic regulatory network of phage λ, and of temperate phage 186 [157]. These two phages have essentially no protein homologies, but have similar regulatory architectures, provided that one identifies repression of a repression with activation, i.e. $-1 \times -1 = +1$. Notice also that CI and CII in 186 are integrated into same regulatory motifs as the evolutionarily unrelated CI and CII in λ. Graph alignment methods can be found in [157, 453, 454]. The networks do not include protein–protein binding, which for λ was measured by [455].

the overall functional response is that presence of s represses whatever is downstream of R.

Formally, in order to truly align two networks, one needs a "scoring" function to estimate the quality of the alignment, as well as quantifying the extent to which the topologies of these two networks are more similar to each other than expected from random chance. This is a more technical issue, that demands some understanding of networks, and in particular what a random network is. We in the next section recapitulate basic lessons on network topology, first ignoring the fact that most biological networks are directed and have signs on their links.

9.2   Connectedness

9.2.1   Links and degree distribution

Networks are a widespread concept from both popular and scientific literature. In general, networks are used to characterize the organization of a system of heterogeneous components, each interacting with a small subset of the other components. Networks are very much about history, as the real-life networks evolved through a sequence of events that took place on a much longer timescale than dynamic processes taking place in the network. The concept of a network may accordingly be useful for systems with:

- Heterogeneity: systems with distinctly different components.
- History: a real network does not appear by random assignment of links, but self-organizes over a long time period. Networks are thus useful for systems

with a separation of timescales, with dynamics on the networks that occur repeatedly and on a fast timescale.

- Distribution and containment of information, energy or material.

Figure 9.2 illustrates these concepts on the large-scale regulatory network of *S. cerevisiae*, a network that consists of many distinctly different proteins that developed with the organism over an evolutionary timescale. This timescale that was billions of times longer than the dynamics of the network, which typically are associated with the two-hour generation time of these organisms. Finally, the light colors in Fig. 9.2 show the response of the network to external perturbation. The response is localized, illustrating that molecular networks not only facilitate information transfer, but also that they confine the information to relevant subsections of the system, a signaling horizon that is also often reflected in the topology of networks in other complex systems [456].

Figures 9.3–9.6 define the basic quantities in network theory. Consider the simplest model for a random network with N nodes, the Erdos–Reynei (ER) network [457]. This can be constructed by connecting each node pair (i,j) with probability p. The expected

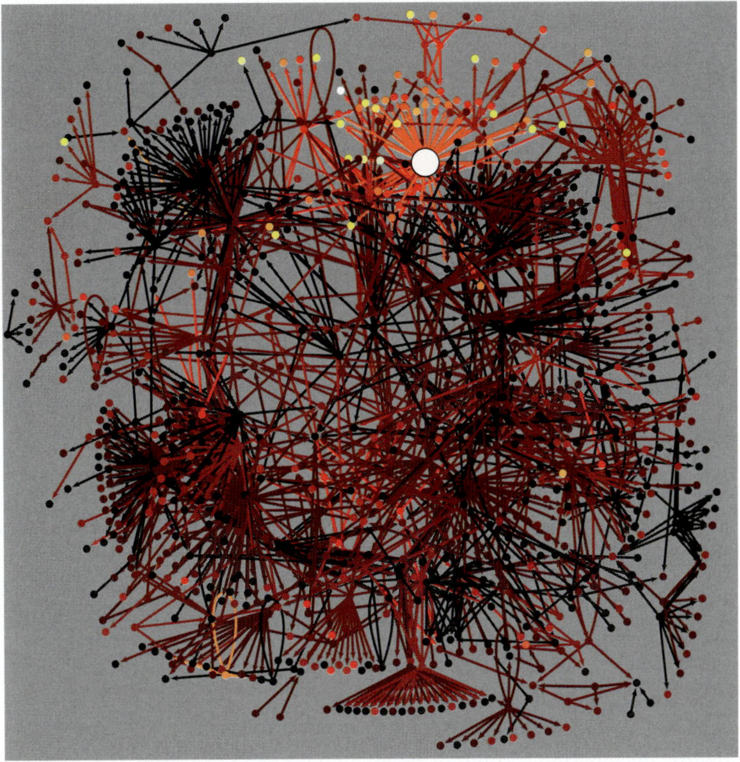

Figure 9.2 Genetic regulatory network in *S. cerevisiae*, from [31]. Notice that the network consists of proteins that are all regulated through the cell nucleus, and does not reflect any "geographical" separation of the proteins. The highlighted nodes reflect genes that change expression in response to a particular external stimulus (amino-acid starvation).

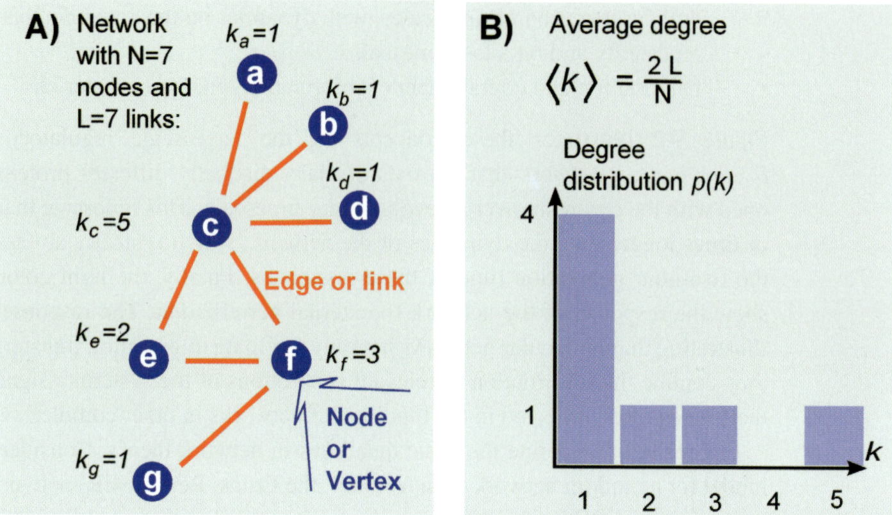

Figure 9.3 (A) Network with nodes and edges: each node characterized by its degree, which is the number of links associated with the node. (B) The degree distribution for the network in panel (A).

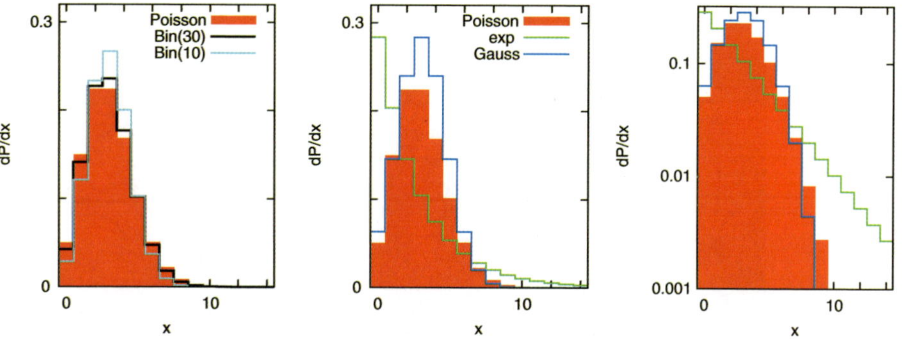

Figure 9.4 Left panel: Comparison between Poisson distribution and binomial distributions both with an average of 3. The binomial distributions are sampled from 10, the other from 30 decisions. Right panel: Comparison of Poisson with exponential and Gaussian distributions. The exponential has the same average, and the Gaussian the same average and standard deviation as the Poisson distribution.

number of edges in the network is then $L = p \cdot N(N-1)/2$. The average degree, defined as number of neighbors per node, is:

$$\langle k \rangle = \frac{2L}{N} \tag{9.2}$$

The degree distribution (probability that a given node has k links to the remaining $N-1$ nodes) is:

$$P(k) = \frac{(N-1)!}{k!(N-1-k)!} \cdot p^k \cdot (1-p)^{(N-1-k)} \tag{9.3}$$

because each node is connected to k specific nodes with probability p^k, and to none of the $N - 1 - k$ other nodes with probability $(1 - p)^{N-1-k}$. The combinatoric prefactor represents the number of ways of selecting the k specific nodes. For large N, $P(k)$ approaches the Poisson distribution (see Fig. 9.4):

$$P(k) = e^{-\langle k \rangle} \frac{\langle k \rangle^k}{k!} \tag{9.4}$$

with an average degree:

$$\langle k \rangle = p \cdot (N - 1) \tag{9.5}$$

and variance $var = \langle k^2 \rangle - \langle k \rangle^2 = \langle k \rangle$, and thus $\langle k^2 \rangle = \langle k \rangle \cdot (\langle k \rangle + 1)$, an equation that will be useful in discussions of signal amplification.

A network is said to be connected if a path exists between any pairs of nodes in the network (see Fig. 9.5A). The distance between two nodes in a network is defined as the minimum number of links that connect them.[2] For a connected network one defines its diameter as the maximum distance between any two nodes.

Imagine a disease that spreads from a node in a random network where all nodes have connectivity of exactly k. At the first step $d = 1$ there are k new neighbors. At the second step, each of the newly visited nodes gives access to $k - 1$ new nodes (see Fig. 9.5B). Assuming that there is no double counting, then the number of visited nodes within distance d from the first node grows as:

$$\text{number}(t) = k \cdot \sum_{d'=1}^{d} (k - 1)^{(d'-1)} \sim k \cdot (k - 1)^{d-1} \tag{9.6}$$

Therefore, the number of visited nodes grows exponentially for any $k > 2$ (see Fig. 9.5). For a more randomized graph, where the degree k may differ between the nodes, the disease will visit the entire network after a number of iterations d, given by:

$$\left(\frac{\langle (k - 1)k \rangle}{\langle k \rangle} \right)^d \approx N \tag{9.7}$$

Importantly, this more complicated equation takes into account that each subsequent node is selected by a probability proportional to its degree (as it sits at the end of a random link). Thus the diameter:

$$\text{Diam} \approx \frac{\log(N)}{\log\left(\frac{\langle k^2 \rangle}{\langle k \rangle} - 1 \right)} \tag{9.8}$$

The main lesson is that the diameter of a random network only grows very slowly with network size N.

---

[2] To calculate the distance from a node i to all other nodes in a connected network, one must first make a list of all the neighboring nodes of i and assign them a distance $d = 1$. Subsequently, one adds new layers of nodes to the list from all the neighbors of the already included nodes, providing that these neighbors are not already on the list. The distance to newly added nodes is calculated from the distance to neighboring nodes already present on the list.

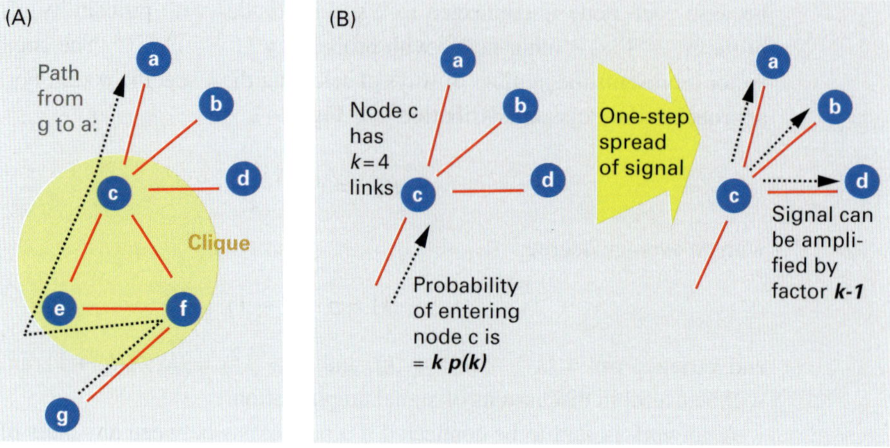

Figure 9.5 (A) Path from node g to node a in a network. Notice that the path shown is longer than the shortest path (g-f-c-a). (B) Amplification of a "virus-like" signal as it enters and spreads across node c with connectivity $k = k_c = 4$. As such, a signal spread subsequently through many nodes, its activity in the network will be multiplied by the connectivity minus one for each node passed.

The above considerations on signal spreading and overall connectedness of a network can be rephrased in terms of the amplification factor $\mathcal{A}$. Consider a perturbation that enters a node, and assume that it is transmitted to all neighbors (see Fig. 9.5B): it is amplified by a factor $k - 1$. However, not all nodes have an equal chance of amplifying signals because the probability of entering a node is itself proportional to its connectivity k. Therefore the amplification in a undirected network [458] is:

$$\mathcal{A} = \frac{\int k(k-1)n(k)\mathrm{d}k}{\int kn(k)\mathrm{d}k} = \frac{\langle k(k-1)\rangle}{\langle k\rangle} = \frac{\langle k^2\rangle}{\langle k\rangle} - 1 \tag{9.9}$$

provided that there are no correlations between the connectivities of neighboring nodes. When $\mathcal{A} > 1$, "disease-like" signals tend to be exponentially amplified, and therefore will spread across the entire network. For $\mathcal{A} = 1$, on the other hand, perturbations will be spread marginally, where some will spread and others will "die out."

To have marginal spreading, one input signal on average should lead to one output through a new link. A network with Poisson-distributed degrees, and $\langle k\rangle = 1$ will have $\langle k^2\rangle = 2$ and $\mathcal{A} = 1$. Such a network will consist of multiple clusters with the power-law distribution of the cluster sizes shown in Fig. 9.9B.

Networks can also be characterized by a cliquishness or clustering coefficient [459, 460, 461]. For each node this is defined as the fraction of cliques in units of the maximum possible number of cliques [461]. For the whole network, the global clustering coefficient is defined as the total number of closed triplets, divided by the total number of triplets in the network (including all sequences of three nodes connected by two subsequent links) [459, 460]. A large clustering coefficient indicates a large locality in the sense that neighbors of a given node tend to be directly connected. A social

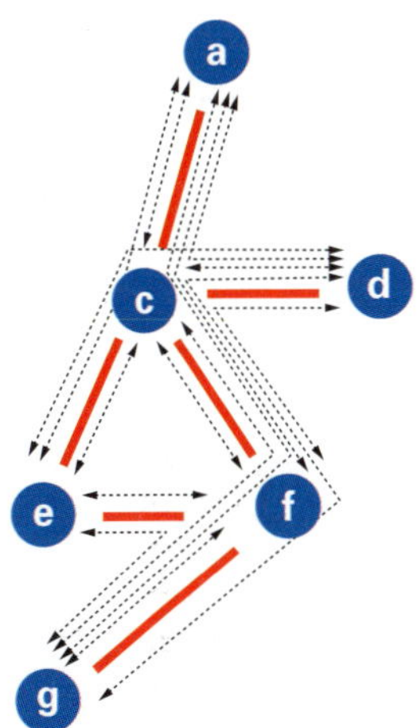

*The betweenness centrality
of a node
is equal to the number
of shortest paths that go
through the node, not including
paths that end at the node.*

Betweenness (a) = 0
Betweenness (d) = 0
Betweenness (c) = 7
Betweenness (e) = 0
Betweenness (f) = 4
Betweenness (g) = 0

Figure 9.6 Illustrating the betweenness of a node, a measure of the communication traffic that goes through the node, assuming that all pairs of nodes send messages to each other, and always use the shortest path between them [462, 463].

network with high clustering reflects a society where nearly everybody has common friends. For the Erdos–Reynei network, the global clustering coefficient is:

$$C = p \sim \frac{\langle k \rangle}{N} \tag{9.10}$$

since the probability that two neighbors of any one node are connected is p. The total number of cliques in such a network is $N \cdot C = \langle k \rangle$, which is independent of the size of the network N.

Another noteworthy concept from characterization of network topologies is the betweenness centrality of nodes, illustrated in Fig. 9.6. The measure ranks nodes according to how centrally they are placed in the network. There are indications that proteins with high betweenness in molecular networks tend to be more important than proteins at more peripheral network locations [464, 465, 466].

It is occasionally useful to represent a network in terms of a matrix, A_{ij}, where a link from node i to node j implies that $A_{ij} = 1$. The absence of a link from i to j implies that $A_{ij} = 0$. A non-directed network is accordingly represented by a symmetric matrix $A_{ij} = A_{ji}$, because a link from node i to j implies that there is also a link from node j to i.

Imagine the population of a disease/virus that is placed on a few of the nodes of a matrix represented by a vector with components v_j, $j = 1, 2, ...N$. The dynamics of disease spreading is a process that allows copying to all neighboring nodes, represented by the update:

$$v_i = \sum_j A_{ij} v_j \tag{9.11}$$

which, in matrix notation, $\mathbf{v}(t+1) = \mathbf{A} \cdot \mathbf{v}(t)$. Applying the matrix multiple times corresponds to applying the infection cycle to neighbors, and next nearest neighbors, and

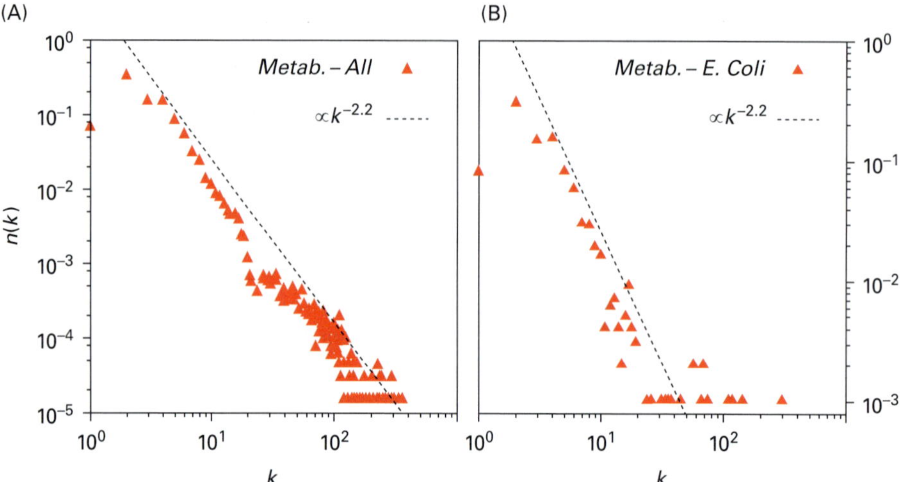

Figure 9.7 Degree distribution of metabolites in networks where two metabolites are linked, one being a substrate and the other a product of the same reaction. The left-hand panel shows the average degree distribution for 107 organisms in the Ma–Zeng database [475]. The right-hand panel shows the degree distribution in the *E. coli* metabolic network. Figure from [476].

back again many times, allowing the disease to present in multiple copies at each node. In the long time limit the product of $\mathbf{A}$ with itself t-times, $\mathbf{A}^t$, $t \rightarrow \infty$ is dominated by the largest eigenvector, and its behavior reflects the region in the network where self-amplification is strongest. Also, in the long time limit, the matrix $\lim_{t\rightarrow\infty} \mathbf{A}^t$ applied to a vector $v_i = \delta(i,j)$ (a vector that is non-zero only at node j), only gives non-zero entries for nodes that are directly or indirectly connected to the node j.

A network can often be subdivided into separate clusters, where all nodes within each cluster are directly or indirectly connected, but where no path exists between different clusters. In matrix notation, these clusters could be mapped into a matrix $\mathbf{A}$ with a block diagonal form (each block corresponding to a cluster). A modular network is one where one allows a few links between clusters, but where links between pairs in different clusters are much less likely than pairs within the same cluster [467, 468, 469, 470]. A modular network corresponds to a nearly block diagonal matrix, where the blocks along the diagonal are supplemented by a few non-zero entries at other places in the matrix.

Notice that the matrix representation allows simple manipulations. The number of triangles in a non-directed network without self-links is:

$$n(\triangle) = \frac{1}{6} \cdot \text{trace}\left(\mathbf{A}^3\right) \tag{9.12}$$

where the factor $6 = 2 \cdot 3$ comes from "going" clockwise, or counter-clockwise around each triangle (a factor of two), and from the three contributions associated with the fact that any of the three nodes in the triangle can give a contribution to the count. Please notice that the adjacency matrix in this case should only include links between nodes, and not any self-links (trace $(\mathbf{A}) = 0$).

The adjacency matrix $\mathbf{A}$ reflects the replication dynamics of the copying activity of the network. Insight into network topology may also be gained from other processes in networks. Of these the most popular is the diffusion-like process, where one follows random walkers in the network [468, 469, 470, 471, 472]. At each timestep the walker takes a random step along one of the (out) links from its current node. The overall random walk behavior is described by a matrix $\mathbf{D}$, which has non-zero values at the same place as the adjacency matrix $\mathbf{A}$, but where each link from a node i is assigned a weight $1/k_i$, where k_i is the total number of links pointing away *from i*.

A random test particle tends to visit sites with high (in) degree more often (see Question 9.25). A variant of this random walk is used for ranking nodes on the famous Google internet site.[3]

9.2.2   "Scale-free" networks

A common feature of many biological, as well as other real networks, is that their degree distribution is very broad [458, 473, 474]. In fact, in real networks the number of nodes with connectivity k may often be approximated by a power law:

$$n(k) \propto \frac{1}{k^{\gamma}} \tag{9.13}$$

with exponent $\gamma \in \,]2.0; \ 2.5[$, a type of distribution shown in Fig. 9.8. Networks with this type of degree distribution are denoted scale-free.[4] For example, for proteins inside living cells, there are many proteins that control only a few other proteins, but there also exist some proteins that control the expression level of many other proteins. Examples of a hub protein is CI in phage λ, whereas CII only regulates a few proteins. Transcription factors in the regulatory networks (see Fig. 9.2), protein–protein interactions (see Fig. 9.12) and metabolites in metabolic networks [474] all have connectivity distributions that are much broader than both the Poisson and the exponential distributions.

One aspect of a broad connectivity distribution is the possibility of a huge amplification $\mathcal{A}$ of disturbances/signals. This feature can be inferred from Eq. (9.9). For broad connectivity distributions, $\mathcal{A}$ typically depends on the node with the highest

---

[3] Google uses a ranking that is proportional to the probability that a walker that starts at a random site visits the node within the above five random steps. This probability is calculated from $\mathbf{D}^5 \cdot \mathbf{1}$ where the vector $\mathbf{1}$ has 1 in all entries. This procedure is easily implemented on a directed network, in practice with the addition that a walker in a node without an exit link is moved to a random other node.

[4] That a distribution $n(k)$ is scale-free is equivalent to $n(k) \propto k^{\gamma}$. To prove this, assume first that the distribution is scale-free:

$$n(ak)/n(k) = n(a)/n(1) \Rightarrow log(n(ak)/n(1)) = log(n(k)/(n(1)) + log(n(a)/n(1)) \Rightarrow f(ak) = f(a) + f(k)$$

where $f(k) = log(n(k)/n(1))$ and $f(1) = 0$. Thus $f(k)$ is a logarithm of k: $log(n(k)/n(1)) = \gamma \cdot log(k) \Rightarrow n(k) \propto k^{\gamma}$. Conversely, if one assumes $n(k) \propto k^{\gamma}$, "scale-freeness" is proven by multiplying the argument k by a factor a, and observing that the frequency n changes with the same factor a^{γ} for all values (scales) of k.

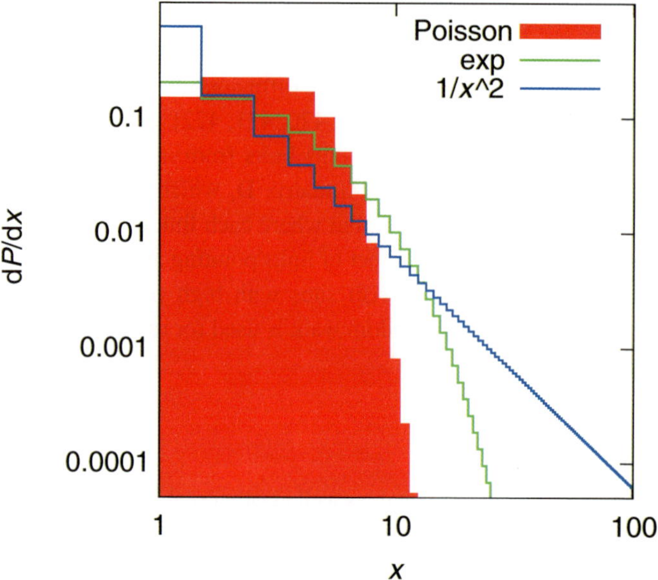

Figure 9.8 Comparison between a power law with exponent 2, i.e. $1/k^2$, the Poisson distribution and an exponential distribution. Notice that both the x- and y-axes are logarithmic.

connectivity. To see this, assume a scale-free network of the form given by Eq. (9.13). Then, from Eq. (9.9):

$$\mathcal{A} = \frac{\int_1^N \frac{k^2 dk}{k^\gamma}}{\int_1^N \frac{k dk}{k^\gamma}} - 1 \sim N^{3-\gamma} \tag{9.14}$$

for $\gamma \in]2; 3[$. In this case the denominator becomes independent of system size N, whereas the nominator increases with N. Thus for $\gamma < 3$, $\mathcal{A}$ is dependent on the upper cut-off in the integral, which represents the node with the highest connectivity. For comparison, most real-world networks have exponents between 2 and 2.5.

9.2.3   Robustness and hubs

$\mathcal{A}$ can also be used to estimate the robustness of the overall connectedness of the network against removal of a fraction f of its nodes.[5] This problem may, for example, have relevance in disease spreading, where f would then be the fraction of people that are vaccinated against a given, potentially epidemic, disease [479, 480, 481].

---

[5] In the literature [477] the derivation was first presented using the average connectivity of a node at the end of links:

$$\kappa = \mathcal{A} + 1 = \frac{\langle k^2 \rangle}{\langle k \rangle} = \frac{Var(k)}{\langle k \rangle} + \langle k \rangle$$

and identify the percolation threshold with this being = 2.

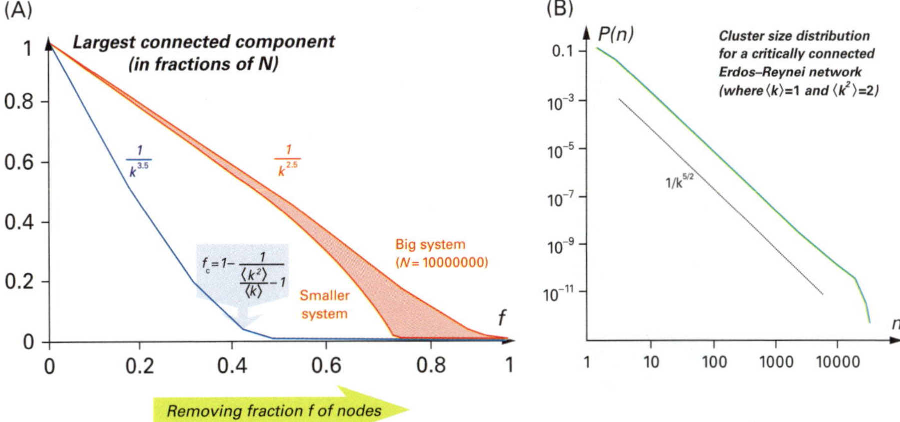

Figure 9.9 (A) Fraction of a system in the largest cluster (F) as a function of the fraction of removed nodes [477]. The blue curve refers to a network with a relatively narrow degree distribution, $1/k^{3.5}$, which exhibits a critical threshold similar to ER networks. The red curves show behavior of networks where $\langle k^2 \rangle$ is dominated by the largest hub in the network. The latter case is also simulated for different system sizes, demonstrating that very large networks remain connected until nearly all the nodes have been removed. (B) Cluster size distribution in a critical ER network $P(n) \propto 1/n^{2.5}$ [478], implying that a random node will be in a cluster with size distribution $n \cdot P(n) \propto 1/n^{1.5}$. This critical distribution is also obtained as one removes random links from a well-connected ER network until $\mathcal{A} = 1$ and $\langle k \rangle = 1$.

The break up of the network after removal of a fraction f of the nodes is determined by the value of f at which $\mathcal{A}$ becomes less than 1. If the initial network has an amplification factor $\mathcal{A}$, then after removal of f nodes, the amplification will be reduced by a factor corresponding to the remaining node fraction, $1 - f$:

$$\mathcal{A} \rightarrow \mathcal{A}' = \mathcal{A} \cdot (1 - f) \tag{9.15}$$

This comes about because each remaining node will lose each of its links with probability $f' = f$.[6]

The network remains super-critical when $\mathcal{A} \cdot (1 - f) > 1$ or:

$$(1 - f) > \frac{1}{\mathcal{A}} = \frac{1}{\langle k^2 \rangle / \langle k \rangle - 1} \tag{9.16}$$

Accordingly the critical fraction [477]:

$$f_c = 1 - \frac{1}{\langle k^2 \rangle / \langle k \rangle - 1} \tag{9.17}$$

This threshold is close to one for scale-free networks with a degree exponent $\gamma < 3$, as also seen in Fig. 9.9A. For more narrow degree distributions, the critical threshold

---

[6] Following a signal that enters into a node, (see Fig. 9.5), each of its remaining $k - 1$ links have a probability $(1 - f)$ of surviving the pruning. Thus, the local amplification $(1 - k) \rightarrow (1 - k) \cdot (1 - f)$ and the global amplification factor $\mathcal{A}$ is reduced by the factor $(1 - f)$.

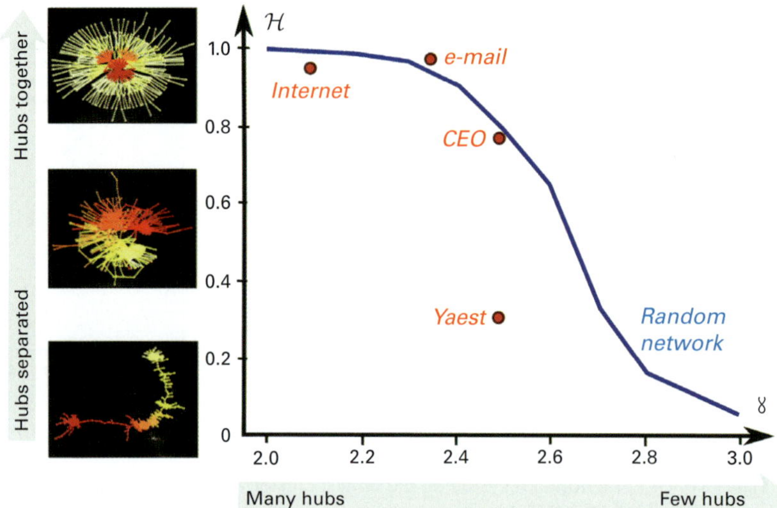

Figure 9.10 A topological hierarchy [482]. The left-hand part of the figure illustrates maximally hierarchical (top) and anti-hierarchical (bottom) networks of size N = 400 with $1/k^{2.5}$ degree distribution. The main figure shows how $\mathcal{H}$ depends on the degree distribution in random scale-free networks with distribution $f(k) \propto 1/k^{\gamma}$. As the degree distribution narrows, the hubs tend to separate and for $\gamma > 3$ the hubs are distributed along a "stringy" network that is dominated by nodes of low degree. The network examples used refer to the hardwired network of internet routers, an email network, the network of board members in American companies and the yeast protein–protein interaction network.

clearly separates from 1, with a value $f_c = 1 - 1/\langle k \rangle$ for ER networks, when one simply uses the Poisson distribution property $\langle k^2 \rangle = \langle k \rangle^2 + \langle k \rangle$.[7]

Another way to illustrate the differences between networks with degree distributions $\sim 1/k^2$ and networks characterized by distributions $\sim 1/k^3$ is to consider links or absence of links between highly connected nodes. For $\gamma \sim 2$ there are so many links in the system that these hubs tend to be directly connected. In contrast, the hubs tend to separate as γ increases towards three. This can be quantified in terms of the topological hierarchy [482, 483], which assigns rank proportional to degree, and assigns a network hierarchy measure = 1 if the shortest path between any pair of nodes follows a hierarchical path. That is, the shortest path has to first go from low to high degree, and then from high to low. The fraction of pairs in the network that are connected by such hierarchical paths is denoted as $\mathcal{H}$.

---

[7] After a fraction f is removed from an ER network, the fraction F in the largest cluster obeys $F = (1-f) \cdot (1-e^{-\langle k \rangle \cdot F})$ [477]: that is, the probability that a given remaining node is not connected to the largest cluster is $\exp(-\langle k \rangle \cdot F)$. Therefore any of the remaining $(1-f)$ nodes will belong to this cluster with probability $(1 - e^{-\langle k \rangle \cdot F})$. At criticality the largest cluster collapses and conforms to the scaling of the other clusters (see Fig. 9.9B).

Figure 9.10 shows $\mathcal{H}$ as a function of γ for randomly generated scale-free networks.[8] The figure also compares a few real-world networks, leaving us again with the challenge of properly defining what a random network actually is; we will do that in Section 9.5. Notice that $\mathcal{H} \sim 1$ for $\gamma \sim 2$, whereas $\mathcal{H}$ decreases to 0 for somewhat larger γ, reflecting that the hubs then have too few links to connect directly to each other. This is also reflected in the behavior of the probability that none of K neighbors has a degree higher than K, $(1 - \int_K^N k^{1-\gamma} dk / \int_1^N k^{1-\gamma} dk)^K \sim \exp(-(\gamma - 2) \cdot K^{3-\gamma})$, see [482].

9.2.4 Questions

9.2.1 What is the minimum number of links needed to connect 100 nodes in one large component (a collection of nodes that are directly or indirectly connected to each other)?

9.2.2 What is the largest diameter one can have in a network with 100 nodes?

9.2.3 How does the diameter of a network scale with the number of nodes N, when these are organized on a square/cubic... grid in d dimensions. Consider, for example, 4096 nodes, organized in one-, two-, and three-dimensional lattices. Determine the number diameter of the 4096 node network when organized in an Erdos–Reynei network with an average connectivity of six (the same number of neighbors as a three- dimensional cubic lattice). Convince yourself that an Erdos–Reynei network has infinite dimensions.

9.2.4 Generate Erdos–Reynei networks with $p = 3/(N - 1)$ (three neighbours per node on average, and not allowing self-interactions) for $N = 10$, $N = 100$ and $N = 1000$, and count how many triangles there are at various network sizes. How does the number of triangles change with N for a fixed average connectivity (fixed connectivity implies that p decreases with system size).

9.2.5 Consider a non-directed network that only consists of one large component. Prove that a random walker, after an infinitely long time will visit each node with a probability that is proportional to its degree. Notice that random walks in networks are at the core of search engines such as Google, although the walkers also complete other moves to deal with the properties of directed networks. Hint: consider the steady-state flux between two connected nodes with different degrees.

9.2.6 Generate a random network of size $N = 200$ with 150 links (average degree $\langle k \rangle = 3$) and monitor the size of the largest component as nodes are subsequently removed. Do the same when removing links subsequently, but maintaining all nodes. Hint: the central part of this question is to make an

---

[8] To generate a network of N nodes with a degree distribution $n(k) \propto k^{-\gamma}$, with a maximum of one link between each pair, one first assigns each node i a degree k_i from this distribution. Subsequently one starts at the node with highest degree and connects it to other nodes, linking it to the node with the next largest degree and subsequently connecting lower nodes until all links for this high-degree node are assigned [482]. Subsequently, lower-degree nodes are assigned neighbors in the same orderly way, until all nodes have their assigned degree. Finally the now constructed network has to be randomized, a procedure that is done using the pairwise link swapping described in Section 9.4.

algorithm that detects all parts of a network that is connected indirectly to a given node.

9.2.7   Use the following equation for the fraction F of the nodes that remain in the largest cluster of an ER network, after a fraction f is removed [477]:

$$F = (1 - f) \cdot (1 - e^{-\langle k \rangle \cdot F}) \tag{9.18}$$

to determine the critical value of f where $F = 0$ and examine F as f approaches this critical point. Hint: expand the exponential in the above equation using $\langle k \rangle F \ll 1$ close to the critical point.

9.2.8   One vaccination strategy is to vaccinate people at the end of links. By vaccinating a fraction f' of the nodes, one removes a fraction:

$$f = f' \cdot \langle k^2 \rangle / \langle k \rangle \tag{9.19}$$

of the links. Argue for this equation, and express the vaccination fraction f needed to stop epidemics in a scale-free network with $N = 10\,000$ nodes and degree distribution $n(k) \propto k^{-2.5}$.

9.3   Large-scale molecular networks

Cells are controlled by the action of molecules upon molecules. Receptor proteins in the outer cell membrane sense the environment and may subsequently induce changes in the states of specific proteins inside the cell. These proteins again interact and convey the signal further to other proteins, and so forth, until some appropriate action is taken. The states of a protein may, for example, be methylation status and phosphorylation or allosteric conformation, as well as subcellular localization. The final action may be transcription regulation, thereby making more of some kinds of proteins, it may be chemical or it may be dynamic. A chemical response would be to change the free concentration of a particular protein by binding to other proteins. A subsequent dynamic response could be the reversal of the directed activity of a motor, as in the chemotaxis of *E. coli*.

There are (at least) four types of molecular network in a cell:

- Regulatory network, where proteins/sRNA regulate production of transcription or translation. Example shown in Fig. 9.2.
- Protein–protein binding networks, sometimes supplemented with catalyzed degradation of one of the proteins. Example in Fig. 9.11.
- Signaling network, where proteins transmit signals by phosphorylating other proteins.
- Metabolic networks, describing the production and breakdown of small molecules.

This distinction into different types of networks is for our use only, in particular reflecting that each experimental method focuses on one type of network. In the cell

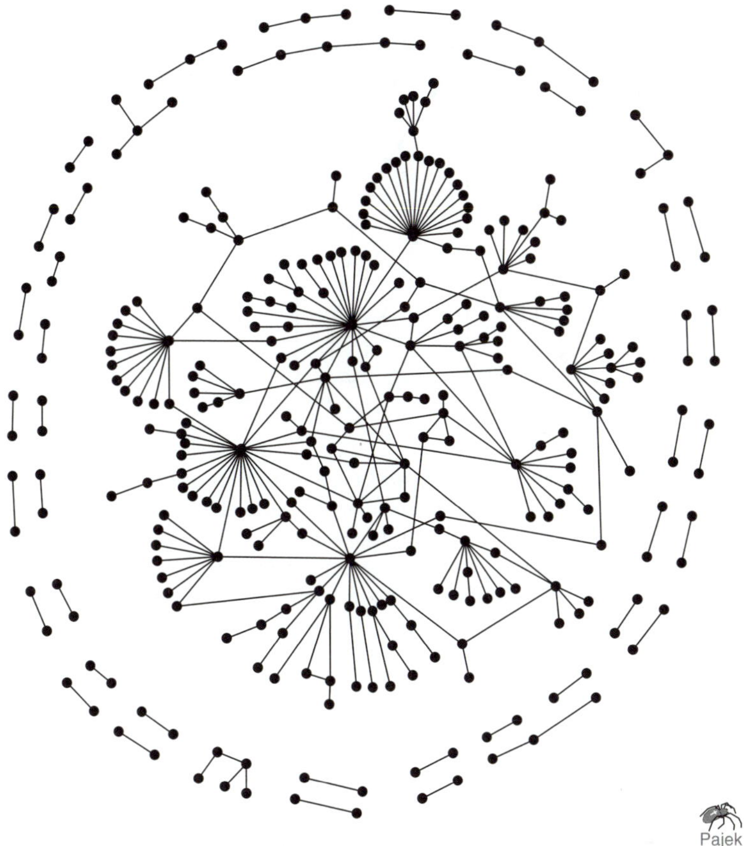

Figure 9.11 A subset of the protein–protein interaction network in yeast [484] obtained from two-hybrid experiments by [485]: yeast is grown under conditions where it needs a particular enzyme, which in turn is controlled by a regulator that contains a DNA-binding domain and an RNAP-binding domain. The regulator is divided into these two domains, and to each domain is fused a protein from the yeast. If the two proteins bind to each other, the regulator will act and the yeast will survive. Surviving yeast cells thus imply that the corresponding protein pair makes a physical interaction.

these networks are coupled to each other, allowing specific response systems at the interface between the different networks.

A common aspect of molecular networks is the wide distribution of directed arrows from individual proteins [487]. There are many proteins that control only a few other proteins, but there also exist some proteins that control the expression level of many other proteins. Not only proteins in the regulatory networks (Fig. 9.2) have this wide variety of connectivity, but also protein–protein interaction networks (Fig. 9.12), as well as metabolic networks (Fig. 9.7).

The known regulatory network of yeast is shown in Fig. 9.2. In this network proteins control the production of other proteins through genetic regulation. We again remind the reader, that, in addition, protein expression can be regulated by mRNA degradation

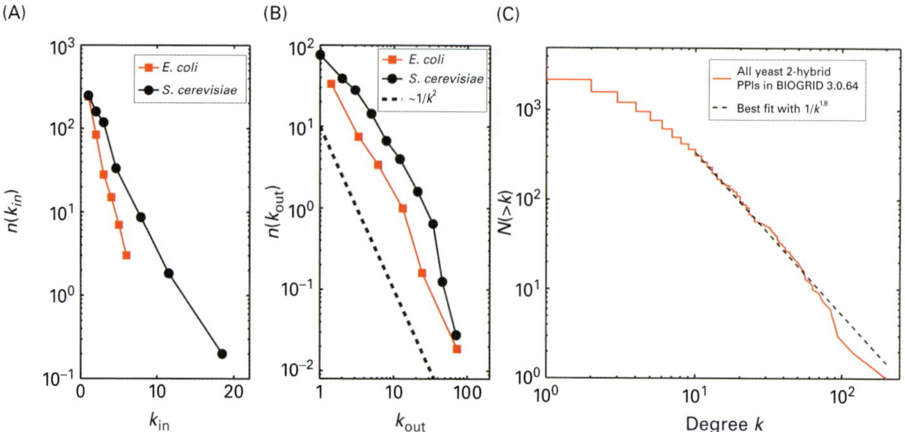

Figure 9.12 (A,B) $n(k)$ for regulatory networks in *S. cerevisiae* and *E. coli*, separated into the in-link and the out-link connectivity distributions, respectively. Notice that the in-degree is exponentially distributed, whereas the out-degree $n(k) \propto 1/k^{1.5 \to 2}$. (C) Connectivity for protein interaction networks from two hybrid measurements from BIOGRID (see also [485, 486]). The figure shows the cumulative distributions $n(> k)$, with $n(> k) \propto 1/k^{1.8}$ corresponding to $n(k) = dN/dk \propto 1/k^{2.8}$. Both figures from S. Maslov.

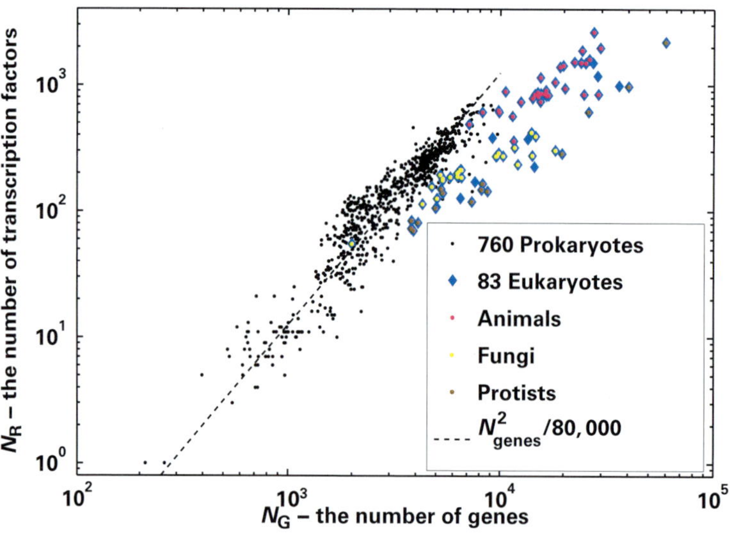

Figure 9.13 Fraction of proteins that regulate other proteins, as a function of the number of genes in the organism, first plotted by [488], but here drawn by S. Maslov. The squares indicate that each added gene should be regulated with respect to all previously added genes. Notice that for single-celled eukaryotes, and multi-cellular eukaryotes, the regulatory overhead only grows linearly, perhaps indicating a more modular organization.

and degradation of the proteins themselves. The transcriptional regulators may also be regulated, for example through binding to other proteins and to various small molecules.

Regulatory networks are essential for epigenetics and thereby multicellular systems, but are not essential for life. In fact, prokaryotes exist with almost no genetic regulation. Figure 9.13 shows the number of regulators as a function of genome size for a number of prokaryotic organisms. One notices that species with a very small genome have almost no regulation. More strikingly, it appears that the number of regulators N_{reg} grows much faster than the number of genes it regulates N [488].

In fact, sampling all prokaryotes (see Fig. 9.13) one finds [488, 489, 490]:

$$N_{reg} = N \cdot N/80000 \quad \text{for prokaryotes} \tag{9.20}$$

Using the relation [489]:

$$\langle k_{in} \rangle \cdot N = \langle k_{out} \rangle \cdot N_{reg}$$
$$\Rightarrow \frac{\langle k_{in} \rangle}{\langle k_{out} \rangle} = \frac{N_{reg}}{N} \propto N \tag{9.21}$$

and therefore that larger prokaryotic genomes tend to have either higher in-degrees for proteins, or, that they have transcription factors that have fewer proteins downstream. Here the option of a higher in-degree would correspond to regulation per gene that increases with genome size, consistent with a model where each added gene becomes regulated with respect to all other genes.

9.4 Analysis of network topologies

9.4.1 Randomization: constructing a proper null model

In order to identify non-trivial topological features of networks one needs to go beyond the single node property defined by the degree distribution. The hope is that this in the end may help us to understand function–topology relationships on larger scales. The key idea in this analysis is to compare the network at hand with a properly randomized version of it.

Aiming to pinpoint patterns one step beyond the degree distribution one should compare the network at hand with a random network with exactly the same degree distribution. The best way to generate such random networks is shown in Fig. 9.14. The idea is to swap links, pair by pair, multiple times until all nodes in the network

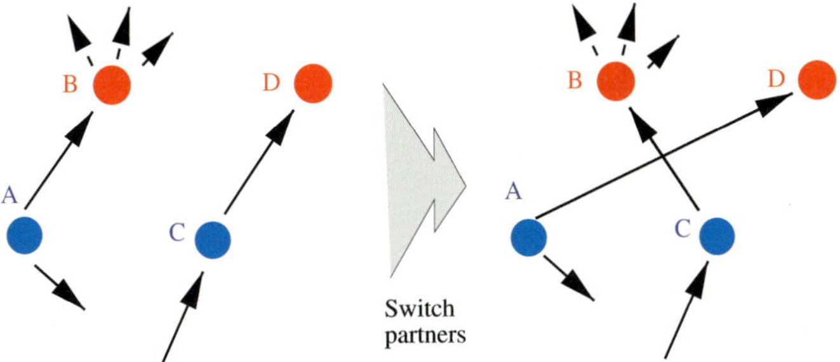

Figure 9.14 One step of the network randomization algorithm [484, 491]. A pair of directed edges A→B and C→D switch connections in such a way that A becomes linked to D, and C to B, provided that none of these edges already exist in the network. An independent random network is obtained when this is repeated a large number of times. The algorithm conserves the in and out connectivity of each individual node.

are assigned new random links [484, 491]. For a system with L links, after t swaps the probability that a given link is not changed is:

$$\text{Fraction (unchanged links)} = \left(1 - \frac{2}{L}\right)^t \approx e^{-2t/L} \tag{9.22}$$

which becomes insignificant when the number of "swaps" t becomes substantially larger than L. Notice, that one cannot allow all random swaps: if there is already a link between two nodes, then an attempted assignment of a second link should be aborted. Also note that one may keep a network connected in one big component by simply only allowing "swaps" that maintain overall connectedness.

Given an adequately randomized network, or better, a sample of about 1000 independent random networks, the significance of any quantifiable measure Q is given by the probability that a random network has the same value of Q as the real network.

Technically the significance of a "*pattern*" is quantified by its Z score:

$$Z(\text{pattern}) = \frac{N(\text{pattern}) - \langle N_{\text{random}}(\text{pattern})\rangle}{\sigma_{\text{random}}(\text{pattern})} \tag{9.23}$$

where $N_{\text{random}}(\text{pattern})$ is the number of times the pattern occurs in the randomized network. Here:

$$\sigma^2_{\text{random}}(\text{pattern}) = \langle N_{\text{random}}(\text{pattern})^2\rangle - \langle N_{\text{random}}(\text{pattern})\rangle^2 \tag{9.24}$$

is the variance among the random networks. If the "Z-score" = 2, the probability of obtaining a comparable pattern in a randomly generated network is 2.5%.

Considering patterns of links between proteins as a function of their degree, [484] reported relatively few links between hubs in both regulatory and protein–protein interaction networks in yeast (see Fig. 9.15). By considering occurrences of higher-order local patterns of control, [382, 493] found some particularly frequent motifs in gene regulation networks. The most abundant three-node motifs are illustrated in Fig. 9.16. [382, 493] emphasized the relative abundance of the feed-forward motif, and [494] suggested that "feed-forward" acts as a noise filter that only allows a signal to pass when

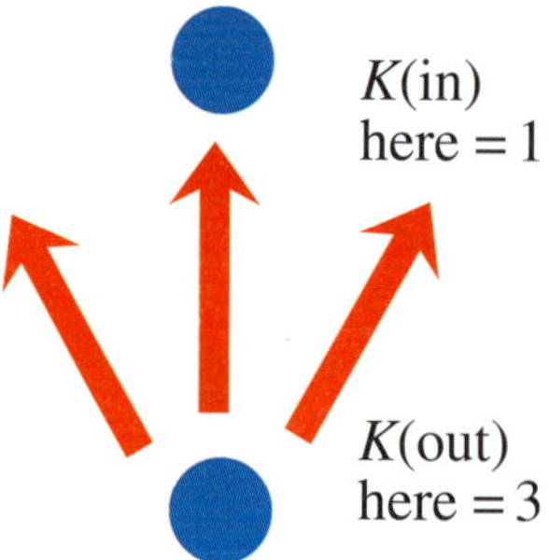

Figure 9.15 Correlation [484] between directly connected proteins in the regulatory network of *S. cerevisiae*, quantified in terms of Z scores. Notice the abundance of high K_{out} controlling single K_{in} proteins. The suppression of direct links between hubs allows these to provide independent input to downstream targets.

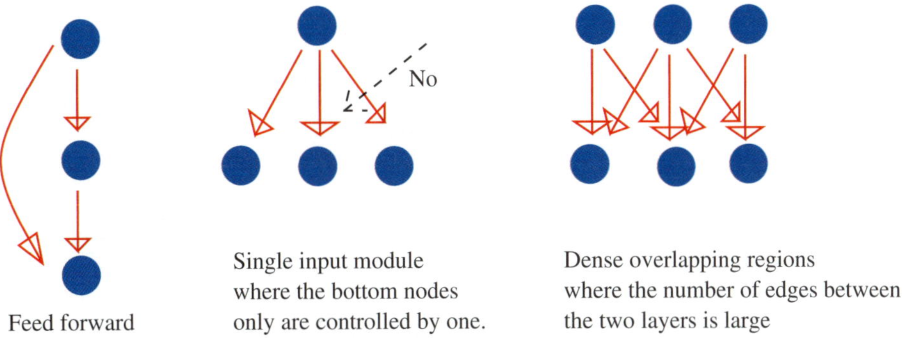

Figure 9.16 Genetic regulatory motifs which are found to be over-represented in regulatory networks of both *E. coli* and *S. cerevisiae* [382]).

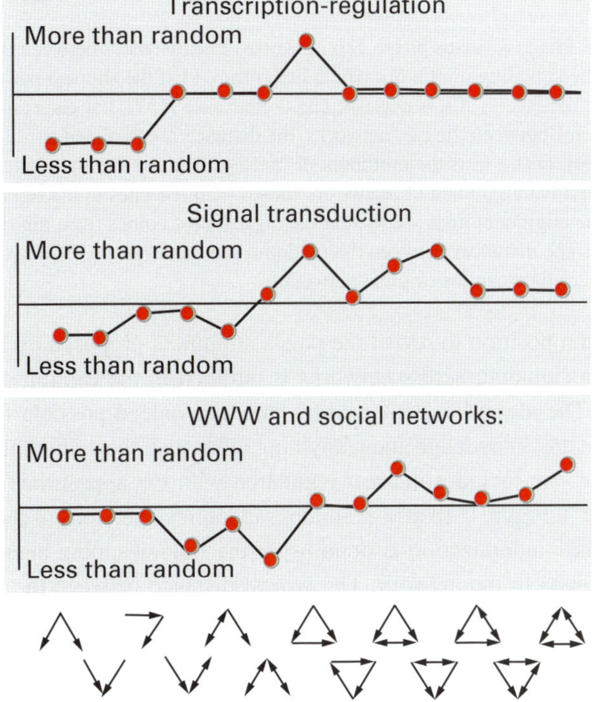

Figure 9.17 Super-families of motifs, as suggested by statistical analysis of different types of networks by Milo *et al.* [492]. One can see that feed-forward is over-represented in transcription and signaling networks. Social networks and the worldwide web favor other three-node motifs. In all cases, triangles in some form are favored, reflecting some three-node clustering.

it is persistent.[9] Fig. 9.17 shows a more elaborate investigation of motifs for different types of network.

---

[9] If a signal goes directly from A to B, and indirectly from A to C to B, then B only gets activated when the signal that passes the longer way is in accordance with the direct signal.

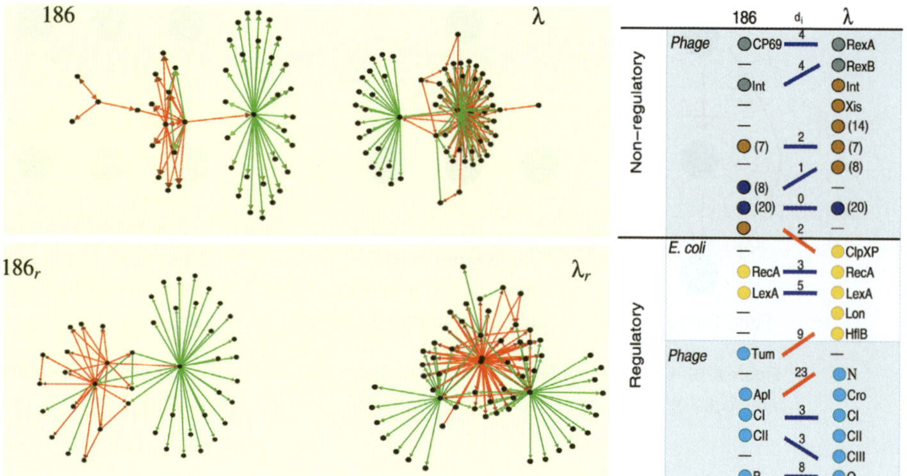

Figure 9.18 Top panel: Phage λ versus phage 186 networks. The distances between the networks are calculated from the signaling matrix S_{ij}, which has entries +1 if the shortest path from i to j results in positive regulation, -1 if it is negative and 0 otherwise [157]. For each attempted identification of proteins between the two networks, the distance is calculated as $\sum_{i,j}|S_{ij}(\lambda) - S_{ij}(186)|$. Dist = 43 is the minimum distance over many attempted identifications [157], reflecting an optimal alignment of regulators shown with red lines in the right-hand panel. The number above the alignment lines shows that the biggest cost comes from misaligning Apl with N. Lower panel: The distances between the randomized counterparts of the two networks are larger than distances between the original networks.

Finally we are in a position to revisit the phage alignment proposed by Fig. 9.1. An algorithm based on signaling across networks is outlined in the caption of Fig. 9.18, taken from [157]. The algorithm demonstrates that it is indeed possible to align networks even without any knowledge about anything other than the sign of all regulatory links, provided that one includes indirect regulation with the appropriate sign (using "+1 = $(-1) \times (-1)$"). Figure 9.18 also examines randomized versions of the two phage networks, where the randomization is done using the link-swapping algorithm [484] that preserves the hubs in the network. The typical distance between the randomized networks is much larger than for the real network, illustrating that the alignment relies on more distinct features than the out degree of regulators.

9.4.2 One hub reflects one function

In biological regulation one repeatedly finds regulator proteins that regulate a few other proteins, but also, in some cases, a hub that regulates the production of many other proteins. Common to the targets of each regulator is that they participate in one fairly well-defined biological process, as, for example, sugar metabolism, or SOS repair. Similarly, Fig. 9.2 provides an example of a particular response pattern in the yeast regulatory network.

One may speculate that the broad degree distribution of molecular networks reflects a broad distribution of the number of proteins needed to do the different tasks in a living

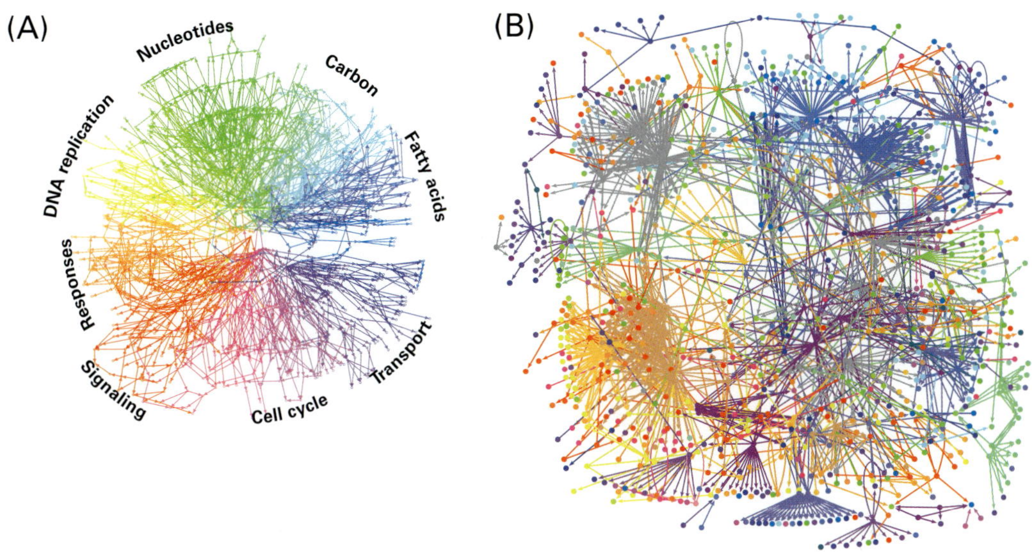

Figure 9.19 (A) Proteins distributed on the GO hierarchy according to the process they participate in. (B) Yeast regulatory network with proteins colored according to their corresponding position in the GO classification. Figure from [31], but the investigation also holds for protein–protein interaction networks, as verified by [495].

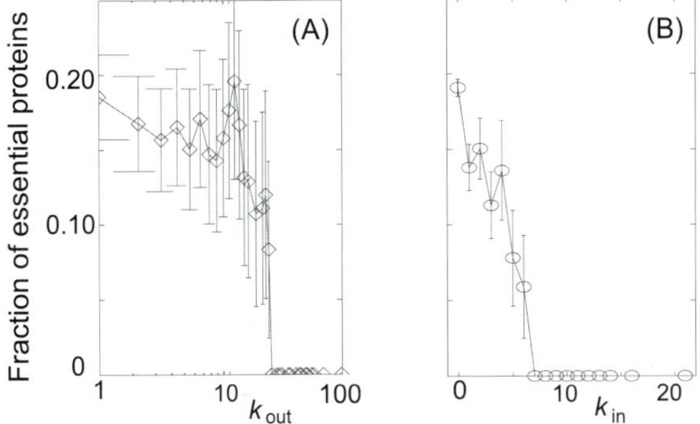

Figure 9.20 Anti-correlation between the likelihood of being essential and in- and out-degrees of a protein in the regulatory network of *S. cerevisiae* [496]. Both panels demonstrate that the "importance" of a protein decreases with its degree in the regulatory network. To improve the statistics, the degree is used cumulatively, plotting the fraction of essential proteins among all tested proteins with the degree above K_{out} (in panel A) or above K_{in} (in panel B).

cell [496, 497]. Some functions simply require many proteins, whereas many functions only require few proteins. This scenario is not only supported by a number of one-system "case histories," but also by Fig. 9.19, which shows that proteins involved in the same cellular functions are indeed co-localized in the same region of the genetic regulatory network [31, 495].

The one hub, one function scenario is also supported by the absence of correlation between the connectivity of a regulatory protein and its likelihood of being essential (see Fig. 9.20). That is, let us assume the opposing scenario, where highly connected proteins were involved in many functions, and also assume that each function had a certain likelihood of being essential. This would imply that this likelihood would grow linearly with the degree of a protein in the network. This is contradicted by the correlations in the yeast transcription network [496]. Accordingly, one again arrives at the one hub, one function type of regulation.

9.4.3   Questions

9.4.1   Construct a network $N = 100$ nodes subdivided into 10 different classes with 10 nodes in each. Generate a random network where each node has about three links, and nodes within the same class have 10 times larger probability of being connected than nodes of different classes. Calculate the number of loops, and compare this with the number of loops when all the links are randomized.

9.4.2   Repeat the above procedure for a network matrix where nodes that are within ± 10 of the diagonal have 10 times larger probability of being connected than nodes that are further away from each other.

9.5   Models for scale-free molecular networks

9.5.1   Threshold networks

A threshold network is one particularly simple model for obtaining scale-free networks [498, 499], aimed at fitting the apparent power law distribution for the number of protein partners in yeast two-hybrid experiments. In one version each protein i is assigned a non-specific binding strength G_i that is selected from an exponential distribution, $P(G) \propto \exp(-G)$. Subsequently, one assigns a link to all node pairs i, j where $G_i + G_j$ is larger than a fixed detection threshold Θ, thus obtaining a network with scaling properties (see Fig. 9.21).

The scaling comes about because a given protein is assigned a G value with probability $\exp(-G)$, and therefore this protein will have a number of binding partners equal to the number of nodes n with $G' > \Theta - G$. This number $n \propto e^{-G'} \propto e^{+G}$. Therefore, there is a probability $P(> G) = \exp(-G)$ of having $n \propto \exp(+G)$ partners:

$$P(> n) = 1/n \Rightarrow \frac{dP}{dn} \propto \frac{1}{n^2} \tag{9.25}$$

Accordingly, threshold networks predict scale-free networks, provided that the threshold Θ is rather large. Noticeably, the threshold model is inherently non-specific: good binders bind to all reasonably strong binders. Presumably, real networks have a specific reason for having a broad connectivity distribution. It seems challenging to make a cell work if some proteins aggregate many other proteins. Therefore it was suggested

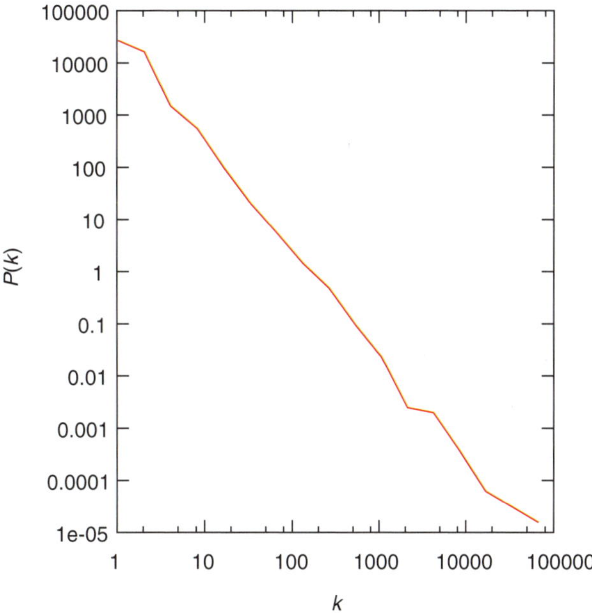

Figure 9.21 Simulation by threshold network, for N = 100 000, each protein i is assigned a non-specific binding $G_i = -ln(r)$ with r being a random number uniformly selected in $[0, 1]$, and the link subsequently assigned between protein pairs i, j where $G_i + G_j > 12$.

[498, 499] that the threshold model may be more appropriate for reproducing link densities for the case where proteins are over-expressed to an extent that allows substantial non-specific binding properties, a situation that often occurs for the experimental setup in two-hybrid experiments [485, 486].

9.5.2   Rewiring of transcription networks

Genetic networks can, in principle, change faster than single point mutations would allow for, using, for example, transpositions. Using transposons, a DNA segment can be copied or moved from one position of the genome to another. Transposition was discovered in maize by Barbara McClintock in the 1950s [9], and was later found in organisms from bacteria [501] to animals [502]. Using large-scale genetic engineering, evolution may proceed in much larger steps than envisioned by single nucleotide changes; for multicellular species, this involves changing the body plan, which at a large scale is governed by organization of the Hox genes [503, 504, 505].

The history of genomic rearrangements can be traced by comparing protein sequences in an organism as it is today. For *S. cerevisiae*, for example, one finds that about 2000 of its 6200 proteins have more than 20% similarity with at least one other yeast protein (see Fig. 9.22). Protein duplication must therefore be a fairly common event on an evolutionary timescale. Two duplicated proteins within the same organism are called

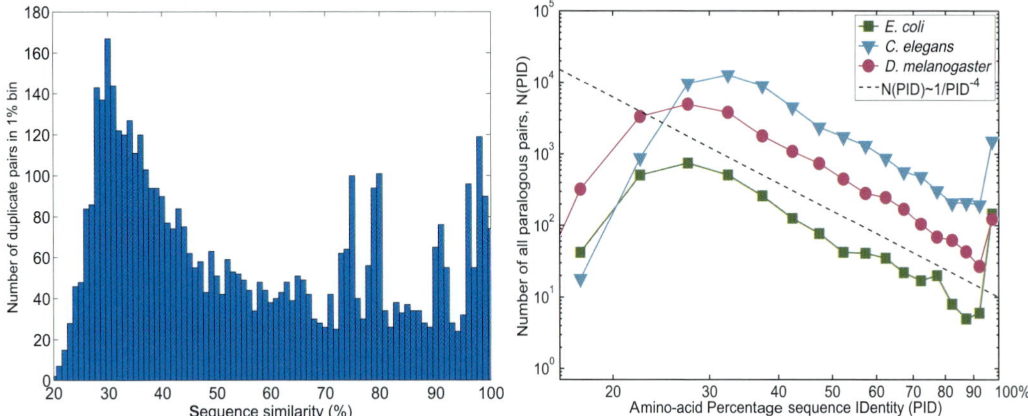

Figure 9.22 Right: Detailed histogram for paralogs in the distribution of evolutionary distances between pairs of protein paralogs in *S. cerevisiae* [500]. Left: Histograms for evolutionary distances between pairs of proteins within each of three organisms. Figure with thanks to S. Maslov. The three histograms can be rescaled to the same behavior by dividing by the total number of protein pairs in the respective organism, $N \cdot (N-1)/2$. Notice the unexplained scaling of the paralog number with distance (protein identity, PID), converging to zero identity at a level where nearly all proteins in the corresponding organism are included.

paralogs. Duplication may be seen as a source of raw material for proteins, which subsequently may be repositioned in the regulatory network to provide the raw material for development of new functions through mutations.

Figure 9.23 illustrates how the overlap in the molecular networks typically diminishes as paralogs diverge. In fact, the overlap changes 3% for each 1% divergence in amino-acid sequence. Thus, the figure demonstrates that rewiring of the genetic network is indeed important.

This type of data [507, 508] has inspired a number of network evolution models based on gene duplication [509, 510, 511, 512, 513, 514, 515]: if a gene is duplicated, then all its network neighbors increase their number of links by one. Therefore the chance that a gene gains by neighbor duplication increases proportionally to its degree. Thus gene duplication can give a scale-free network. However, the duplication model also predicts that neighbors of a gene should be related [31, 513, 514, 515], in particular that all genes directed from the same hub should be descendants of the same ancestor. This turns out not to be true. In fact, proteins in the same hub are only slightly more related than random proteins in a given organism [31, 514]. Accordingly, rewiring and recombinations after duplication is so large that the duplication in itself should have little impact on the degree distribution.

9.5.3   Models for scale-free networks that do something: the toolbox model

There have been very few attempts to understand the broad degree distribution using the fact that molecular networks serve a function. Here we consider the part of the

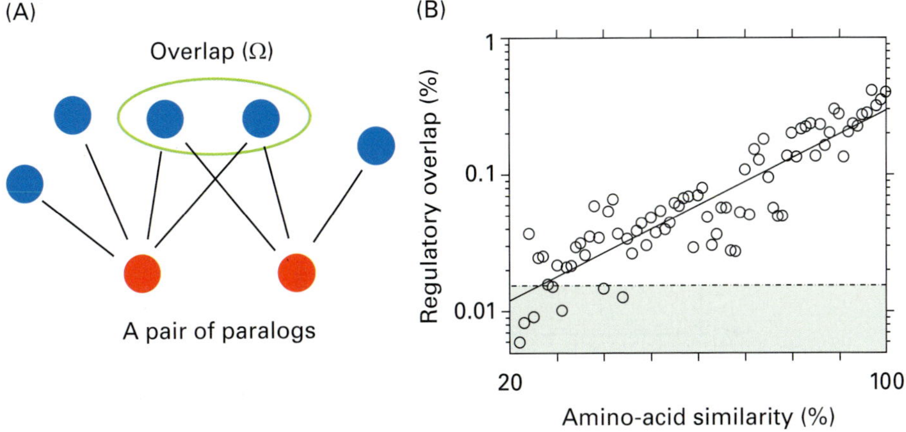

Figure 9.23 (A) Overlap between the network neighborhood of two paralog proteins. (B) Decay of overlap of upstream transcription regulation (for yeast), with decreasing similarity between the proteins [500]. The binding of proteins to DNA is done by ChIP-on-chip [506].[10] Notice that even straight after duplication, most of the overlap is gone, indicating that survival of a duplication event demands very fast integration into a new functional context. The dashed line is overlap in a randomly rewired network.

regulatory networks that controls metabolic networks [497], and use this to derive the scaling relation:

$$N/N_R = 80000/N \qquad (9.26)$$

which implies that there are fewer genes assigned per added regulator for a larger genome.

The reason for focusing on regulation of metabolism is the fact that this constitutes a large fraction of the prokaryotic genome, and that the size of a metabolic network is believed to be correlated to the number of different environments that a given prokaryote can live in. For *E. coli*, for example, about half of all its transcription factors have a binding site for a small molecule [422, 518]. The regulatory out-degree of *E. coli* (Fig. 9.24, left) will be modeled in terms of out-degrees needed to regulate enzymes in particular metabolic pathways (Fig. 9.24, right). Pathways that we here identify by their end points, i.e. that each new food source requires a new set of reactions in order to be connected to the already existing metabolism. Such a complete set of reactions would be very difficult to evolve in eukaryotes, but could be added in prokaryotes by their extensive use of horizontal gene transfer with a global microbial pan-genome [519, 520, 521, 522].

If each new food substrate requires one new transcription regulator, the "broad" degree distribution of regulators R may reflect the varying length of the metabolic

---

[10] ChiP-on chip is a method where the protein of interest binds to DNA *in vivo* by adding a small amount of formaldehyde. Subsequently the cell is lysed, the protein extracted by an antibody, and the associated DNA fragment is isolated. Finally the DNA is identified using DNA micro-array.

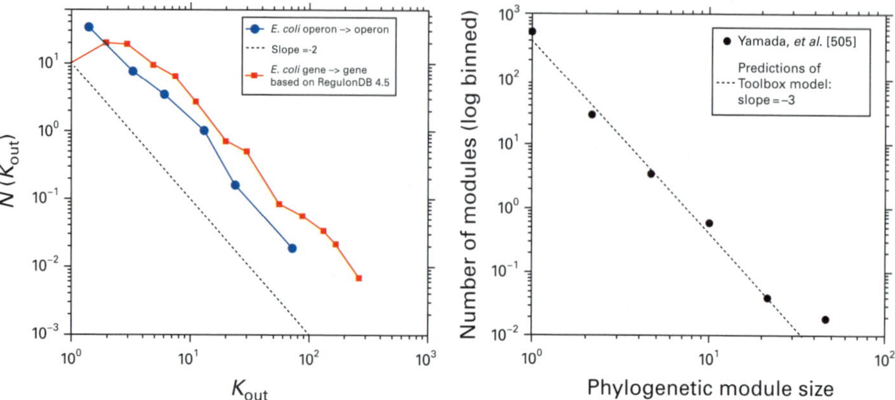

Figure 9.24 Left: The size of operons in *E. coli*, as well as the out-degree distribution for transcription factors, both distributed as $\sim 1/K_{out}^2$. Right: The size of metabolic pathways in *E. coli*, counted by number of enzymes that are simultaneously present from bacterial genomes [516]. The $1/K^3$ distribution is steeper than the out degree $1/K_{out}^2$, reflecting some limitation in the proposed identification of metabolic paths with single transcription factors. Both figures from S. Maslov, private communication [517].

pathways: some food molecules can be broken down by one enzyme, whereas other metabolites need many enzymes to be broken down, a fact visualized by the broad size distribution of metabolic pathways in Fig. 9.24. To mimic this, we suggest that long pathways with many metabolic reactions are associated with the first food molecules an evolving bacteria is exposed to. Subsequent metabolites need shorter pathways to link up with the earlier pathways, and the number of enzymes added per food source = per regulator, will therefore be smaller for larger genomes.

Consider a particular metabolic network with $N = N_{\text{enzymes}}$ enzymes and a comparable amount of metabolites, both representing a subset of all possible enzymes/metabolites $N_{\text{all}} = N_{\text{all-possible-enzymes}}$ among all available organisms (the available pan-genome [519, 520, 521, 522]). Adding one new food source with one new regulator, requires the addition of about $dN/dN_R = N_{\text{all}}/N \propto 1/N$ new enzymes/reactions to make connections to the existing metabolism.[11] Thus the number of regulators:

$$\frac{dN}{dN_R} \propto \frac{1}{N} \Rightarrow N_R \propto N^2 \tag{9.27}$$

This in fact resembles the observed quadratic scaling for the number of regulators among prokaryotes (see Fig. 9.13). The model outlined in Fig. 9.25 also includes pathway elimination, thus allowing steady-state sampling of genomes of various sizes. The obtained scaling turns out to be maintained with this important addition, resembling a steady state of the ongoing dynamics within the microbial pan-genome [519, 520, 521, 522].

---

[11] $dN/dN_R = N_{\text{all}}/N$ assumes a random sampling between the number of enzymes and the sequence of enzymes that is needed to connect a given reaction to the set that is already in the genome.

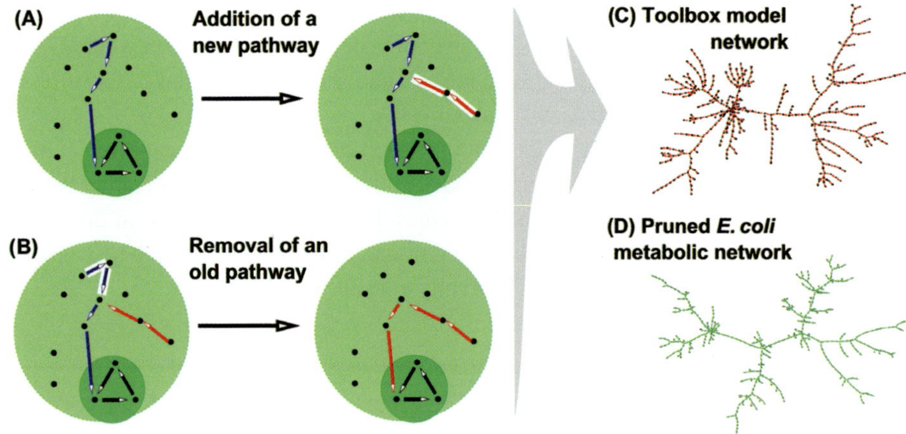

Figure 9.25 "Toolbox model" for evolving metabolic networks [497]. (A) Addition of a new metabolite requires its end point to be connected by a new metabolic pathway (red) to an existing pathway (blue). All paths ultimately link up to the core (black). (B) Removal of a random nutrient is implemented by removing the pathway partway down to the point where it becomes required for metabolism of another nutrient. The union of all black dots denotes all metabolites in the microbial pan-genome. A regulatory network can be assigned by assuming that each end metabolite is associated with one regulator [497], that then has an out degree that is at least equal to the branch length. Right panel: A model network from connecting 400 metabolites out of a pan-genome of 1800 metabolites (numbers from [523], see also [524, 525]).

The model also predicts a broad distribution of sizes of metabolic pathways: pathways with size $> K$ must be added at stages where there are less than $N = N_{\text{all}}/K$ enzymes in the network. At such an early stage there are relatively few metabolic branches and even fewer regulators $N_R \propto N^2 \propto 1/K^2$. The number of pathways that are longer than K is $P(> K) \propto 1/K^2$, giving the size distribution $1/K^3$ (see Fig. 9.24). Again, this result is robust to the inclusion of ongoing addition and removal of pathways [497]. We stress that the $1/K^3$ scaling in the right-hand panel is substantially steeper than the size distribution of operons. This difference is not explained by the simplest one-to-one identification between the larger operons and the relatively smaller metabolic paths.

Notice that the assumption of linear pathways in the above toolbox model may be relaxed, and the model may be extended to use all metabolic reactions needed to utilize a given new food source. The formal way to do this is based on the formalism of [524, 525], implemented as described by [526]. In any case, the model presents a possible scenario for how "one hub, one function" is consistent with scaling laws in gene regulatory networks.

9.5.4 Questions

9.5.1 Simulate a threshold network for a $N = 1000$ system. With G_i assigned as in the text, good results should be obtained with a threshold of 7.

9.5.2 Simulate a threshold network for a $N = 1000$ system where G_i is selected from Gaussian with mean 0 and standard deviation 1. The threshold $\theta = 3.5$ should give reasonable results.

9.5.3 Derive Eq. (9.27) with pre-factors, as well as the expected number of metabolites in each added pathway using random stepping between available reactions/enzymes. Reiterate the derivation of the scale-free distribution of the added number of enzymes for each new metabolite in the growth model. Eventually consult ref. [497].

9.6   Summary

- Networks consist of nodes, each with specific properties and with specific connections to a few other nodes.
- Real networks are formed by some dynamic process, taking place on much longer timescales than the dynamics associated with signaling or transport across networks.
- Networks often have broad degree distributions, $N(k) \propto \frac{1}{k^\gamma}$ with $\gamma \sim 2 \to 2.5$. Such scale-free networks can be efficient in transmitting diseases because of the large value of the amplification factor:

$$\mathcal{A} = \frac{\int k^2 n(k)\,\mathrm{d}k}{\int k n(k)\,\mathrm{d}k} - 1 \tag{9.28}$$

- Signaling sign and logic are central to the functions of regulatory networks.
- The broad connectivity distributions in signaling and regulative molecular networks reflects the different numbers of proteins needed to deal with the different functions of the cell.

10 Signaling and metabolic networks

10.1 Signaling across networks

10.1.1 Strong components

Cells need to send signals, between each other (see 8.5 and [552]), between compartments inside a cell and between different protein types within a cell. Signals are sent in response to environmental conditions to allow the cell to take appropriate action. Computation by the regulatory system makes sure that the right signals are sent at the right times to specific proteins. Signals are facilitated by reactions of various types, for example enzymatic reactions or transcriptional regulations, as outlined in Fig. 10.1.

As an introduction, we return to *E. coli* as a model system, including all known regulatory reactions, as analyzed in [527] and shown in Fig. 10.2. To address the large-scale network properties, the first question is to what extent signals can be propagated on or across this network. This depends on the ability to transmit a signal across each link, and also on the logic and sharpness of signal integration at each molecular reaction. Signals that can send multiple links across a network have been seen in the induction of lysis in phage λ (see Fig. 4.6).

Figure 10.2 shows the network of proteins and all the known reactions in *E. coli* in which these proteins take part. It includes transcription regulation, protein complex formations and enzymatic reactions, but the network does not include the metabolites themselves. Further, in this figure, we have chosen a finegrained bipartite representation, with separate nodes for the proteins and for the reactions. Each protein has an arrow towards all reactions in which it is a reactant. Further, the protein receives an arrow from every reaction in which it is a product. Similarly, every reaction has an arrow towards each of its products, and receives an arrow from each of its reactants.

For example, a reaction node associated with equilibrium binding will have bi-directional arrows to both incoming and outgoing proteins/protein-complexes. Another example is a transcriptional reaction, where each transcription factor has a directed arrow to the reaction and there will be a directed arrow from the reaction to the proteins downstream of the regulated promoter.

The directed, bipartite graph representation in Fig. 10.2A consists of 2846 protein nodes and 2774 reactions (of which 848 are transcription reactions, 980 are irreversible and 1794 are reversible reactions). Already, with such a representation, there are four degree distributions: the in and out degrees for proteins and the in and out degrees for

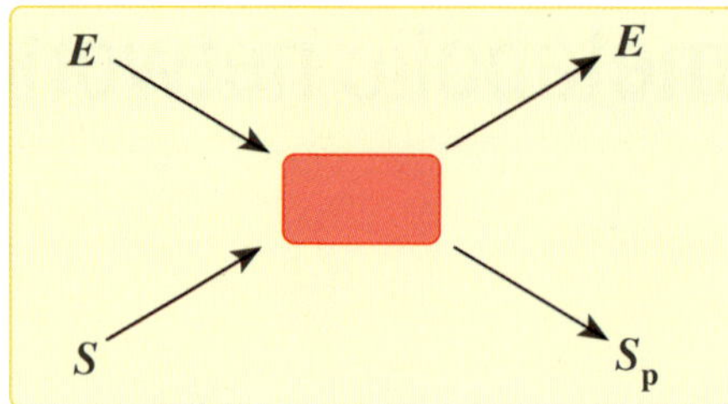
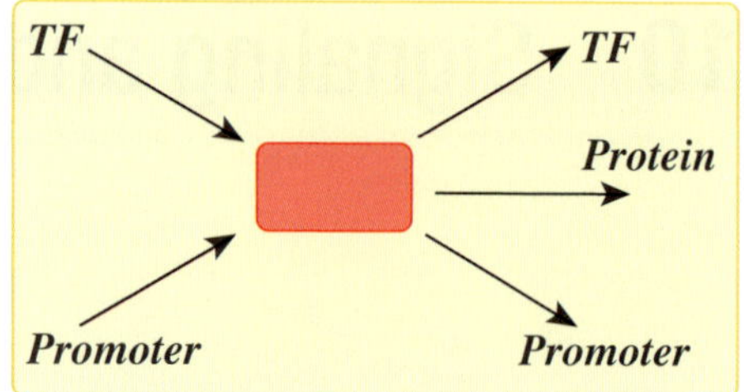

Figure 10.1 Basic information processing in molecular networks. On the left we see an enzyme phosphorylating a substrate **S**, on the right a standard transcription factor (TF) influencing the production of a protein. Other reactions would be degradations, typically done by particular proteases, and also sequestering reactions where some proteins bind others and thereby keep them from reacting elsewhere. The reaction nodes plus the reactant nodes together form a bipartite network, which we visualize on a larger scale in Fig. 10.2.

the reaction nodes, all shown in Fig. 10.2D,E. Of all these four distributions, only the out-degree distribution from proteins is broad and possibly scale-free. The in-degree distributions are both narrow.

Another large-scale design feature is the tendency of transcription reactions (cyan, in Fig. 10.2A) to be at the center of the network. The shell-like structure, apparent in Fig. 10.2A, with reactions alternating with protein nodes is in part due to the bipartite nature and in part due to the higher interconnectedness of the core of the network, consisting mostly of transcription factors.

Figure 10.2B analyzes the directed network in terms of strong components. A strong component is defined as a collection of nodes (here reactions, proteins or protein complexes) where everybody can reach everybody else by following links in the forward direction. Thus in a strong component everybody is both below and above everybody else.

In Fig. 10.3 we visualize the definition of strong components, and show how the coarse graining in Fig. 10.2B was done.

The network in Fig. 10.2A is composed of a large number of small strong components. Thus, although the non-directed version of the network from Fig. 10.2A is connected into one large component, the strong components do not allow for signals to be propagated across the network. The fragmentation of the network into separate strong connected modules instead demonstrates that the allowed signaling is substantially below the percolation threshold.

Figure 10.2C shows a network [484, 491] obtained by randomization with link swapping, for both reaction nodes and protein nodes, thereby preserving the number of both in and out links for all nodes. The giant strong component confirms that there are enough links in the system to put it substantially above the percolation threshold. Comparison between its randomized counterpart and the real *E. coli* protein accordingly emphasizes the modular nature of the real network.

The simplest aspect of the network topology related to signaling is the number of nodes downstream of any given starting node. The possible signals from the starting node are obviously limited to reach only these nodes. In the random network most nodes

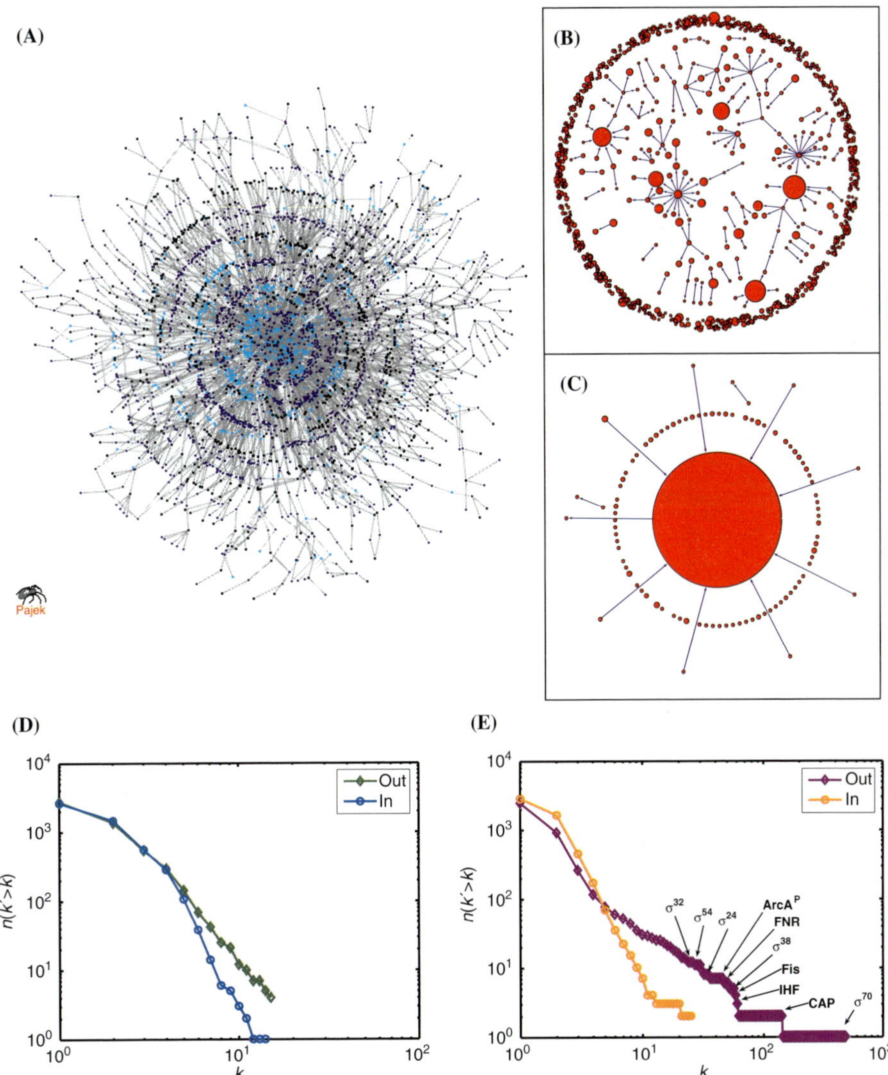

Figure 10.2 *E. coli* network (left top), from [527]. The graph is bipartite, consisting of protein and reaction nodes. The largest hubs, σ_{70} and *CRP*, have been removed for clarity. Inset top right: a condensed graph of the network where nodes reflect the size of strong components, and arrows indicate paths connecting two strong components. Inset middle right: a similar condensed graph for the randomized version of the *E. coli* network. Two bottom panels show the degree distribution for reaction nodes and reactant nodes respectively. The network does not include metabolic reactions.

can reach almost all other nodes, whereas each protein in the real network has a much smaller number of downstream targets (see Fig. 10.4).

It is often speculated that the apparent constraints of the real network serve as functional compartmentalization that will be weakened if metabolic reactions are included, as small metabolites provide additional feedback from the "bottom" layer of transport

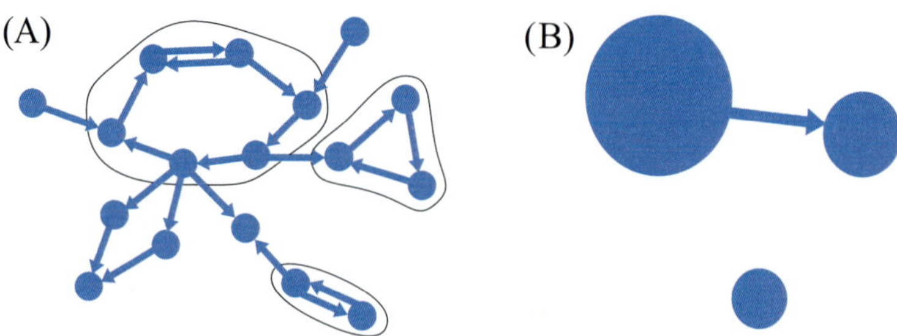

Figure 10.3 A strong component, and how the coarse graining in the previous figure was done in practice [527]. (A) A network and its decomposition in strong components. (B) How the network on the left is visualized in terms of the coarse-grained representation of its strong components.

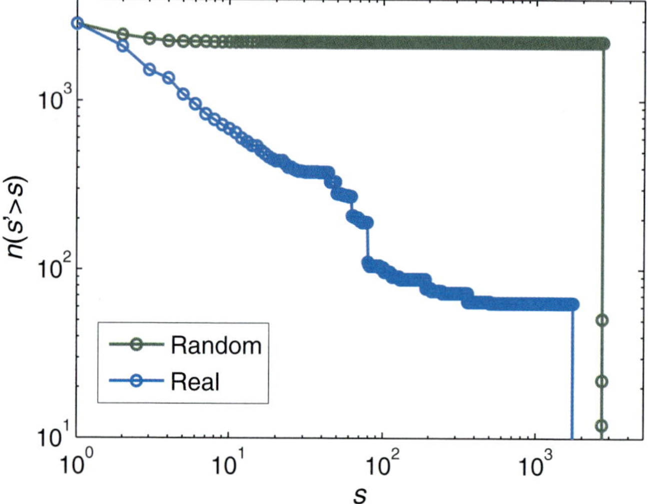

Figure 10.4 Distribution of number of downstream targets for proteins in real, and randomized networks [527]. The giant strong component in the randomized network implies that most proteins have all other proteins as downstream targets. In the real network, however, many proteins only have a limited downstream sphere of influence.

proteins and enzymes to the top transcription factors [353, 355, 380, 425]. One such connection is through the highly connected transcriptional regulator CRP (also called CAP) that is placed in the center of the network in Fig. 10.2A [528].

10.1.2   Signal transmission across a promoter

Signal processing often involves integration of more than one input for each output. For example, a reaction will often rely on the presence of several reactants, as envisioned in Fig. 10.1. For chemical reactions, each reaction typically demands the presence of all reactants, implying an effective "AND" gate between reactants, for the propagation of the signal.

Promoters are often regulated by more than one transcription factor. When changing the concentration of one of the inputs, it is therefore not certain that the activity of the promoter changes. The sensitivity to input depends on the logic of the regulatory input [529, 530, 531, 532, 533], which may well go beyond the AND gate for standard chemical reactions.

Following [532] we here consider a gene that is regulated by transcription factors A and B. If both A and B separately activate the gene, then if A is present enough to fully activate the gene, then further regulation by B is over-ruled. On the other hand, if both need to be present to activate the gene, then the expression of the gene will become dependent on B, if A is already present. This dependency is encoded in the way the transcription factor binds to the regulatory DNA around the promoter.

One important aspect of DNA looping is that it makes it possible to influence a given promoter from many different operators. This allows extensive combinatorial regulation, which does indeed seem typical in eukaryotes [530, 534, 535]. Figure 10.5 shows possible mechanisms [532] for logical output as a function of two proteins A and B. In all cases one monitors the output in the form of RNAP binding to the promoter or not. In practice, the Boolean nature of the logic is limited by both the "sigmoidal shape" of any binding curve, and by the basal transcription of the downstream gene (for quantitative models see the questions).

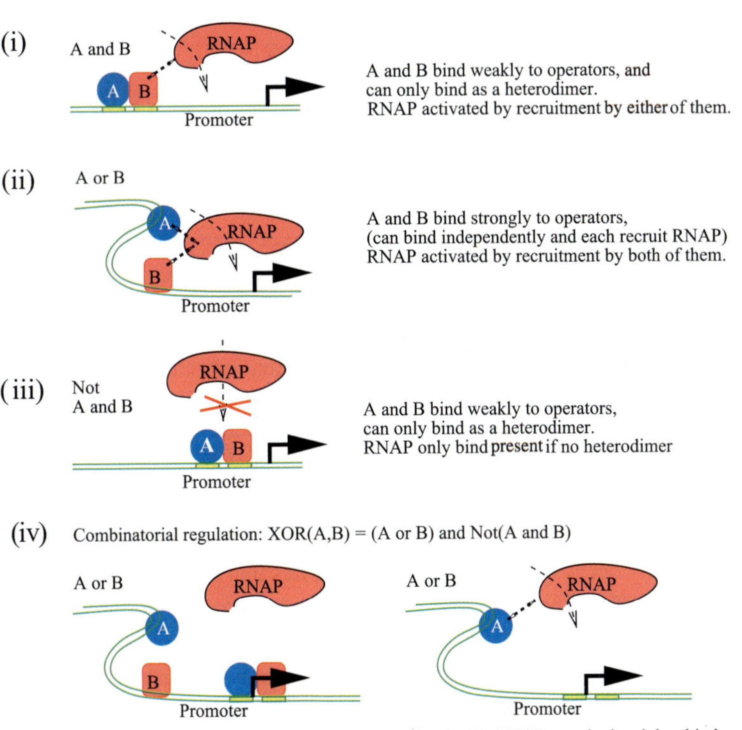

Figure 10.5 Combinatorial transcription logic. Protein A and protein B determine the activation of a promoter. The RNAP binding is regulated by either recruiting it to the promoter, or by preventing it from binding to the promoter. In the upper two panels RNAP needs either A or B in order to bind to the promoter. In the third panel RNAP is assumed to bind without help, provided there is no AB complex on the promoter. In panels (i–ii,iv) the RNAP can only bind by recruitment.

Input logic is expected to be important for signal propagation in the so-called feed-forward loop motif [382]. For A, which activates B and C, and B, which activates C, the "AND-gate" at C would imply that its activation requires a persistent signal from A (such that the signal has time to pass around B as well). On the other hand, if C is already "on," it would then be closed down on the fast timescale associated with removal of the direct signal from A.

An overall important lesson is that more input links to a given promoter makes it more difficult to transmit a signal across this node, in the sense that it will demand all inputs be in certain relations before the output actually depends on one specific input.

10.1.3   Signaling and protein–protein binding

Protein–protein networks are widely available, in part due to the two-hybrid binding assay [485, 536, 537], which allows a genome-scale screen of pairwise protein–protein bindings (see Fig. 9.11). Following [538], this section investigates how well signals in protein–protein networks propagate. The propagating signal is quantified in terms of the relative change in the free concentration of the various proteins $F \rightarrow F + \Delta F$.

The main equation for equilibrium interactions is:

$$F_i F_j = k_{ij} D_{ij} \qquad (10.1)$$

which expresses the free concentration for two proteins i and j in terms of the concentration of their heterodimer D_{ij}. Each free concentration F_i, is in turn related to its total concentration C_i by:

$$C_i = F_i + \sum_{\text{dimers}} D_{ij} = F_i + \sum_{\text{dimers}} \frac{F_i F_j}{k_{ij}} \qquad (10.2)$$

where the sum runs over all links in the protein–protein network. Thus, the sum takes into account all dimers in the network. The above equations can be solved iteratively [538]:

$$\{F_i\} \rightarrow F_i = \frac{C_i}{1 + \sum_{\text{dimers}} F_j / k_{ij}} \qquad (10.3)$$

starting with $F_i = C_i$ for all i. This analysis is done for the protein–protein network in yeast shown in Fig. 10.6. Overall, one can see that the effect of a perturbation at node 0 rapidly decays with distance. This implies that protein–protein bindings typically are inefficient for more than one-step signal propagation, an observation that is not at odds with the fact that protein–protein interations typically are used in motifs where an abundant protein sequesters its low-abundance transcriptional regulator.

10.1.4   Questions

10.1.1   Calculate (plot as two-dimensional surfaces) promoter activities for AND, OR, NAND and XOR logic as functions of A and B concentrations between 0.001 and 100. Set all binding constants equal to 1, A and B binding with co-operativity equal to 0.01 when they bind to neighboring operators. Disregard any leaking, and assume perfect recruitment of A or B or AB, when needed.

10.1.2 Repeat the AND, NAND and XOR gates above, using an AB complex in solution, with binding constant 0.01.

10.1.3 Consider the network of three proteins on a line $0-1-2$, with total initial concentrations $C_0 = C_1 = C_2 = 1$. Calculate the free concentration of all proteins for $K = 1$. Investigate the steady-state change $\Delta F/F$ for all proteins, as the total concentration of the first protein is increased to 10 (F denotes free protein concentration).

10.1.4 Repeat the above question with the stronger binding $K = 0.1$.

10.1.5 Repeat Question 10.1.4, but assume that total concentrations are $C_0 = 10$, $C_1 = 3$ and $C_2 = 1$, changing to $C_0 = 100$, $C_1 = 3$ and $C_2 = 1$. Finally, repeat 10.1.3, but with $C_0 = 1$, $C_1 = 2$ and $C_2 = 10$ changing to $C_0 = 10$, $C_1 = 3$ and $C_2 = 10$.

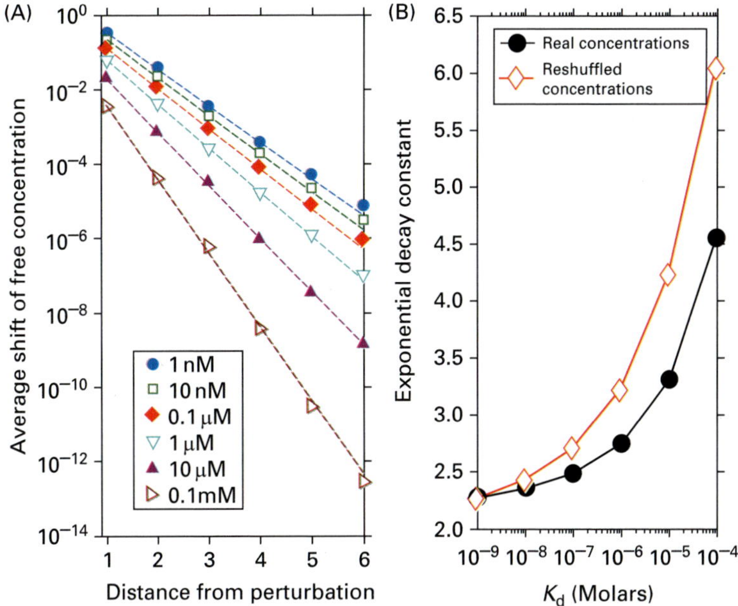

Figure 10.6 Theoretical investigations [538] of spreading of perturbations in measured protein–protein binding networks [539]. The perturbation is measured in terms of change in freely available protein, as a function of network distance to a changed concentration of an input protein. The different curves to the right refer to increasing value of dissociation constants, with minimum $K_d = 1\,nM$ corresponding to 34 molecules per yeast cell. The right curve shows that the exponential decay from panel (A) becomes steeper if concentrations of various proteins are reshuffled away from their observed values (protein abundance in yeast is from [540]).

10.2   Enzyme kinetics and phosphorylation

Consider an enzyme F that catalyzes the process:

$$X + F \rightleftharpoons XF \rightarrow X_{\mathrm{P}} + F \tag{10.4}$$

with $(F - [XF]) \cdot (X - [XF]) = K_F[XF]$. When the complex is at relatively low concentration, i.e. $[XF]^2 << [X][F]$, this can be approximated to (see also Fig. 10.7):

$$[XF] = \frac{[F] \cdot [X]}{K_F + [X] + [F]} \tag{10.5}$$

Note that a vanishing complex concentration $[XF]^2 << [X][F]$ corresponds to a fast reaction, implying that the intermediate state only exists transiently.

To simplify the analysis one often assumes that external enzyme concentrations I and F are always much smaller than intrinsic enzyme concentrations ($[F] << [X]$). Thereby all external reactions are expressed in the Michaelis–Menten form:

$$[\mathrm{XF}] = \frac{[\mathrm{F}] \cdot [\mathrm{X}]}{K_{\mathrm{F}} + [\mathrm{X}]} \tag{10.6}$$

an often-used equation that in fact is identical to our equation for a relatively abundant transcription factor (X) acting on a single operator site (F). This equation implies that the reaction rate becomes insensitive to [X] when $[X] >> K_{\mathrm{F}}$, i.e. when binding is strong.

The Michaelis-Menten equation (MM) for an enzyme reaction is:

$$\text{Reactions per second} = \frac{k_{\mathrm{cat}} \cdot [\text{enzyme}] \cdot [\text{reactant}]}{K + [\text{reactant}]} \tag{10.7}$$

This equation emphasizes that the number of reactions does not increase unlimited with the number of reactant molecules. Enzymatic reactions are limited by the availability of enzymes, and take place at a typical rate set by k_{cat}.

10.2.1   Goldbeter–Koshland phosphorylation cascade

Signals can be propagated by enzymatic addition of a phosphor group (PO_4^{3-} molecule) to certain proteins, performed by enzymes called kinases [541]. In a classic paper from 1981 Goldbeter and Koshland [27, 363] investigated the transmission of a signal in the form of state change of a protein Z, to a phosphorylated state Z_{p}. In effect, it is the enzyme E that catalyzes this reaction to initiate the signal, resulting in the changed phosphorylation status of Z, in spite of an ongoing constant dephosphorylation of Z_{p} due to another enzyme (denoted F). The kinematics of catalyzed phosphorylation is outlined in Fig. 10.7. Following [27, 363] we consider a "push–pull" combination of phosphorylation and de-phosphorylation as particularly efficient in signaling.

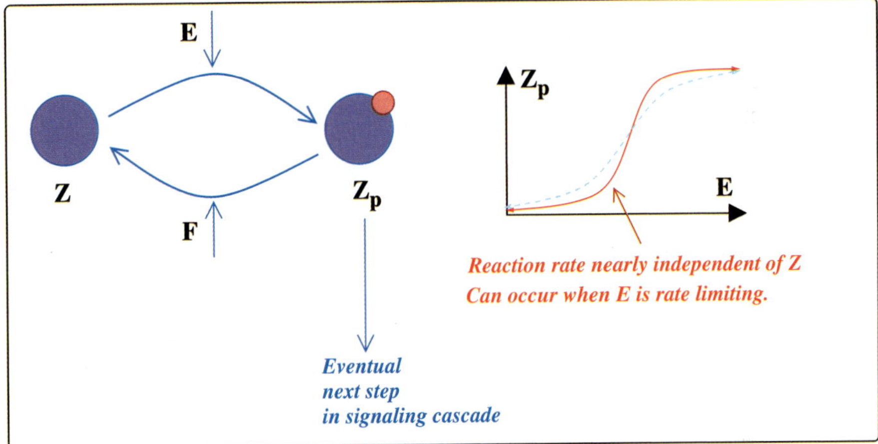

Figure 10.7 The top panel investigates in detail the criteria for such substrate independence, demonstrating that this can occur when K is small and $[E] << [Z]$. As ultra-sensitivity requires that the enzyme is limited both ways, in fact it requires that $Z_{total} >> E, F$. The lower panel shows Goldbeter–Koshland reaction kinetics, giving ultra-sensitivity when reaction rates become independent of substrate density (Z and Z_p, with $Z_{total} = Z + Z_p$ kept constant).

Push–pull reactions allow a switch-like response. Thus, when reaction rates become independent of Z and Z_p[1] the two competing conversions will result in a net drift:[2]

$$\frac{dZ_p}{dt} \approx E - F \tag{10.8}$$

---

[1] Using the Michaelis–Menten approximation, which towards $\rightarrow Z_p$ gives a rate proportional to $[ZE] \approx E \cdot Z/(K+Z)$. It comes naturally as the next to last approximation in Fig. 10.7, setting $K + E \approx K$.

[2] For simplicity we here avoid the reaction rate constants k_{cat} for the catalytic reactions in both directions. Thus we have simplified $\frac{dZ_p}{dt} \approx k_{cat}^{\rightarrow}E - k_{cat}^{\leftarrow}F$ to $\frac{dZ_p}{dt} \approx E - F$, with the understanding that these constants in fact should be introduced for a full description.

that proceeds until there is none of the minority species left, i.e. the drift cannot continue into negative concentrations. Therefore:

$$Z \approx Z_{\text{total}} \text{ and } Z_p \approx 0 \quad \text{for } F > E$$

$$Z \approx 0 \text{ and } Z_p \approx Z_{\text{total}} \quad \text{for } E > F \tag{10.9}$$

This equation states a sharp switch as E increases above some critical value that is set by the amount of the opposing enzyme F. The equation is only true when only the enzymes define how many Z proteins is converted per second. As a result of this extreme assumption, the switch from one state to the other becomes infinitely sharp.

In reality, at some point the reaction rate becomes sensitive to the vanishing concentration of one of the substrates (Z or Z_p) and approximately (in the limit where the complex EZ is very short lived, see Fig. 10.7 for details):

$$R_\rightarrow = R(Z \rightarrow Z_p) = \frac{[Z][E]}{[Z] + [E] + K_E} \tag{10.10}$$

$$R_\leftarrow = R(Z_p \rightarrow Z) = \frac{[Z_p][F]}{[Z_p] + [F] + K_F} \tag{10.11}$$

where we still abstain from multiplying by the k_{cat} rates. This leads to a more accurate version of Eq. (10.8):

$$\frac{\mathrm{d}Z_p}{\mathrm{d}t} = R_\rightarrow - R_\leftarrow$$

$$= \frac{[Z][E]}{[Z] + [E] + K_E} - \frac{[Z_p][F]}{[Z_p] + [F] + K_F} \tag{10.12}$$

When Z or Z_p becomes low, the rates:

$$R_\rightarrow \sim \frac{[Z][E]}{[E] + K_E}, \quad \text{for } [Z] < [E] + K_E$$

$$R_\rightarrow \sim \frac{[Z_p][F]}{[F] + K_F} \quad \text{for } [Z_p] < [F] + K_F \tag{10.13}$$

approach zero proportional to the minority species. This causes a steady state Z, Z_p that is not exactly zero and accordingly gives a softer transition than that outlined by Eq. (10.9).

One may consider a sequence of push–pull reactions, where a first step could be that Z_p catalyzes a new substrate $X \rightarrow X_p$. Provided that one maintains a hierarchy of subsequent substrate limitations, $E << Z << X << \ldots$ one may in fact get a subsequent increase in the sharpness of the transition. The increase will put still higher demand on actual reaction rates in order to be feasible, and also the subsequent steps come with an amplification of noise [543, 544]. But anyway, it is remarkable that such push–pull reactions allow for signal transmission over several steps in a signaling chain.

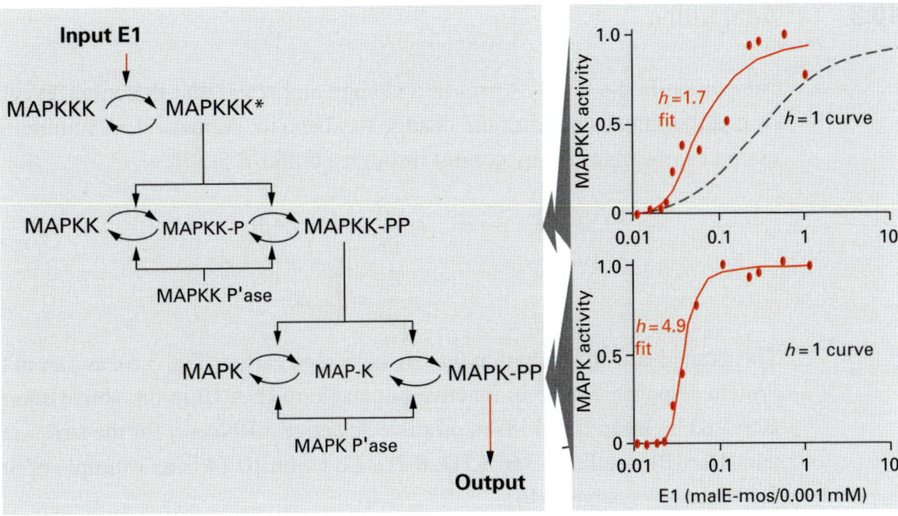

Figure 10.8 Increased ultra-sensitivity of activity as measure input-output relation along the MAPK-signalling pathway. Figure reproduced from ref. [542].

The MAPK signaling cascade is illustrated in Fig. 10.8. This sequence of enzymatic reactions is involved in decisions about cell division.

10.2.2 Questions

10.2.1 Compare the exact expression for [EZ] with [EZ] = [E], with $[EZ] = \frac{[E][Z]}{K+[Z]}$ and with $[EZ] = \frac{[E][Z]}{K+[E]+[Z]}$ by plotting the expressions as a function of [E] $\in [0:10]$ for fixed [Z] = 1. Use $K = 1$. See what happens when $K = 10$ and $K = 0.1$.

10.2.2 Solve the Goldbeter–Koshland system from lower panel of Fig. 10.7 using $[Z] + [Z_p] = 100$, and plotting [Z] as function of E between 0.5 and 1.5 when $F = 1$ is fixed. Hint: simulate $\frac{d^z p}{dt}$ until steady state for each [E] value. First consider the case of strong binding $K_E = K_F = 0.1$. Subsequently consider $K_E = K_F = 10$.

10.2.3 Consider again the Goldbeter–Koshland system from Fig. 10.7 but now for $E = 1$ and $E = 0.9$ and $K_E = K_F = 0.1$ and simulate dynamics using the Gillespie algorithm (that is, one has to simulate, say, at least 1000 events, each being one of the two competing reactions $Z_p \to Z_p + 1$ and $Z \to Z + 1$. Each event is selected according to rates $= k_p \cdot [ZE]$, respectively $= k \cdot [Z_p F]$ where concentrations of complexes can be expressed from the equations in Fig. 10.7. Set $k_p = k = 1\,\text{s}^{-1}$.

10.2.4 Construct logical gates based on phosphorylation processes, with the output being counted in terms of the phosphorylation status of one protein. Hint: allow for complexes of enzymes, with the possibility that these complexes may be active, or passive, respectively.

10.3   Adaptation

Adaptation is the ability to sense a change and react, and thereafter to return to more or less the same state after the change has become permanent. A minimum adaptation circuit can be obtained by control through a buffer B [119]:

$$\frac{dB}{dt} = 1 - L \cdot B$$

$$\frac{dy}{dt} = L \cdot B - y \tag{10.14}$$

This is an idealized adaptation that we have seen earlier (Fig. 3.6), as part of the unfolded protein response where an inactive but stable mRNA (B in the above equation) can be activated by Ire1p (L) [118] to produce an active mRNA (y) for the stress-relieving protein (Hac1P), see Figs. 3.6, 8.1D, 8.2D. Equation 10.14 is an example of an incoherent feed-forward system [546].

The idealized Eqs. (10.14) support a steady-state level of the output variable $y = 1$, independent of the level of the input L. At the same time, changing the input L to say $L \rightarrow 100 \cdot L$ makes a rapid increase in y which is then later counteracted by a decreased value of the buffer B. Figure 10.9 shows a simulation of the adaptation of this idealized circuit. The rest of this subsection discusses how adaptation can be obtained from coupled enzymatic reactions.

Figure 10.10 shows a combination of proteins that can perform adaptation, in a network where each node consists of a passive and an inactive version of an enzyme. Figure 10.10 presents a negative feedback system (from [545]):

$$\frac{dB}{dt} = \frac{k_{CB} \cdot C \cdot (B_T - B)}{(B_T - B) + C + K_{CB}} - \frac{k_{FB} \cdot F \cdot B}{B + F + K_{FB}} \tag{10.15}$$

$$\frac{dC}{dt} = \frac{k_{LC} \cdot L \cdot (1 - C)}{(1 - C) + L + K_{LC}} - \frac{k_{BC} \cdot B \cdot C}{B + C + K_{BC}} \tag{10.16}$$

which can be simplified by assuming that $L << 1 - C + K_{LC}$ and $F << K + K_{FB}$, as shown in the figure. In this system a sudden increase in input L makes C increase until it

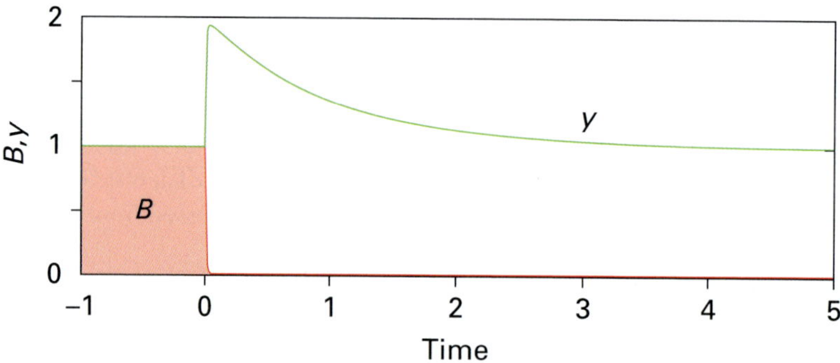

Figure 10.9 Minimum adaptation circuit, from Eq. (10.14) simulated for $L = 1 \rightarrow L = 100$ change in input for time $t > 0$.

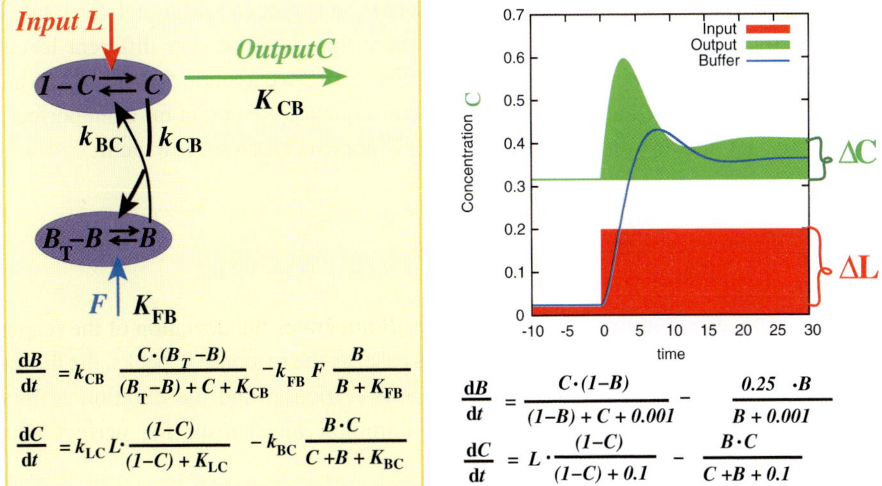

Figure 10.10 System of two nodes that can perform adaptation, from [545]. Here each node consists of a protein that can be in two states, active or inactive. These states may for example, differ by a phosphate group (PO_4^{3-}). The active state may catalyze the transition in other proteins, and furthermore some of the transitions may be catalyzed by outside forces F. The network transforms an input L to an output C, which for optimal adaptation exhibits a short strong pulse upon a sudden shift in L, and thereafter returns to its value before shock. Here B is a buffer node.

induces a delayed increase in the buffer B. When B increases and approaches its higher steady state, it does so in balance with a subsequent decrease in C, which is decreased due to the term $\sim -k_{BC} \cdot B \cdot C$.

L and F can have a huge impact, through their reaction constants. In the right-hand panel of Fig. 10.10 we have absorbed these reaction constants into L and F, while still assuming that their actual concentrations can be ignored when compared to the proteins they act on.

The motif in Fig. 10.10 exhibits adaptation, defined as the ability to show an output C with a big initial response to change in input $L = 0.02 \rightarrow 0.2$, and thereafter to relax back to a value close to its original value. Perfect adaptation is characterized by a large, but short, change in C, followed by a perfect return to the earlier state ($\Delta C = 0$, see Fig. 10.10).

Consider now the network in Fig. 10.10, in the limit of large amounts of the protein B, i.e. where $(B_T - B) > K_{CB}$, $B > K_{FB}$ (saturation), as well as $B_T - B > C$ and $B > C$ (substrate limited) then:

$$\frac{dB}{dt} = C \cdot k_{CB} - F \cdot k_{FB} \tag{10.17}$$

with steady state

$$C^* = F \cdot k_{FB}/k_{CB} \tag{10.18}$$

which is independent of the input L. Thus, for a big reservoir of B, $B_T - B > C$ the output C can return to a value that is independent of input L: it can adapt perfectly.

Notice that we, in Eqs. (10.15),(10.16) use a concentration of 1 for total C, but B_T for total B, thereby emphasizing that there may well be very different levels of the two proteins. In the right-hand panel of Fig. 10.10 we assume that $B_T = 1$, which instead is similar to the level of C_T, and as a consequence we obtain far from perfect adaptation.

An interesting aspect of Eq. (10.17) appears when we rewrite it:

$$\frac{dB}{dt} = (C - C^*) \times k_{CB} \tag{10.19}$$

$$B = B_0 + \int_0^t (C - C^*)dt \tag{10.20}$$

This equation implies that the buffer B integrates the deviation of the response function C from its steady-state value. In this case, as the response C peaks, the buffer B increases to an extent that is given by the peak response, and the duration of this peak. This so-called "integral control" is a common mechanism for perfect adaptation, see [366, 546, 547, 548].

10.3.1   Adaptation and scale-invariant greed

A particularly classical adaptation system relevant to bacterial chemotaxis [549, 550, 551, 553, 554, 555] was suggested by Barkai and Leibler [366], based on studies of adaptation in the *E. coli* chemotaxis network.

As in other adaptation motifs, the Barkai–Leibler adaptation is associated with the phosphorylation state of some enzyme Z_p that transiently increases when the ligand L increases. Adaptation is secured by a steady state of Z_p that is independent of L as long as $Z_u = 1 - Z - Z_p$ is large. In stylized form, the equations from Fig. 10.11 read [546]:

$$\frac{dZ_p}{dt} \approx L \cdot Z - Z_p, \quad \text{and} \quad \frac{dZ}{dt} \approx -L \cdot Z + \text{constant} \tag{10.21}$$

thereby employing the mathemathical structure of Eq. 10.14 to obtain adaptation, using Z (and implicitly $Z_u = 1 - Z_p - Z$) as the buffer.

The purpose of adaptation in chemotaxis is to allow the bacterium to sense food gradients, both when the food level is low and when it is much higher (see Fig. 10.12). The chemotaxis works by measuring the decline or increase of food L as the bacteria moves, and thereby deciding whether the bacteria should change direction or not (see Fig. 10.12). In fact, using near to perfect adaptation, the *E. coli* can perform chemotaxis over more than a 10 000-fold change in food levels [550]. A scale-invariant greed that needs more than the one buffer shown in Fig. 10.9 to function across multiple scales in food concentration. The full system also involves clusters of receptors on the bacterial surface [551, 553, 555].

10.3.2   Questions

10.3.1   A pedagogical example of adaptation could be our intestines, occupied in part by bacteria B, which consumes the available food F. Argue for and identify variables in the following model:

$$\frac{dB}{dt} = F \cdot B - B$$

$$\frac{dF}{dt} = E - F \cdot B - C \cdot F \qquad (10.22)$$

Assume that host consumption $C = 0.5$ and simulate the system for a switch from steady state at $E = 1$ to a steady state where the host eats twice as much, i.e. $E = 2$. What could make this adaptation less perfect?

10.3.2 Consider the simulation in the right-panel of Fig. 10.10, but replacing the assumed total concentration of $B_T = 1$ with a $B_T = 10$. Examine the equations as $B_T \to \infty$ (i.e. that the buffer is in huge excess). Demonstrate that the output C obtains perfect adaptation ($\Delta C \sim 0$). Hint: set $L = 0.02$ and solve the equations by simulating a trajectory with small time increments ($dt = 0.01$) until steady, then change L abruptly to $L = 0.2$ and plot the response of output C.

10.3.3 Investigate Barkai–Leibler chemotaxis adaptation in Fig. 10.11 numerically, in particular with respect to increasing the rates between Z and Z_p by a factor of 10. Normally these reactions are assumed to be fast, whereas methylation reactions are assumed to be slower for this system.

10.3.4 Vary the saturation condition for methylation by CheR for the model in Fig. 10.11 (change the constant 0.001 to 1). Investigate what happens if one changes the CheB rate limits $(Z_p/(Z_p + 1)) \to Z_p/(Z_p + 0.1)$ in one of the terms for dZ/dt).

10.4 Metabolic fluxes

Metabolic networks describe the conversion of small molecules to other small molecules; for example, they may describe the synthesis of amino acids. The metabolic networks also characterize the breakdown of some small molecules to basic elements that can be used for the building blocks of their metabolites.

As most reactions are associated with specific well-characterized enzymes (proteins), the allowed metabolic reactions in an organism can be read off from its genome. As a consequence, the metabolic networks are well known for a range of organisms (for their scale-free aspects see [474, 556, 557, 558, 559, 560]). Figure 10.13 shows the (pruned) metabolic network of *E. coli*, illustrating a particular geometry with long "production" lines of particular metabolites. In *E. coli* the number of reactions is of the order of 1000, representing less than one-third of all known allowed metabolic reactions in the living world [523].

Ref. [561] proposed a useful way to model metabolic networks, using the fact that these networks describe the flux of small molecules as these are used as building blocks for other small molecules. There are now good textbooks [562] and tutorials [563, 564]

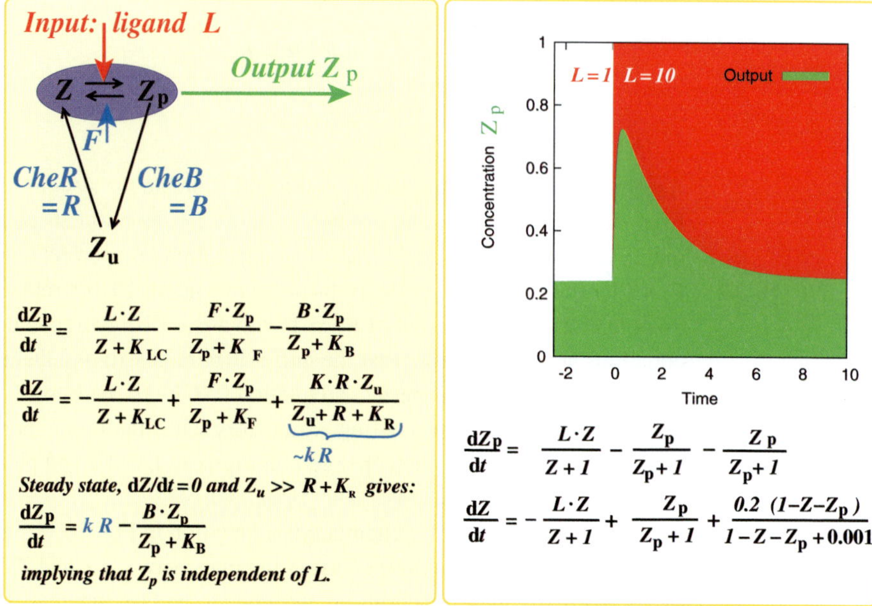

Figure 10.11 Adaptation motif for chemotaxis suggested by Barkai and Leibler [366, 546, 548]. This motif uses a combination of two modifications on Z, methylation (CH_3) and phosphorylation (PO_4^{-3}). The methylated receptor Z is sensitive to some external ligand (food molecules) L. The presence of L favors phosphorylation $Z \to Z_p$. Z_p can be de-phosphorylated directly (through F), or Z_p can be de-methylated by the enzyme CheB. When Z is de-methylated (to Z_u), it also loses its phosphate group. Z_u is converted back to the non-phosphorylated but methylated Z by the enzyme CheR (which is rate limiting). With a change in L, the output Z_p shows adaptation, with near complete restoration of the old level after a transient response, (provided that Z_u is large!!).

on flux balance analysis, so we here only provide a brief account of the basic concepts of this useful matrix formulation of interconnected assembly/dissasembly lines.

Figure 10.14 shows a metabolic model system consisting of five metabolites and seven fluxes. The metabolites are Gluc6P, Fruc6P, Fluc1-6P2, ADP and ATP+ADP, whereas glucose is only an input, parameterized solely through v_1.

The production of each metabolite demands a number of other metabolites, specified by the stoichiometry of the reaction. For example, consider the production of one Gluc6P molecule, which requires one glucose molecule and is associated with the conversion of one ATP to one ADP molecule. Accordingly, if one assigns the reaction:

$$\text{glucose} + \text{ATP} \to \text{Gluc6P} + \text{ADP} \tag{10.23}$$

with the rate v_1, then glucose $\to$ Gluc6P is assigned the rate v_1 and transitions from ATP to ADP should in addition, include a rate v_1. This is indeed included in the first, fourth and fifth metabolic equations in Fig. 10.14. Notice also that rates, in principle,

If there is no adaptation, tumbling will become less and less frequent and *E. coli* will end in a "random walk" without bias:

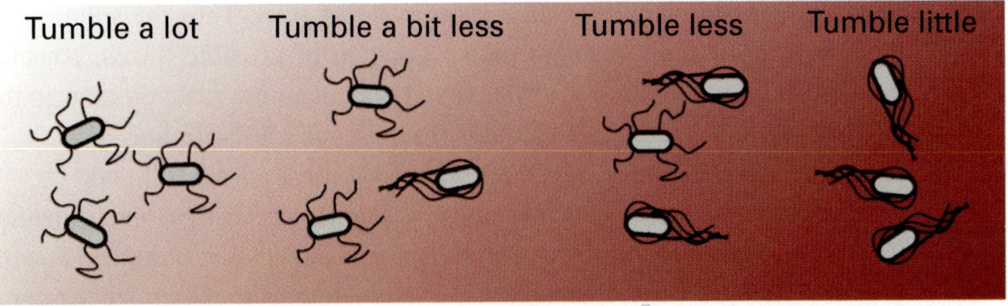

With adaptation, the tumbling frequency can reset ( ) to each new concentration, and its random walk can be persistently biased.

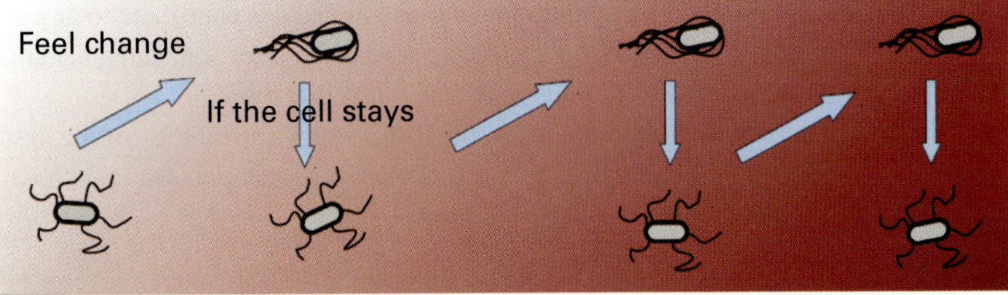

Figure 10.12 *E. coli* have flagella that rotate. If they rotate one way it propels the bacteria forward, while it tumbles and changes direction when the motor reverses [550]. The likelihood of reversing depends on signals from a receptor for food molecules. When an *E. coli* moves towards a food source, the signal from the receptor increases and the tumbling is repressed, thereby making the bacterium perform a directed random walk [550]. When the receptor adapts to the outside signal, the *E. coli* can maintain the directed walk over a large range of concentrations (figure with courtesy of N. Mitarai).

could be negative, in which case they represent a net reaction in the opposite direction to what is specified. Irreversible reactions will always have only one sign.

Given all the reactions, one constructs the stoichiometric matrix [564, 565], describing the relation between changes in metabolite concentrations and the fluxes between these metabolites:

$$\frac{dX}{dt} = \mathcal{S} \cdot v \tag{10.24}$$

where X is the vector of metabolites, and v the currents between the metabolites. $\mathcal{S}$ typically has more columns than rows, reflecting the fact that in general there are more fluxes than metabolites. In the above example of ATP removal by Gluc6P production, the inclusion of v_1 as loss and gain in the fourth and fifth metabolic equations is reflected in -1 and $+1$ in the first positions of the last two rows of the $\mathcal{S}$ matrix, respectively.

Flux balance is built on the assumption that for any particular metabolite, the steady-state current in and out of that metabolite is equal. Thus one assumes that $dX/dt = 0$ for all metabolites X. This gives a relatively simple matrix equation:

$$\mathcal{S} \cdot v = 0 \tag{10.25}$$

which in general has more variables than constraints. Thus the solution space will be "infinite," i.e. it will consist of some polygon of possible fluxes, constrained by the bounds set by the input to the system and the maximum reaction rates in the system. In Fig. 10.14 we illustrate this solution space by random sampling of v_2, v_5 and v_7, only accepting solutions where some pre-defined constraints are fulfilled.

Finally, flux-balance analysis is often supplemented by a maximization principle. That is, one assumes that some combination of enzymes is produced at a maximal rate. It is, for example, often assumed that *E. coli* attempts to maximize the combinations of amino acids, nucleic acids and lipids that represent the average composition of the cell.

The overall procedure for flux-balance analysis is:

- Find matrix S by counting how many molecules contribute to any given metabolite. Notice that fluxes are assigned a direction, and a negative flux then just corresponds to the opposite net flow direction.
- Determine the constraints of the fluxes set by the biology of the system.
- Find what fluxes solve $S \cdot v = 0$. These fluxes are consistent with Kirshoff's law for metabolites.

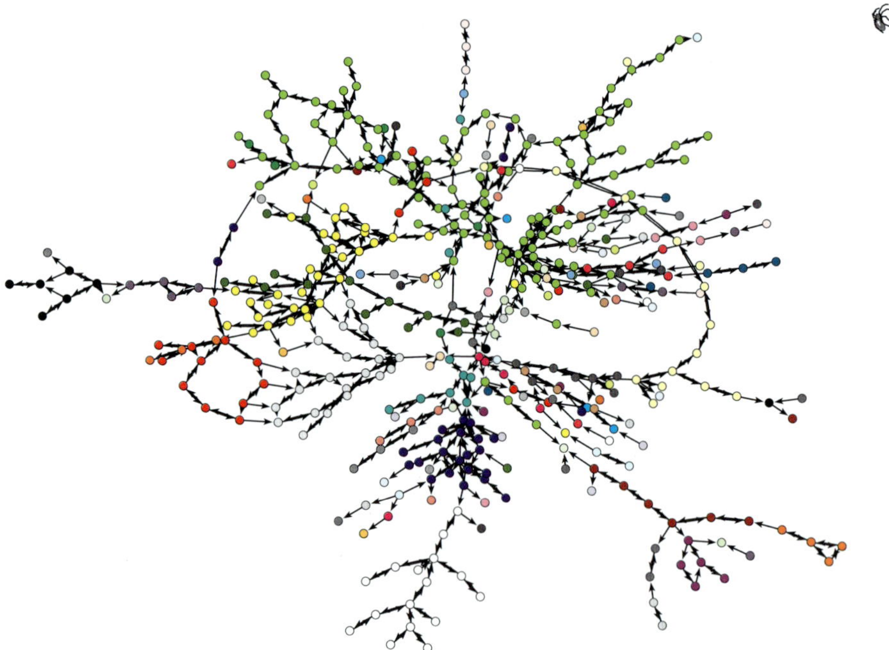

Figure 10.13 *E. coli* metabolic network [497] that is pruned by removing highly connected metabolites. Each node is a metabolite, and each arrow a metabolic reaction. Many highly reacting metabolites are not included, as they are involved in so many reactions that they obscure the visualization of the main pathways.

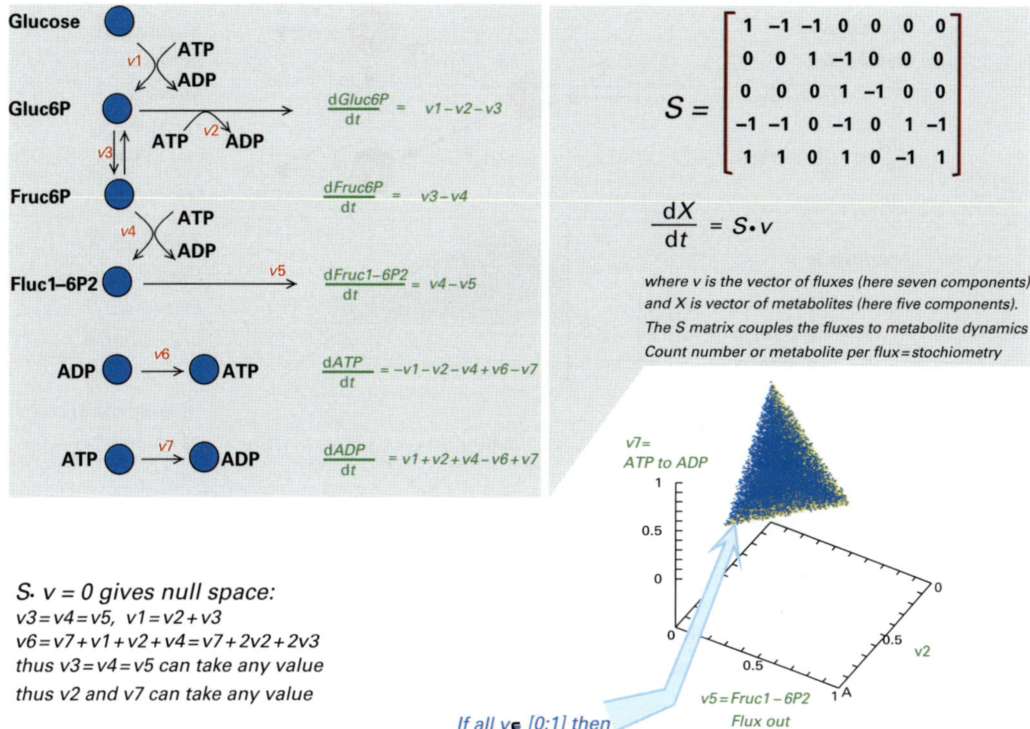

Figure 10.14 Tutorial of flux balance analysis using a classical glucose example. The method works on metabolic networks, drawn such that each node is a metabolite, and each link between two metabolites represents how much reactant is needed to make the product in the end of the link. If one, for example, wanted to maximize the output of Fluc1-6P2 in the above network, one would have to search for the maximum value of v_5, given whatever constraint the biology puts on the other reactions. In the three-dimensional figure we examine solutions constrained by having all $v_1, \ldots v_7 \in [0; 1]$. Network from E. Klipp, *Systems Biology in Practice* Wiley-VCH (2005).

- Define an objective target function, for example what one thinks the organism wants to achieve with the fluxes. A reasonable guess is that it wants to generate the maximum biomass with a composition equal to the composition of metabolites in the organism.
- Find which fluxes satisfy the constraints, $S \cdot v = 0$ and the maximization. The last often uses linear programming, see, for example, [566].

For real metabolic networks the search for optimal solutions is a complicated scanning of high-dimensional polygons using so-called linear programming (for a review see [564]). For small systems, one can obtain a good insight by sampling solutions randomly, as shown by the many blue dots in the three-dimensional plots in Figs. 10.14 and 10.15. The yellow planes show projections of solutions on the horizontal and vertical planes. The optimum solution will then be one of the "blue dots" that maximize the target function within the constraints set by the polygon. Therefore, the maximum will

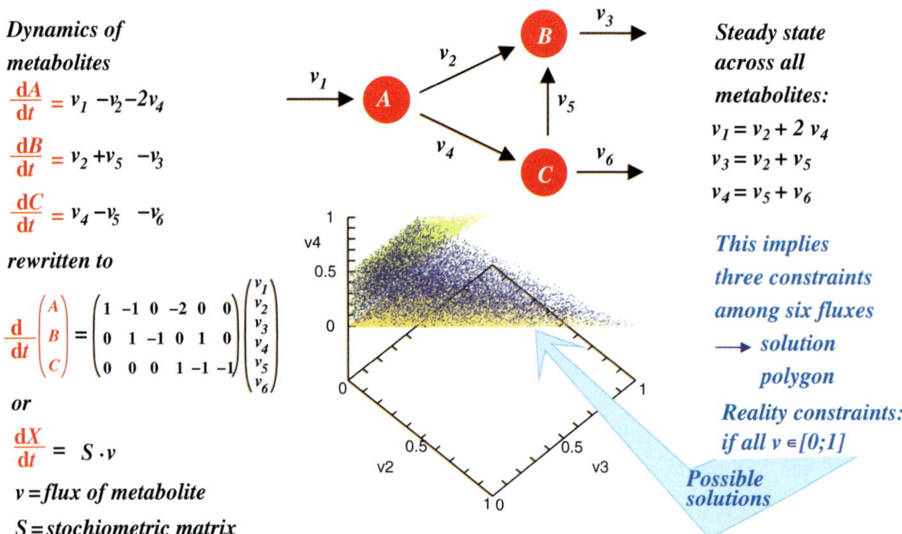

Figure 10.15 Small network of metabolites A, B and C that are converted to each other through fluxes $v(i)$, $i = 1, 2 \ldots 6$. In this example, production of one C needs two A molecules, whereas B needs one A or one C. If all $v(i)$ are limited to the interval $[0; 1]$, the allowed polygon of solutions is illustrated by random sampling in the three-dimensional figure. A similar example is found in [18].

always be found on a side, edge or corner of the polygon. If the maximum flux solution is not found on a corner, there is more than one solution.

10.4.1 Questions

10.4.1 Examine possible solutions for the glucolysis network in Fig. 10.14, assuming that all v are in the interval $[0,1]$. Which solutions maximize production of Fluc1-6P2. What are values of all fluxes in that state? Explain the solution in words?. Hint: sample solutions by selecting v_2, v_3 and v_7 randomly in the chosen interval and obtain other vs from the equation for the null space. Investigate the subset of solutions that fulfill all constraints.

10.4.2 Repeat question 10.4.1, allowing v_6 and v_7 to take any value, i.e. allowing transitions between ATP and ADP to be very fast.

10.4.3 Repeat the random sampling in Fig. 10.15, showing instead the allowed values of v_1, v_3 and v_6.

10.4.4 Examine the metabolic model network in Fig. 10.15, and the sample solution space, where we assume that all $v \in [0; 2]$. Find solutions where $v_3 + v_6$ is maximal.

10.5   Summary

- Signal propagation across networks requires fine-tuning of parameters. Protein–protein binding in itself only allows the signal to pass from the more-abundant to the less-abundant protein. Transcription factors allow information transfer in the opposite direction.
- Phosphorylation allows efficient boolean signaling. Central to obtaining on–off behavior is the push–pull reaction:

$$\frac{d \, output}{dt} = Input - threshold \tag{10.26}$$

  where *threshold* is set by the pull reaction.
- Adaptation is the ability to react strongly, but only transiently, to a changed external input. Networks exhibiting adaptation often include a buffer B, for example $dB/dt = 1 - L \cdot B$, whereas the output $\propto L \cdot B$.
- Metabolic networks can be viewed as carriers of a current in analogy with Kirshoff's law for electric circuits.

11 Agent-based models of signaling and selection

11.1 Excitable agents as cells in a tissue

Here we introduce agent-based models (ABMs) by demonstrating how they can be applied to to the inflammation propagation model from Chapter 8. The ABM was originally introduced by Von Neumann [567] to deal with system properties of many identical and relatively simple agents that repeatedly use specified rules of mutual engagement [567, 568, 569, 570, 571]. ABMs are accordingly suited for adressing self-organization and emergent phenomena, and have been used to study the properties of living, and in particular, social systems, including segregation [568], traffic jams, evacuation behavior [570], social insect organization [572] and stock-market dynamics, as well as dynamic pattern formation, with ABMs taking the form of cellular automata [567, 569, 573, 574].

Chapter 8 introduced an excitable medium as a model for waves of signaling molecules that facilitate signaling across tissues. Such wave propagation can in fact be rephrased in terms of a simple ABM with agents placed along a line [573, 574, 575]. A cellular automaton is defined in terms of a site variable $s(x)$ that takes a value $0, 1, 2, \ldots \sigma$ or takes a value identified as the excited state f. At each timestep all sites $x = 1, 2, \ldots L$ are updated synchronously:

- For all sites x where $s(x) < \sigma$ then $s(x) \rightarrow s(x) + 1$.
- For all sites where $s(x) = f$ then nearest-neighbor sites are set to f if they already have value $s = \sigma$. Subsequently, the firing site is reset to zero, $s(x) = f \rightarrow s(x) = 0$.
- A constant inflammation source is modeled by always maintaining the site $s(L/2) = f$.

The model is defined in terms of a single variable σ, which sets the refractory time τ that was an essential parameter for wave propagation in the model from Fig. 8.13. Figure 11.1A shows the results of a simulation with $\sigma = 5$. The ABM can also be reformulated to take into account stochastic effects, implemented by using a random update of the $s(x) < \sigma$ states. Figure 11.1B examines a model where at each timestep, each site is updated $s(x) \rightarrow s(x) + 1$ with probability $p = 0.5$.

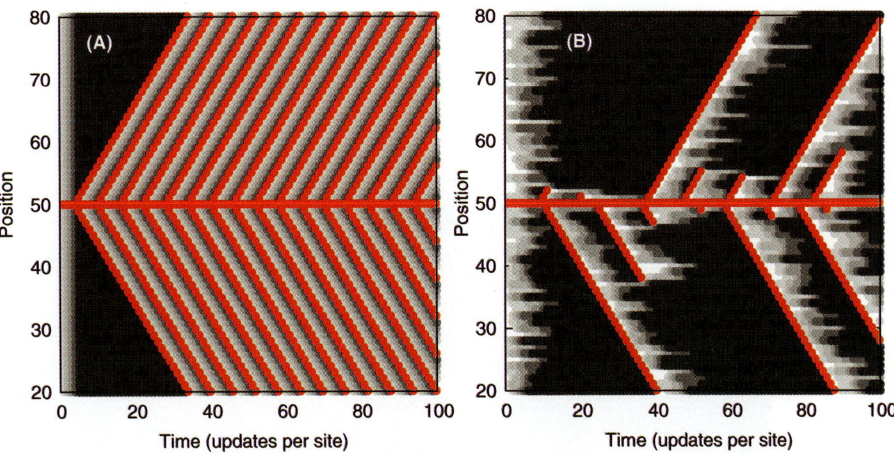

Figure 11.1 Cellular automaton model of a one-dimensional excitable media. The left-hand panel shows a deterministic model, whereas the right-hand panel updates the progress toward the excitable modes more slowly and randomly. In both cases each site has seven possible states, including the excited state f, which can activate neighbor states, provided that these have $s = 5$. After a site in the excited state has attempted to update a neighbor state, it is reset to state $s = 0$. In (A), at each time unit all sites with $s < 5$ are updated by $s \rightarrow s + 1$. In (B) sites with $s < 5$ are updated stochastically with probability $p = 0.5$. The plots show the $\sigma = f$ state in red, and lower s values in a grayscale.

11.1.1 Questions

11.1.1 Simulate the excitable media cellular automata with $\sigma = 10$ and $\sigma = 20$, for an $N = 100$ system with site number 50 always being excited.

11.1.2 Simulate the excitable media cellular automaton with $\sigma = 10$, and size $N = 100$ where a random site becoming excited for every system update. Also consider a random excitation for every 10 system updates.

11.1.3 Simulate an excitable medium in two-dimensions using stochastic updating with $p = 0.5$ and $\sigma = 5$.

11.2 Information spreading on social scales

11.2.1 Information can spread like infection waves

Information gathering is also important on larger scales, allowing individual organisms to optimize their survival and proliferation (see Fig. 11.2). For social animals, information is collected by communication, involving some sort of language. Interestingly, language and communication within our own species also seems to follow rules that result in wave-like behavior.

Figure 11.2 Information gathering is central to managing in a complex world, and strategies for obtaining it are manifold. Here is a picture of a mountain animal gathering large-scale information from a suitable location.

It has long been observed that linguistic features, like some diseases, spread outward from an originating center. A beautiful example is the geographical distribution of the word "snail" in Japan. This was investigated by Yanagita [576], who found that ancient forms of the word still existed in the southern and northern parts of the country, but not in the middle. He concluded, using his wave theory, that this reflected the strong influence of Kyoto, Japan's old capital.

Following [577], consider the dynamics of culture spreading around a strong pulsating cultural center. As a proxy for the spreading of cultural traits, one may use the spreading of words where the key feature is that new words are more prone to be adopted than old:

Information is sorted, using that **New** is better than **Old**.

Figure 11.3 shows the geographical distribution of swearwords in Japan. There are about 20 words present in the map, and the overall trend is that old words are found far away from Kyoto, whereas new words are close by. The drawn circles show their center of mass distribution with respect to the absolute distance from Kyoto. The data also shows that the gaps between two adjacent circles are not uniform, but tend to grow with increasing distance from Kyoto. From old records of when the word first appeared in Kyoto, the speed of swearword propagation has been estimated to be $v_{\mathrm{word}} = 1\,\mathrm{km\,y^{-1}}$ $(0.5\text{--}2\,\mathrm{km\,y^{-1}})$. Accordingly, the words in the northern and southern parts of Japan are found to be about 500 years old.

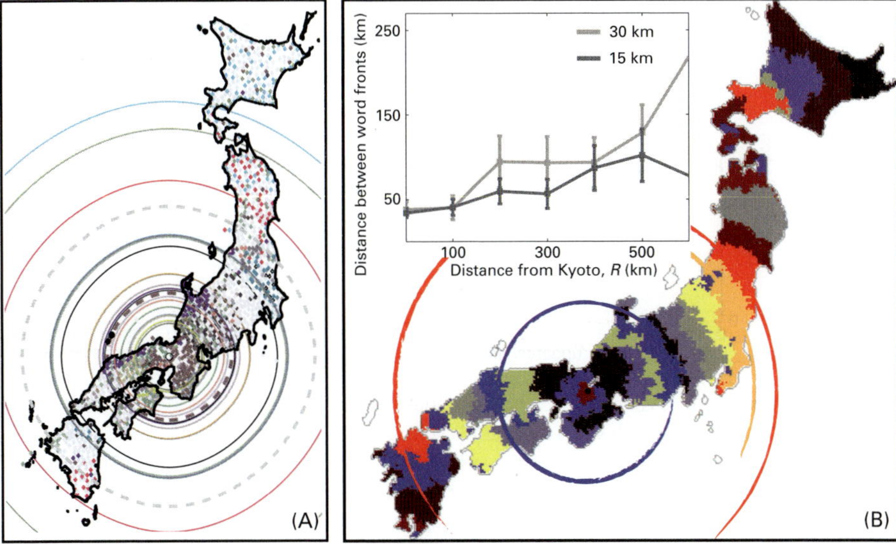

Figure 11.3   The left-hand panel shows distribution of swearwords, as measured in the 1980s. The geographical distribution of concentric circles around Kyoto are the result of 600 years of history. The right-hand panel shows a snapshot of a simulation of the spatial dynamics of word spreading over the Japanese mainland. Blue and red circles show two examples where the same word forms are found symmetrically on either side of Kyoto. The graph in the upper left corner shows the mean distance between two adjacent fronts (averaged over many runs) as a function of distance from Kyoto. The orange broken circle belongs to a word that is only present on Kyoto's east side. The probability that a word co-exists on both sides decays with distance from Kyoto in a way shown in the inset. Figure from [577].

The ABM for "word" spreading is defined on a two-dimensional lattice on which words spread, after being coined in the cultural center. The N agents of the simulation are placed on a square lattice. The model aims to capture the ongoing spread of new words that originate stochastically at the center with a frequency f_{word}:

- At each update a new word is initiated at the center, with probability f_{rmword}/N, and assigned a birth time according to a time counter.
- At each update a lattice site is chosen, and its word is communicated to a randomly chosen neighbor site on the two-dimensional square lattice.
- If the word is younger than that already at this site, it overwrites the old word at this site. If the word is transmitted to a place where an even newer version exists, the older word is ignored.

As words spread, they always retain their original birth time, assigned at origination at the center. In Fig. 11.3 the two-dimensional lattice is constrained within the land borders of Japan, thus allowing us to include the simplest geographic features.

The simulations depend on the frequency of new words f_{word} from Kyoto and also on the coarse graining of space. With larger, but fewer, patches, fluctuations increase

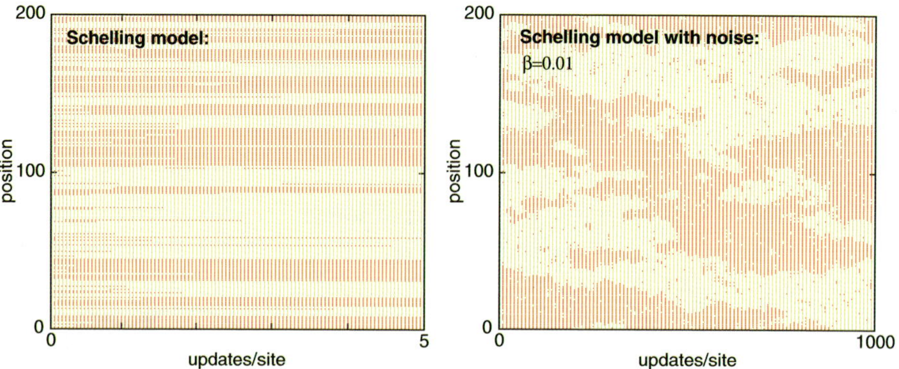

Figure 11.4 Dynamics of Schelling model, giving segregation of agents with two colors "red" or "yellow". At each step one agent is selected and attempts to replace position with another randomly selected agent in the system. The attempted move is accepted if the number of opposite color agents within the nearest four neighbors is reduced. Left panel shows how the agent segregates within a few updates per agent in the system but then freezes as no further movements are possible. Right panel shows that accepting suboptimal moves with some small probability $\beta = 0.01$ predicts large scale segregation.

and the likelihood that a word dies out becomes larger. The graph shows the case where each "agent" represents a lattice square with $\Delta = 30$ km across, and where the word frequency is such that about 20 words remain simultaneously on Hunshu island, as required by the data. The size of Δ is adjusted to fit the increasing distances between words as one moves out from the center, see Fig. 11.3. Inportantly, the words are quite different from the center to the perephery of Japan. Therefore culture may also diverge, a divergence that would be diminished with the addition of shortcuts, or information highways. One may speculate that the success of, for example, the Roman Empire was built on their extensive road networks, securing such social coherence.

The language-spreading model with the basic assumption that "'new" over-rules "old" resembles a minimal disease-spreading model, where people get infected, and are subsequently immune to each disease, a process that allows subsequent waves of emerging new diseases [578, 579].

11.2.2   Segregation and self-organized communities

Information spreading and boundary formation present two opposing processes in many biological systems. Boundaries and segregation may most easily be modeled by using the famous ABM model proposed by T.C. Schelling [568], suggested as a description of spontaneus segregation of white and black neighborhoods in American cities. In a simple formulation, the Schelling model consists of two agents, one "red" and one "yellow." Initially, one assumes equal numbers of each color, and positions the agents randomly along a one-dimensional line $x = 1, 2, ...L$. At each step:

Figure 11.5 Segregation and social structure may emerge as a consequence of preferences in communication, here visualized by some that communicate and some that do not.

- Select a pair of agents, and attempt to swap the positions of these two agents.
- The attempted move is accepted if it would lead to a reduction in the number of opposite color agents within the nearest four neighbors. The attempted move is abandoned if the number of opposite color agents is not decreased.

Notice that since there are only two colors, it is enough to check whether the neigborhood of one of the potentially reshuffled agents "improves." A simulation of the model can be seen in Fig. 11.4A. One can observe that the system freezes relatively quickly into a pattern of patches with a similar color, a pattern that no agents can improve by further pairwise reshuffling. Noticeably, the model may be reformulated to take noise into account, which in fact allows the system to develop steady-state dynamics, where typical patch sizes are larger than obtained without noise.

Segregation models have been suggested to be important for a range of biological problems, ranging from the evolution of colisin-producing bacteria [580] to spatial sorting of different cell types into tissues, using the differential adhesion hypothesis [581, 582], i.e. assuming that similar types of cells attract each other more strongly than cells from different tissues. Noticeably, however, segregation does not require the pre-imposed differences in color as envisioned by Schelling. It may in fact emerge spontaneously through an interplay with preferences in communication, perhaps as illustrated among the three parrots in Fig. 11.5.

11.2.3 Information spreading versus social fragmentation

Hierarchy formation is important for stability and function of colonies of many social animals. For ants, their hierarchical organization allows for the prioritization of

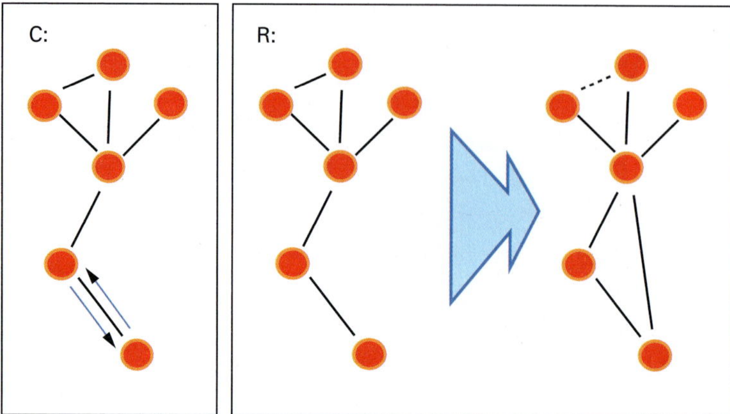

Figure 11.6 An ABM for social networking, consisting of a communication step (C) and a rewiring step (R). During communication, agents share information with nearest neighbors on the network, allowing new information about the current position of each agent to spread in a process mimicking the swearwords spreading across Japan. During the rewiring, a new link is formed to a friend of a friend, mimicking social climbing towards the source of new information [585, 586].

resources and work. The hierarchy in human society may, in addition, be coupled to flow of information between its members.

Simple agent-based modeling of spontaneous hierarchy formation typically emulates positive feedback between winning and the future chance of winning [583, 584]. Interestingly, social hierarchies are also predicted by ABMs acting through information gathering and networking [585, 586]. We will see that a hierarchy around a central node is a natural consequence of information gathering of individual agents, because of a positive feedback between being centrally placed and having access to newer information.

The ABM in [586] considers the dynamics of a social network consisting of N agents that are connected via a fixed number of L links (see Fig. 11.6). Each agent is assigned an individual memory that includes three vectors: (1a) N pointers to each of the $N - 1$ other agents in the system that show which friend[1] provided information about the other agent, (1b) the age of information pinpointing the direction to each of the $N - 1$ other agents[2] and (2) a priority vector that allocates space proportional to the interest in each of the other agents in the system. The memory in (1a) and (1b) is a map that is updated based on the principle of new is better than old information, whereas the priority in (2) is only used in the segregation version of the model.

The network model is executed in timesteps, each consisting of one of the two events (see Fig. 11.6):

---

[1] Friends are agents that the agent in question has communicated with and thus use in its contemporary map of pointers to nodes further away in its surrounding network.

[2] The information age that is used to compare quality, and subsequently update the information when communicating with neighbor agents.

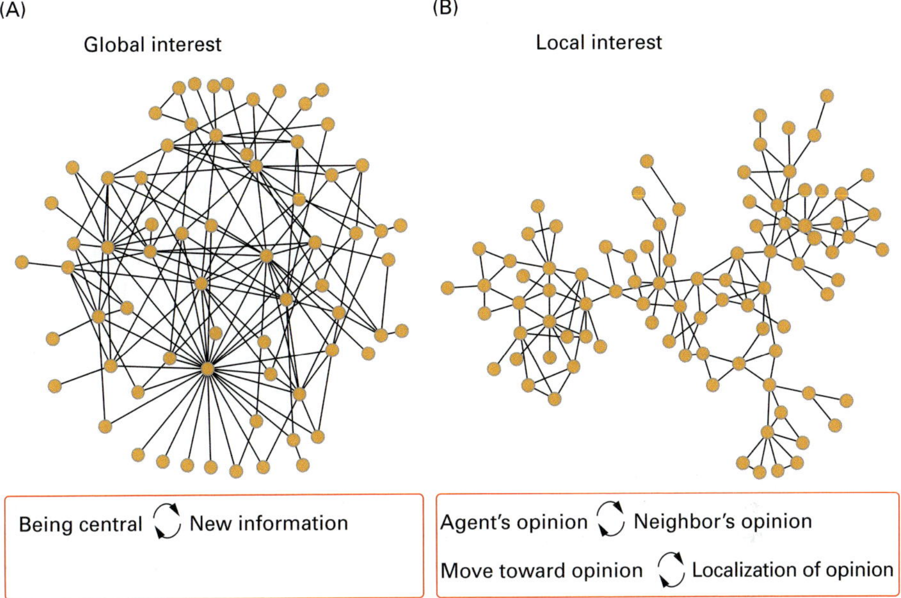

Figure 11.7 An ABM where agents create links to neighbors of neighbors, and remove random links. (A) A network snapshot in the steady state of an evolving social network where everybody attempts to climb towards each other in an ongoing hunt for information about each other. For details see [585]. (B) Snapshot of a segregated network, where the social climbing of agents is limited by the agents' prioritization of recent experiences, leading to self-organization of a limited information horizon [586]. The lower panels emphasize the main positive feedback that drives the topology of the communication networks shown.

- **Communication, C:** Choose a random link and let the two agents connected by the link communicate about a third agent selected by one of them. The two agents also update their information about each other.[3]
- **Social climbing, R:** Let a random agent use the local information to form a link to a friend's friend to shorten its distance to a selected third agent. Subsequently a random agent loses one of its links.

Communication is executed much more frequently than rewiring, thereby allowing each agent to build a reliable contemporary map of where other agents are. Rewiring (R) defines network dynamics where individuals change their neigborhood by gradual social climbing to friends of friends.[4]

The ABM is first simulated such that agents prioritize each other equally all the time. This mimics the case where broad-minded agents simply hunt for new information about

---

[3] When two agents communicate about a third agent, they first decide which of them has the newest information. This information is considered the most reliable, and the agent with the older information updates its information about its position by use of the agent with the newer information about the location of the third agent. Thereby each agent builds a map of the contemporary social network, which will be reliable if communication is much more frequent than rewiring [585].

[4] If an agent is completely disconnected from the system, it reconnects on the basis of a weighted choice from its own prioritized list.

everybody. The network then develops the hierarchical structure shown in Fig. 11.7A, which reflects the positive feedback between being central and having access to new information about other agents. That is, agents with new information are attractive in the R-move of other agents. Therefore central agents tend to gain links and networking reinforces a hierarchy based on information access.

The behavior of the ABM changes dramatically when agents prioritize their "information hunt" based on what they are used to hearing (see Fig. 11.7B). Following Spencer's [587] conjecture of proportionality between interest and previous experience, this version of the ABM lets an agent's priority memory be filled with other agents' names, proportional to their occurrence in recent gossiping. In practice, the priority memory is updated during the communication step, where one of the communicating agents uses the prioritized memory to select the agent of interest to "talk" about. Subsequently both agents increase the memory slot about this discussed agent at the cost of reducing some other memory.

The two versions of the ABM for dynamic social networks illustrates: (1) That agent-based models may indeed reflect common experiences from social networking, (2) that information spreading or containment may act in a close positive feedback loop with contemporary social organization and (3) that social segregation may also emerge among identical agents due to positive feedback involving reinforced interest in local neigborhoods.

11.2.4 Questions

11.2.1 Simulate a Schelling-like model with three colors, where all agents want four nearest neighbors to be the same color as themselves. At each step, select two agents. If both agents either win or at least do not lose in the exchange, then switch positions. Simulate an $N = 100$ system for 2000 updates per agent in the system.

11.2.2 Simulate and visualize the spread of signals along a one-dimensional line, with new words appearing at position $x = n/2$ with high frequency (for example, each time each agent has been involved in one word exchange). At each step, select two neighbors, and let the youngest word spread to replace the oldest word.

11.2.3 A classical way to obtain scale-free behavior is "the rich get richer dynamics" [473, 588, 589]. For networks this is formulated in terms of a growth model where each node links to already present nodes by linking up to the nodes at the end of random selected links in the network [473, 589]. Start with one node at time $t = 0$ and let $n(k, t)$ be the number of nodes with connectivity k at time t. Show that [590]:

$$n(k, t+1) - n(k, t) = \frac{(k-1) \cdot n(k-1, t) - k \cdot n(k, t)}{\sum kn(k)} \quad for \ \ k > 1$$

$$(11.1)$$

Each added node is associated with one link, which means two edge ends, and thus $\sum_k kn(k,t) = 2 \cdot t$.

Argue that:

$$\frac{dn}{dt} = -\frac{1}{2t} \cdot \frac{d(k \cdot n)}{dk} \tag{11.2}$$

Use the "ansatz" $n(k,t) = f(t) \cdot N(k)$ to prove that $N(k) \propto 1/k^3$.

11.1.4 Consider the distribution [588, 591, 592]

$$p(s) \propto 1/s^\tau \tag{11.3}$$

as a distribution for wealth in human society. Ague that $\tau \leq 2$ is fundamentally different society than $\tau > 2$. Notice that $\tau = 2$ is the famous Zipf distribution observed, for example, for word frequencies in books [593].

11.3 Tragedy of the commons

A classical problem in both social behavior and work sharing between microbes is how to combat cheaters or free riders: when multiple individuals depend on a shared and limited resource, it is mutually beneficial that the resource is not depleted. However, it is in the short-term interest of each individual to consume resources fast to gain a competitive advantage. This dilemma, coined as "the tragedy of the commons" [595], illustrates the need for restrictions on the use of limited resources. However, in microbial systems there is no law reinforcement, and therefore restrictions must arise as a stable and sustainable consequence of evolution. To illustrate how this may occur we here consider an ABM model for a cyclic predation between three states: the rock–paper–scissors (RPS) game.

In RPS none is superior, and increased predation by one species will negatively feedback on its own population, because it represses its predator's predator. In the limit where one species is very slow, then an RPS can behave as a robust variant of an excitable media with the excited state being the one that is both preyed upon and predating fast.[5] This is illustrated in Fig. 11.8, where an extreme separation of growth rates is reflected by an extreme separation of length scales.

11.3.1 Tragedy of the commons in rock–paper–scissors

The rock–paper–scissors relation mimics the ongoing cyclic turnover that is found in many ecosystems, ranging from the terrestrial and aquatic to the microbial [596, 597,

---

[5] With the noticeable difference that the refractory period in RBS cannot become excitable by itself, but needs to be predated by the prey of the excited state. This encoded sequence of events secures the ongoing activity of recycling between the three states, even when the predation speeds are all similar.

Figure 11.8 The spatial structure of rock-paper-scissors ecology when one species (green) is much slower than the two other species (Figure from [594]). The ecology is simulated on a $12\,000 \times 12\,000$ lattice with a speed contrast of factor 100. With higher contrast, the red predator will go extinct, and the slow species will take over the system. In terms of an excitable media analogy the red species is the excited state, the green the excitable state and its slow invasion of the blue reflects a refractory period. The obtained contrast will be obtained when two species are allowed to mutate, thus both racing for faster predation. If all three species can mutate, they all increase in speed more or less synchronously.

598, 599, 600, 601, 602]. The rock–paper–scissors relation also partially reflects the ongoing war between a virulent phage and its host, with the phage facing extinction from the absence of host [598, 599]. Other microbial examples include a bacteria that excretes a colisin,[6] a colisin-sensitive bacteria and a colicin-resistant bacteria [604].

This section explores evolution in an idealized rock–paper–scissors game [596, 597, 598, 605], where three species compete for space. One species is allowed to mutate and thereby increase or decrease the speed at which it invades its prey. Following [596, 605], we will show that group selection in a spatial setting removes predators with the fastest growth rates, thereby stabilizing overall co-existence.

---

[6] Colisins are bacterial toxins, which bacteria use to fight each other. Bacteria that cary a colisin also carry an antidote for it, and can therefore preferentially kill competing neighboring bacteria. An interesting microbial ecosystem model with many different colisins is introduced by [603].

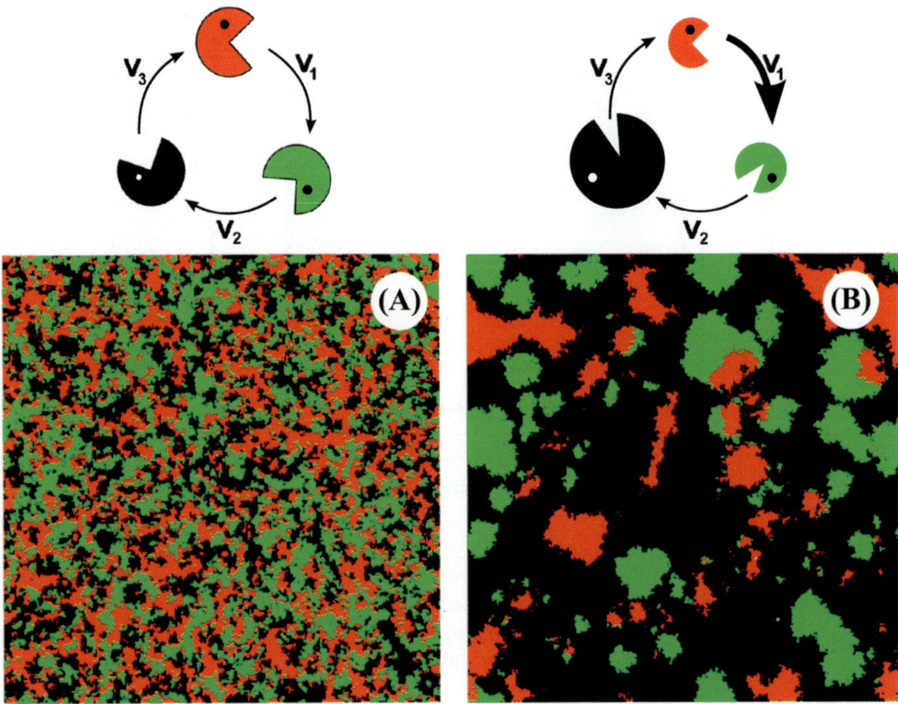

Figure 11.9 Model system of three species with cyclic interactions. Species 1 overgrows species 2, which overgrows species 3, which overgrows species 1. Individuals may grow at different rates. The mean growth rate for species 1, 2, and 3 are denoted v_1, v_2 and v_3. (A) When all species grow at the same mean rate, they will be equally abundant. Members of same species will self-organize to form clusters on the lattice. (B) When species 1 grows faster than 2 and 3, species 3 will be more abundant than species 1 and 2. Here $v_2 = v_3 = v_1/5$. Fig from ref. [605].

Consider a community of three species where species 1 overgrows species 2, which overgrows species 3, which, in turn, overgrows species 1 (see Fig. 11.9a). A key property of such cyclic interactions is that growing fast does not help a species to gain biomass – it will help its predator! By growing too fast, the species will weaken the population of its prey thereby improving conditions for its predator (see Fig. 11.9b–c). Therefore, for the species as a whole, it is advantageous to grow slowly, whereas each individual of the species will gain a competitive advantage from growing fast. It has been found that such a species easily succumb to the tragedy of the commons when interacting globally, but it has also been found that spatially structured populations tend to support biodiversity [598].

11.3.2   An agent-based model

Let each site of an $L \times L$ square lattice be occupied by a member from one of the species, 1, 2 or 3. Initially, all individuals grow at the same rate. At each timestep the following actions take place:

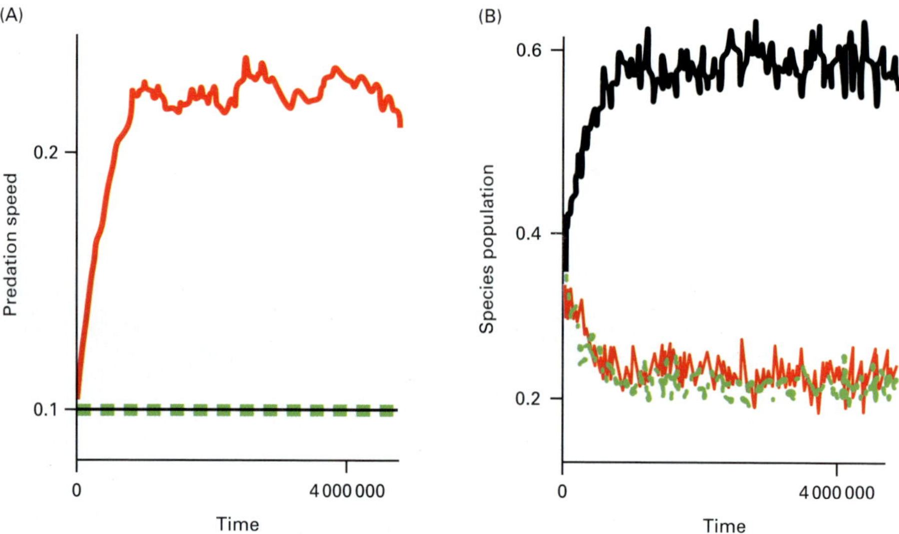

Figure 11.10 Species 1 (red) is allowed to mutate. As a consequence, its average speed accelerates to a growth rate fluctuating around 2.4 times faster than species 2 and 3. Consequently, species 3 (black) grows to become 2.6 times more abundant than species 1 and 2. Simulation with system size $L = 200$, $\gamma = 1.02$, and $p_{mutate} = 5 \cdot 10^{-5}$. Figure from [605]

- A random node i and one of its four neighbors j are selected. If i can overgrow j it does so with a probability v_i, set by the growth rate at position i, thereby, j becomes occupied by the same species as i.
- When i overgrows j, it may mutate by a small probability p_{mutate}. Hereby, j will either become faster growing than i, $v_j = (1 + \gamma)v_i$, or slower growing, $v_j = v_i/(1 + \gamma)$, where γ is a constant that sets the velocity changes.

Assume that the species grow at rates (v_1, v_2, v_3) and that their densities are (p_1, p_2, p_3). In the mean field approximation we achieve a steady state when $v_1 p_1 p_2 = v_2 p_2 p_3 = v_3 p_3 p_1$, leading to [596]:

$$(p_1, p_2, p_3) = \frac{1}{v_1 + v_2 + v_3}(v_2, v_3, v_1) \Rightarrow \frac{p_1}{v_2} = \frac{p_2}{v_3} = \frac{p_3}{v_1} \tag{11.4}$$

Thus the biomass of any given species is proportional to the speed at which its prey grows! For a species, it does not pay to be more efficient, although any individual will gain by spreading its offspring faster.

When only species 1 is allowed to mutate, it will evolve to a certain mean growth rate $v_1 = (2.4 \pm 0.1) \cdot v_2$ [596]. At the same time, species 3 will grow in size to $p_3 = (2.6 \pm 0.1) \cdot p_1$, while the relative sizes of species 1 and 2 will remain about equal (see Fig. 11.10).

Using a simple, one-dimensional argument, one can understand the relative growth rates at steady state. Locally, a faster mutant will always have a competitive advantage over a slower mutant. However, each mutant will be localized [606]. The faster mutants are therefore at risk of exhausting their neighborhood of prey, leaving their offspring in an isolated cluster surrounded by predators (see Fig. 11.11a–c). To avoid this, the

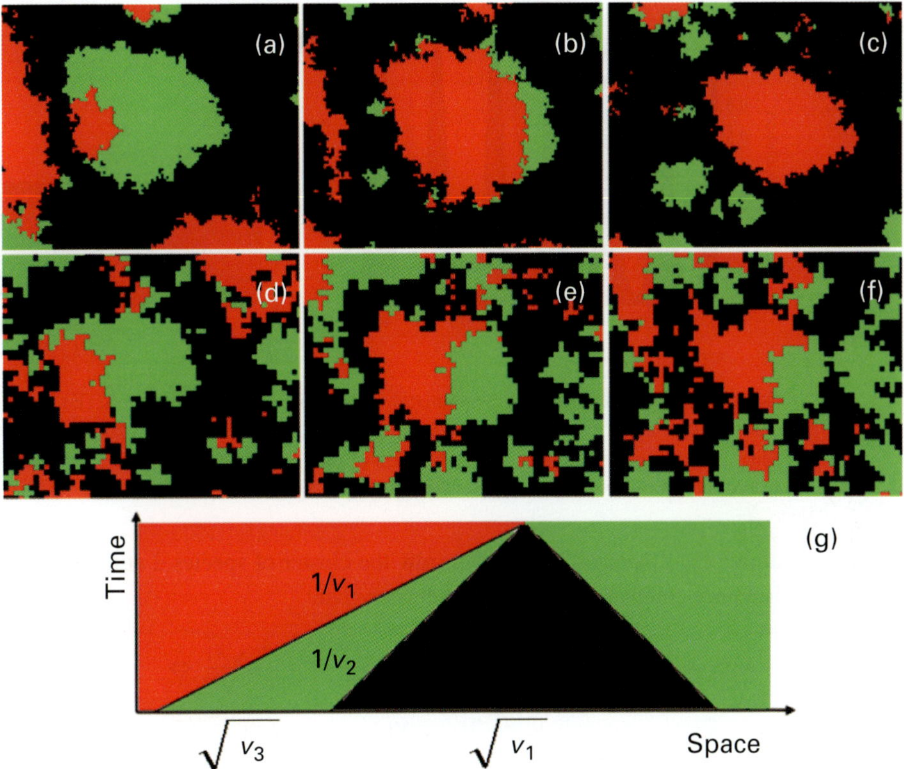

Figure 11.11   (a–c) Simulation when red is five times faster than the other species. Red invades a cluster of green and quickly overgrows it all, leaving itself surrounded by its own black predator. (d–f) Simulation when red grows 2.4 times faster than the other species. Before red has overgrown a cluster of green, this has connected to other clusters of green. (g) Clusters of species 1, 2, 3 and 2 arranged on a line. In the space–time diagram, species 1 grows with slope $1/v_1$ and species 2 grows with slope $1/v_2$. A typical cluster length of species 2 is $\lambda_2 \propto \sqrt{v_3}$ and for species 3 it is $\lambda_3 \propto \sqrt{v_1}$. The maximum speed for species 1 follows by demanding that a typical cluster of species 2 is overgrown just when it connects to a new cluster of species 2. Figure from ref. [605]

survival of a fast mutant requires that it allows time for its prey to grow sufficiently to make a connection to a new cluster of prey (see Fig. 11.11d–f). At the maximum growth rate, a species will grow through a typical cluster of prey, just when this becomes connected to a new cluster of prey. The maximum growth required is obtained when the prey constantly provides a marginally percolating food source (see Fig. 11.11g).

Since the abundances of the three species are given by Eq. (11.4), one expects the typical cluster diameters to scale like:

$$(\lambda_1, \lambda_2, \lambda_3) \propto (\sqrt{v_2}, \sqrt{v_3}, \sqrt{v_1}) \tag{11.5}$$

Imagine clusters of species 1, 2, 3 and 2 arranged on a one-dimensional line (see Fig. 11.11g). As time passes, the two clusters of species 2 will overgrow species 3

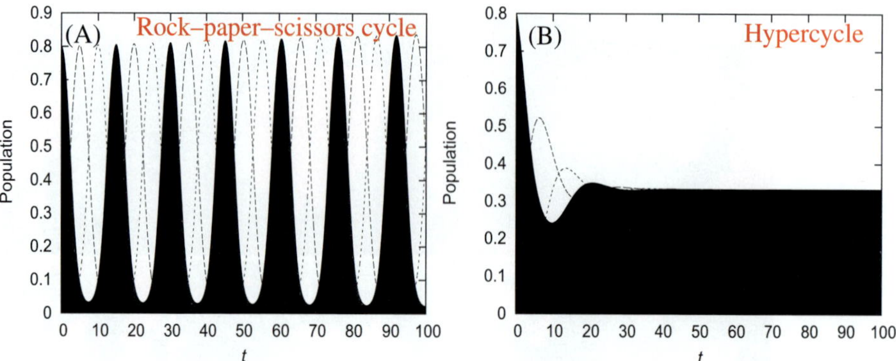

Figure 11.12 (A) Three-cycle deterministic rock–paper–scissor simulation. (B) Three-species deterministic hypercycle. In a cycle the species "eat" each other, in the hypercycle the species "eat" from the common pool. The depletion of the "common pool" occurs in catalytic reactions where the overall biomass is conserved by supply and dilution.

at rate v_2. If species 1 is to overgrow the cluster of species 2 at the exact time that this becomes connected to its second cluster, then:

$$\frac{1}{v_1}\left(\sqrt{v_3} + \frac{\sqrt{v_1}}{2}\right) = \frac{1}{v_2}\frac{\sqrt{v_1}}{2} \Leftrightarrow$$

$$\frac{v_1}{v_2} = 1 + 2 \cdot \left(\frac{v_1}{v_3}\right)^{-1/2} \tag{11.6}$$

When $v_2 = v_3$, the solution $v_1/v_2 = v_1/v_3 = 2.3$ agrees with the simulated value of 2.4 ± 0.1.

The spatial structure of the system is essential for the obtained group selection. In a well-mixed system, the species would instead start a race to extinction and overall collapse of the ecosystem.

In phage–bacteria ecology, experimental evolution of the T4 phage in the lab has demonstrated that unrestricted mixing explicitly favors evolution of efficient killers [607]. Thus, long-term survival of prudent pathogens/mediocre killers demands space [172, 607], which allows selection on a group level. The role of space in maintaining microbial diversity has also been tested experimentally using microbial model systems [599]. Studies that emphasize the stabilizing effects of space in other ecology systems can be found in [596, 597, 598, 608, 609].

11.3.3   Questions

11.3.1   The deterministic counterpart of the non-spatial model can be given in terms of the coupling matrix Γ through the population dynamics of any species i,

$$\frac{dp_i}{dt} = \sum_i \Gamma(i,j) \cdot p_i \cdot p_j - \sum_j \Gamma(j,i) \cdot p_i \cdot p_j \tag{11.7}$$

where the first sum overruns over all prey species of i, i.e. where $\Gamma(i,j) = 1$, and the second sum is over all predators of j (species where $\Gamma(j,i) = 1$). Simulate a three cycle using this equation, and subsequently add an additional species that only prey on one of the cycle species, but do not give anything back (a parasite).

11.3.2 Repeat the above simulation for a five-cycle system.

11.3.3 The cyclic relationship from the previous questions includes a direct predation cost on the prey, which differs from Eigen and Schusters hypercycle [610], where the growth of one species happens at the cost of all other species. The hypercycle equation: $\frac{dp_i}{dt} = p_i \cdot p_{mod(i+1,n)} - \Omega \cdot p_i$ where $\Omega = \sum p_i \cdot p_{mod(i+1,n)}$ describes catalytic chemical equations in a chemostat, where the sum of all species is maintained at $\sum p_i = 1$. Simulate a three-cycle and a five-cycle and compare with corresponding predator–prey simulations. Hint: see Fig. 11.12.

11.3.4 Social dilemmas are most often modeled in the form of the classical two-player prisoners' dilemma [611]. In this, each player decides independently whether to co-operate or defect. A given player subsequently is given a score of 1 if the opponent co-operates and 0 otherwise. Further the player is assigned an additional score $-c$ (i.e. penalty $c < 1$) if he collaborate. Assume that the opponent randomly, with probability r, chooses to co-operate; what is your optimal strategy? Consider now an iterated game where players play multiple times with each other. Assume that the opponent always mimics your last choice; what then is your optimal long-term strategy as a function of c?

11.4 Summary

- Excitable media can be modeled as a forest fire, where the excited state (fire) can propagate when there has been a sufficient refractary period to allow neighboring trees to grow.
- Large-scale structure can emerge as a consequence of many iterations of simple update rules between pairs of agents.
- Agent-based models open for the study of social dilemmas and altruism.
- Without space, "cheaters" cannot be isolated and their growth may lead to system collapse. With space, selection on a clonal level limits the success of the "cheaters" and stabilizes co-existence of evolving species.

12 Competition and diversity

12.1 Phage worlds

Earlier chapters specialized in a few model organisms, including the λ phage and its host *E. coli*. This choice was a modeling choice that reflected a historical contingency in the scientific community starting in the early 1950s. However, in spite of its random origin and positive social feedback [612] the scientific focus had advantages. A precise description of one model system may well have been more informative about the universal mechanism of gene regulation than a listing of various properties of different organisms. However, a prominent and astounding feature of life is its diversity; a diversity of species, strategies and molecular mechanisms.

12.1.1 Phage predation strategies

The approximately $10^{31}-10^{32}$ phages in the world [129, 133] come with many forms, shapes and strategies (see Fig. 12.1 for an attempted family tree). The hierarchical relationships implied by this tree may not reflect the evolutionary relationship between all genes in the respective phage, since phages exchange genetic material extensively, both with each other [614], and between their respective hosts [616, 617]. This non-hierarchical relationship with gene transfer of groups of related genes is emphasized in Fig. 12.2.

In the upper middle branch of Fig. 12.1 one finds the λ phage, which has multiple close relatives, e.g. the $\phi 80$ phage [618] and the 434 phage are both similar to λ. The λ-like phages represent a substantial section of the temperate phages that infect *E. coli*. In fact, temperate phages that infect *E. coli* typically belong to one of the two main classes: the λ-type (lambdoid phages) or the P2 type (e.g. the 186 phage), which are both highlighted in Fig. 12.2. For further information on phage diversity, see [613, 614, 615, 619, 620, 621, 622, 623].

Many temperate phages use regulated randomness[1] to decide whether they kill or live with their host after infection. However, temperate phages exist with simpler strategies, always allowing the host to survive, but at the same time producing an ongoing flux of phages from the infected host. These filamentous phages include the *fd*, *f*1 and the M13

---

[1] However, other phages may not. For example, phage 186 seems to choose lysogeny with a frequency of about 10% that does not depend on multiplicity of infection (I. Dodd private communication).

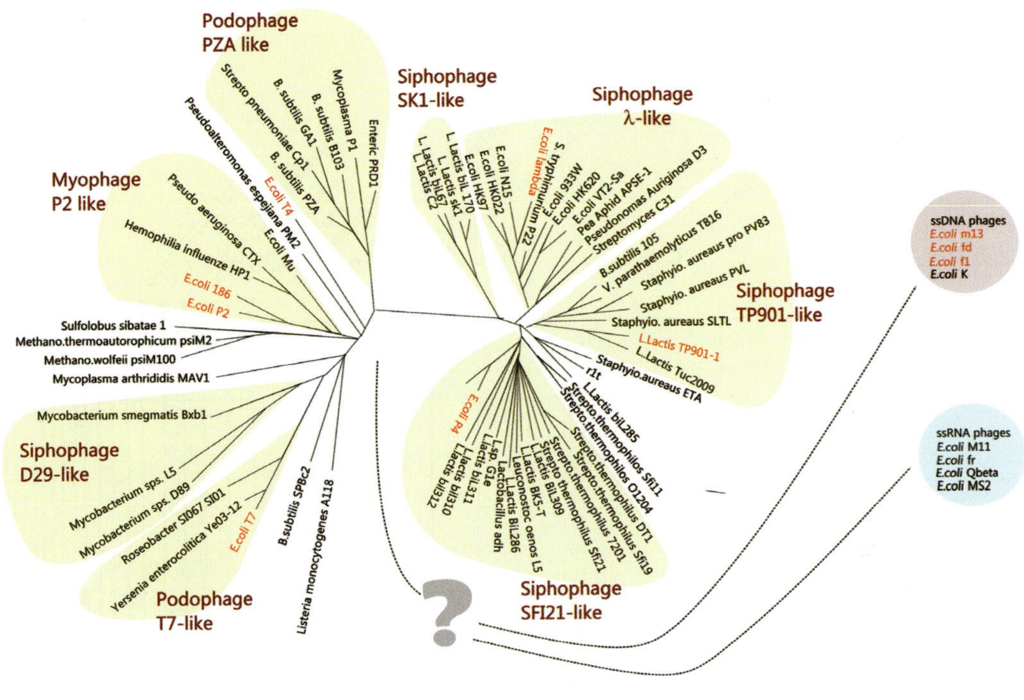

Figure 12.1 Redrawn from [613] with the focus on relationships between double-stranded DNA (dsDNA) phages. The evolutionary tree was generated by comparing protein sequences across the different phage groups [613]. The tree should be taken with many grains of salt, as there are huge horizontal transfers between different phages [614, 615]. Notice that the phages λ and 186, mentioned earlier, are evolutionarily distant from each other. ssDNA phages are often filamentous and have small genomes, about 8 kb. ssRNA phages have small genomes, for $Q\beta$ or MS2 about 4kb. A systematic method for dividing phages into categories based on their contents of protein families [615] found no common families between ssRNA, ssDNA and dsDNA phages. Phages mentioned in the text are highlighted in red.

phages [624]. The M13 phage contains only about 8 kb of DNA, and does not integrate into the host chromosome. The infected bacteria release about 100 M13 phages per generation, at the cost of doubling its generation time. Each released phage is $\sim 800\,\text{nm}$ long, but only 8 nm in diameter. This filamentous design is presumably constrained by the need to leave the *E. coli* cell without destroying it.

In contrast to temperate phages, there is a group of persistently virulent phages, which for *E. coli* include T1, T2, T3, T4, T5, T6 and T7. Temperate and virulent phages can, in principle, be closely related, as a temperate phage easily mutates into a lytic one by losing a few regulatory sites (like λ vir), or the repressor (as evidenced by the clear plaques in spontaneous lysis experiments, corresponding to a spontaneous mutation to virulence that is between 10^{-6} and 10^{-7} per cell generation). However in practice almost no virulent phages outside the lab environment bear any resemblance to temperate phages, nor do they have any trace of an immunity region (the DNA region from OL to PRE in

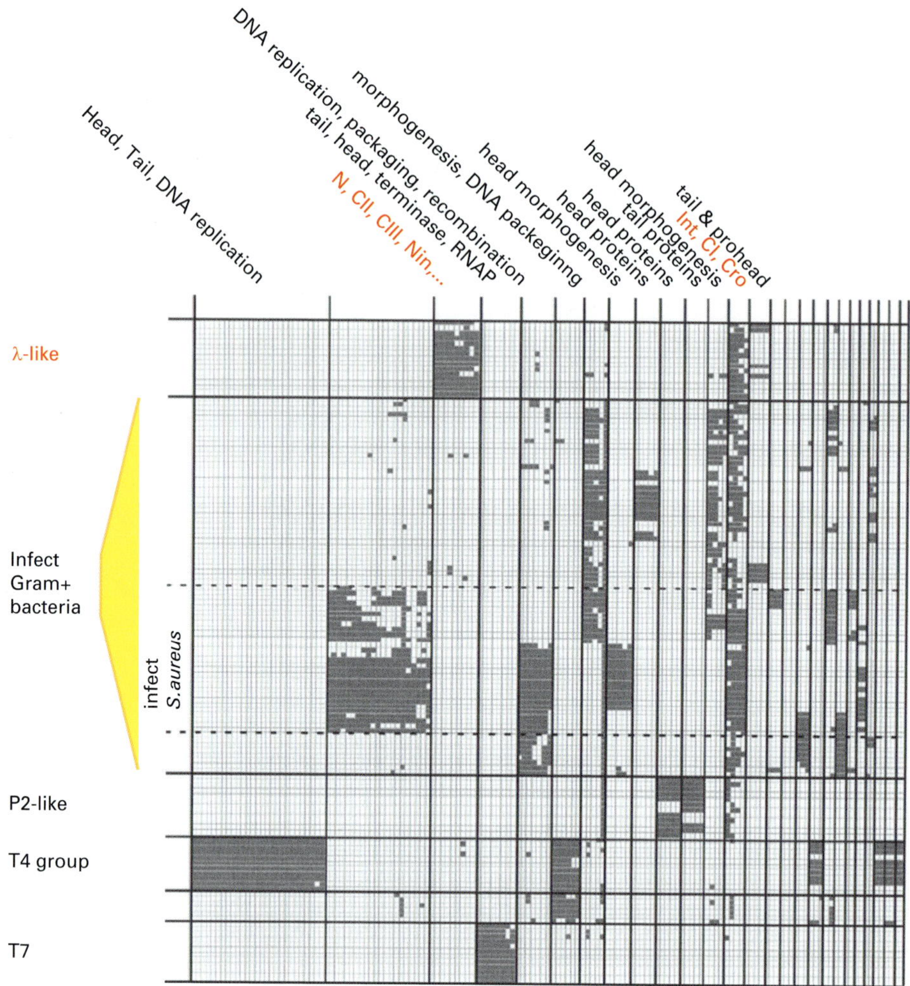

Figure 12.2 Relationships between phages quantified in terms of overlap between functional groups of proteins, as visualized by [615]. The λ-like phages and their two main groups of coherently transferred proteins are marked in red. Noticeably, N, CII and CII co-occur distinctly together, indicating that these three proteins also work closely together. Importantly, the temperate phages (the ones with CI, Cro and int) have very little overlap with the virulent phages (the existing overlap is believed to be associated with horizontal gene transfer, where temperate phages also facilitate gene transfer between distant virulent phage groups [615]). Figure reproduced from [603], by permission of Oxford University Press

phage λ). Thus, survival as a virulent phage must require more than just being able to replicate and lyse their host.

Typically phage burst size is measured as an average over a population (examples are listed later in Fig. 12.11), but it is remarkable that bursting of a given phage in a given type of host is also highly heterogeneous. This heterogenety in killing and predation strategy is illustrated by the very broad distribution of the number of phages that escape infected bacteria (see Fig. 12.3). Perhaps such variability again refects a hedging strategy, or alternately, then occasional small bursts may allow the

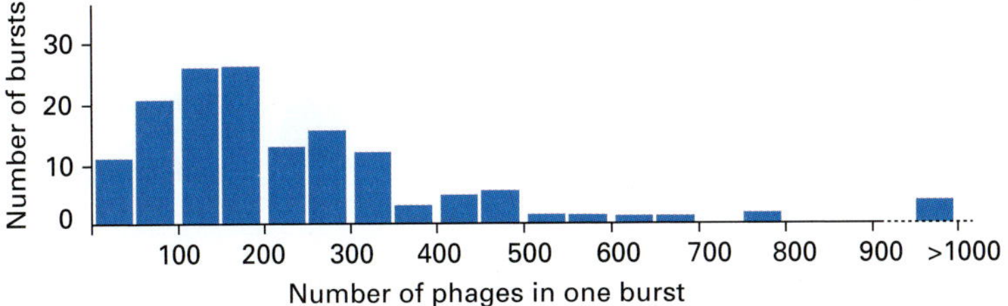

Figure 12.3 Burst size distribution from some unknown virulent phage, as measured by Delbruck in 1945. The distribution is extremely broad, suggesting that perhaps broad distributions and huge heterogeneity is one way to hedge options for a virulent phage in an unpredictable environment.

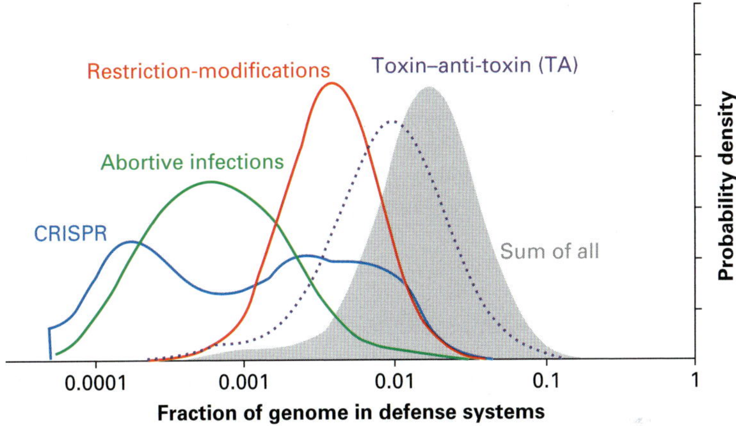

Figure 12.4 Distribution of four different types of defense among a large sample of prokaryotes [625]. The RM systems give immunity to foreign DNA (innate immune system), CRISPR to previously encountered DNA (adaptive immunity), the abortive infections (phage exclusion) leads to senesence or cell death upon infection, and toxin–antitoxin systems provide senesence and may thereby delay phage release. The TA systems are shown here with dotted line, indicating that their role as a main player in phage defence is uncertain. Together, the four types of system can comprise up to 10% of a bacterial genome (gray shaded area).

host to survive locally under conditions where well-mixed co-existence would be marginal [172].

12.1.2 Bacterial defense strategies

An overview of the widespread distribution of bacterial defense systems against phages is found in Fig. 12.4, which is redrawn from [625]. We now briefly describe these systems, and subsequently emphasize the co-evolved attack and defense situations between phages and their host. Common for the defense systems is that they all contain a toxic element that attacks the phage, and often also some antidote that prevents the toxin from attacking the bacteria; a push–pull design that is poised to react when the situation demands it.

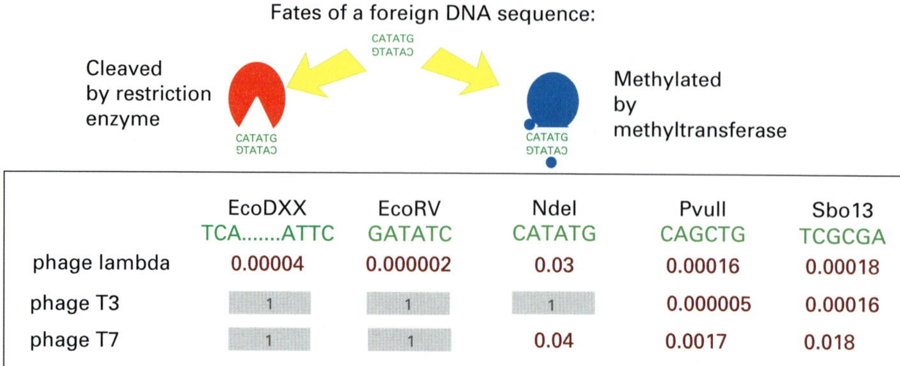

	EcoDXX TCA.......ATTC	EcoRV GATATC	Ndel CATATG	Pvull CAGCTG	Sbo13 TCGCGA
phage lambda	0.00004	0.000002	0.03	0.00016	0.00018
phage T3	1	1	1	0.000005	0.00016
phage T7	1	1	0.04	0.0017	0.018

Figure 12.5 Restriction-modification systems each act on their specific DNA sequences (see second row), with a bias towards cleavage if they are un-methylated. As a result they do not provide perfect immunity. The table shows the probability of three different unmethylated phages successfully infecting an *E. coli* cell that contains one of five different restriction-modification systems [626]. Noticeably, when the phage has successfully completed one lysis cycle, its descendants will be methylated on potential cleavage sites and can therefore complete the next lysis cycle on the same host type with probability close to 1 [626]. The gray shading refers to lack of any immunity [626]. There are at least 200 different recognition sequences among known RM systems [627].

12.1.2.1 Restriction-modification (RM) systems

One way a bacterium can reduce predation by non-specific phages is to use RM systems [628, 629]. Each such system consists of two enzymatic functions that both recognize and act on a certain DNA sequence (see Fig. 12.5). One of the enzymes cleaves the sequence if it is not methylated, whereas the other enzyme adds methyl groups to the sequence. Thereby new and non-methylated DNA from an invading phage will be preferentially cleaved. This provides immunity to infection, as quantified by the examples in Figure 12.5. Noticeably, the immunities of RM systems are not perfect, and phages that survive the first infection cycle[2] will be methylated and therefore able to bypass the restriction enzymes next time it infects the bacterium [626, 628, 630].[3]

12.1.2.2 CRISPR

(**C**lustered **R**egularly **I**nterspaced **S**hort **P**alindromic **R**epeats) A widespread[4] defense mechanism for bacteria is the so-called CRISPR immune system [631, 632, 633, 634], consisting of a group of proteins that can "capture" short DNA sequences of an invading phage and integrate them into the host genome. Subsequently the CRISPR system can

---

[2] Dependent on relative speed of methylation versus DNA cleavage, some foreign DNA may not be cleaved before it is methylated.

[3] Phages can also mutate or evolve to lose restriction recognition sites, as, for example, T7 lacks EcoRV sites.

[4] CRISPR systems are found in about half of all species of bacteria, and in nearly all of the thermophiles [625].

recognize new infections by phages bearing that sequence, and abort their infections. CRISPR systems appear Lamarckian in nature, as the host evolves to deal with the "threat" when exposed to it [635, 636, 637]. However, the Lamarckian adaptation comes with its own imperfection: although the hosts "learn" on the fast timescale of one infection cycle, the chance that a foreign DNA sequence is integrated into the genome of a surviving bacterium is small, estimated to be of the order 10^{-6} [634]. However, when carrying a matching spacer, the corresponding phage will only have 10^{-5} probability of successfully infecting a bacterium [632].

Noticeably, CRISPR only eliminates phages that the host has experienced earlier in its history. Therefore one should expect CRISPR to be co-evolving in an ongoing arms race with the local phage population; an arms race that will make a bacterium with CRISPR more likely to be immune to phages in its own environments than to foreign phages [639, 640].

12.1.2.3 Toxin–antitoxin (TA) systems

Recently, there has been evidence for phage exclusion by TA systems [643, 644, 645]. Such systems are found in most prokaryotes. *E. coli*, for example, has 11 such systems, and *Mycobacterium tuberculosis* contains about 90 TA systems [127]. In nearly all cases, TA systems act through a long-lived toxin that in its free form inhibits cell growth, but which can be inhibited by binding to a short-lived anti-toxin [127]. One possible role for this type of innate immunity could be, for example, to act against a phage that prevents transcription of bacterial genes. The phage would then indirectly activate the long-lived toxin and thereby prevent further translation of its other proteins. The infecting phage may also stress the cell and thereby activate the protease Lon, which in turn is known to degrade the anti-toxin. Remarkably, phage T7 has a counter-defense against this type of bacterial defense system. T7 can inactivate Lon [645], thereby increasing the lifetime of the anti-toxin to a degree that renders the toxin harmless.

12.1.2.4 Abortive infection

Abortive infection (or phage exclusion) [634, 646] systems typically result in the death of both bacterium and phage. These systems work as particularly strong TA systems or occasionally by coupling to more standard TA systems. A classic phage exclusion system is the RexA:RexB system in phage λ, which causes the death of both bacterium and phage when infected by phage T4rII. In this case RexA:RexB is activated, and the membrane-anchored protein RexB acts as an ion channel that causes a subsequent drop in the membrane potential, and thereby a sharp drop in ATP synthesis [647].

12.1.3 Co-evolved strategies

The billions-of-years-old "war" between phage and bacteria [634, 638, 653, 654] have led to multiple attack, defense, and counter-attack mechanisms between the predators

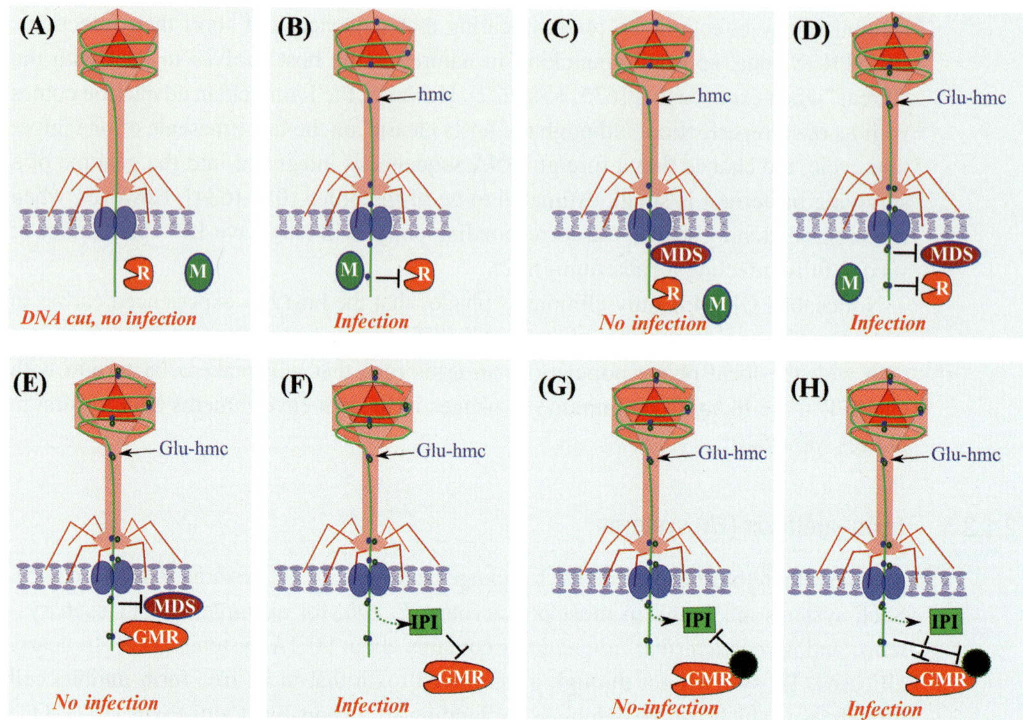

Figure 12.6 Phage T4 infects an *E. coli* that, in the absence of defense, would be sensitive, redrawn from Labrie *et al.* [634]. (A) Phage T4 fails to infect a host containing a restriction-modification system (RM) that cuts phage DNA at specific sites. (B) The genome of T4 contains hydroxymethylcytosine (hmc) and can be methylated as the host DNA. It can therefore avoid the RM system. (C) Some bacteria contain a system that exclusively cleaves hmc-containing DNA, thereby preventing infection. (D) Phage T4 acquired a resistance to MDSs through the glucosylation of hmc residues (glu-hmc). (E) A few *E. coli* cells have aquired the glucose-modified restriction (GMR) system that cleaves glu-hmc-modified phage genomes, thereby blocking infection. (F) Some T4 phages have a protein IPI that hinders the GMR system. (G) Some *E. coli* have a modified GMR system where a translational fusion between some GMR proteins stops the IPI protein from working. (H) Finally, this mechanism is again bypassed by some T4 phage strains. See also review by [638]. Figure redrawn with permission from the publisher.

and their hosts. There are examples of up to eight layers of attack and defense mechanisms for some mutants of T4 that attack *E. coli* (see Fig. 12.6). For example, RM systems can be counteracted by self glucosylation of potential restriction sites on the phage DNA [655], as done by phage T2, T4 and T6 [626, 656]. More aggressively, there are also RM systems that specifically cleave particular sequences of methylated DNA, thereby targeting phages that can methylate themselves or have been methylated in bacteria by standard RM systems [656, 657]. Figure 12.6, in addition, includes examples of defense systems associated with the production of molecules that mask receptors, including production of a polysaccharide coat, which in turn can be degraded by some phages. Remarkably, the phage may also encode parts of CRISPR as an anti-CRISPR strategy, allowing it to bypass this adaptive immunity system [658].

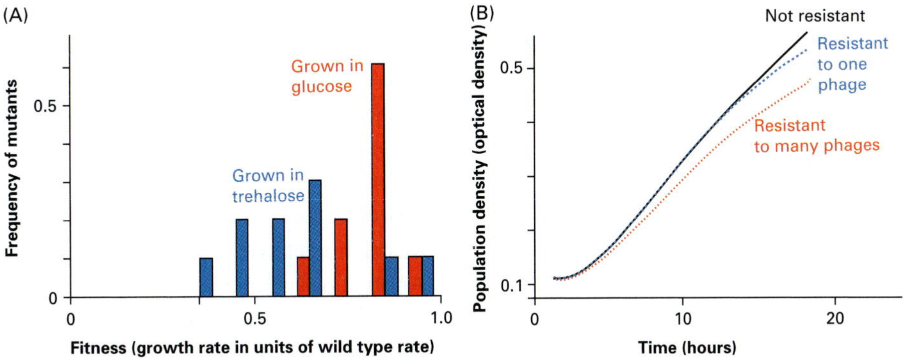

Figure 12.7 Left: Experiments [641] that test the replication rate of bacteria when they evolve immunity to the T4 phage. One sees that there is a mutation-dependent reduction in fitness when the bacteria is grown afterwards without the presence of phages. The reduction in growth rate also appears to depend on the available nutrients. Right panel: Experiment [642] where growth of bacterial strains that are resistant to one, and to many different phages are investigated in the absence of phages. More resistant bacteria grow more slowly, reflecting a sizeable cost in maintaining the acquired immunity.

Rapidly acquired immunity often comes at a substantial cost in growth rate, and thus a fitness cost in environments lacking phages. This has been demonstrated in several laboratory environments [641, 642, 659] (see Fig 12.7). Interestingly, phage predation on bacteria may indirectly kill the winner [660], thereby preventing development of "superbugs." This predation provides negative feedback on any peaks in fitness landscape formed by other bacterial properties. This "fitness landscape flattening" makes more mutations neutral in contemporary environments, even if a priori they lead to less efficient bacteria [661].

Phages interfere with each other using much the same systems as bacteria use in defending against phages (see Table 12.1). In fact, immunity casettes are presumably being transfered between bacteria by phages and other transmittable systems, such as plasmids and episomes (the latter being plasmids that can also integrate into the bacterial chromosome). Many RM systems and TA systems are, for example, encoded in plasmids where they may well support continued propagation by killing the host if it loses the plasmid.[5]

One particularly famous plasmid is the "*F*" factor [662, 663], with *F* standing for fertility because it allows the bacteria to connect and have "sex" by transmitting "*F*," and occasionally additional genetic material from the bacteria. At the same time, "*F*" conveys resistance to some phages, but is needed for entry of other phages [664]. Thus, overall, "*F*" acts as an especially benign phage that provides both advantages and disadvantages for the host, with a gain or cost that depends a lot on the environment.

---

[5] As toxin is long-lived and anti-toxin short-lived, accidentally losing the production of both will effectively leave the bacteria with only the toxin. Therefore, TA systems can prevent the "cheaters" from proliferating and support their continued presence in the gene pool.

Phage	Phage target	Mechanism
λ	T4 mutants	RexA,B induce suicide of *E. coli*
P4	P2	P4 lyse a P2 lysogen and "steal" its head [648, 649]
P2	λ	P2 prevent λ from entering
P2	T4, T5, T7	P2 prevent infection
P1	"Others"	Restriction enzymes + recognize methylated DNA
P7	"Others"	Restriction enzymes + recognize methylated DNA
933W	HK97	Abort superinfection [650]
P1,P10	"Others"	Proteins prevent superinfection [651, 652]
λ^{434}	P22	Hex exclude superinfection of P22
HK022	λ	NUN of HK022 prevent N anti-termination in λ
T4	T2, T6	T4 encodes Imm that block DNA injection.
T4	T6	T4 derepress T6 (and also 30 other phages)
T6	T2	Weak prevention of infection?
RB69	T2, T4, T6	RB69 prevent infection

Table 12.1 Anecdotal phage interactions, with the phages in the first column influencing the phages in the second column. P4 is particularly specialized, acting as a parasite of P2. Some phages are very similar, e.g. the lambdoid phages that include λ, phi80, 434, HK97, 21, 82, 933W, HK022 and H19-B. Another is the P2 group, which includes phage 186, HP1, L-413C and K139 (infecting *Vibrio cholerae*), PSP3 (*Salmonella*) and ϕCTX (*Pseudomonas aeruginosa*). P2 infects several types of host, including *E. coli, Shigella, Serratia, Kelbsiella* and *Yersinia* (I. Dodd, private communication). Yet another phage, P22, is known for its ability to pack alternative DNA segments upon lysis. Among virulent phages, the T4 group is particularly large and includes RB69, RB49, Aeh1 and 44RR2.8t.

Closely related to *F* are the family of "*r*" episomes, which carry multiple drug-resistant genes [665].

12.1.4   Phage–bacteria networks

An ecological characterization of phage–bacteria networks remains elusive, with the best characterized network dating back to a journey across the Atlantic in 1981 [666]. This network was recently rediscovered and analyzed extensively by Flores *et al.* [667]. The full network includes 215 phage types and 286 host strains with 1332 interactions (see Fig. 12.8). Remarkably, phages and bacteria were found to interact more when they were found relatively close to each other in the ocean [666]. Quantitatively a phage infects hosts from its own location with probability 16%, whereas it infects hosts from locations 100–2000 km away with a probability that declines from 4% to $\sim$2%. One interpretation of this observation is that phage propagating on susceptible hosts increases the local phage population and thereby makes it more likely to see phages when their hosts are present. Consequently, bacteria appear to be more resistant to phages they have not experienced before, than to phages in their closer surroundings, an overall trend that is also corroborated by [669, 670].

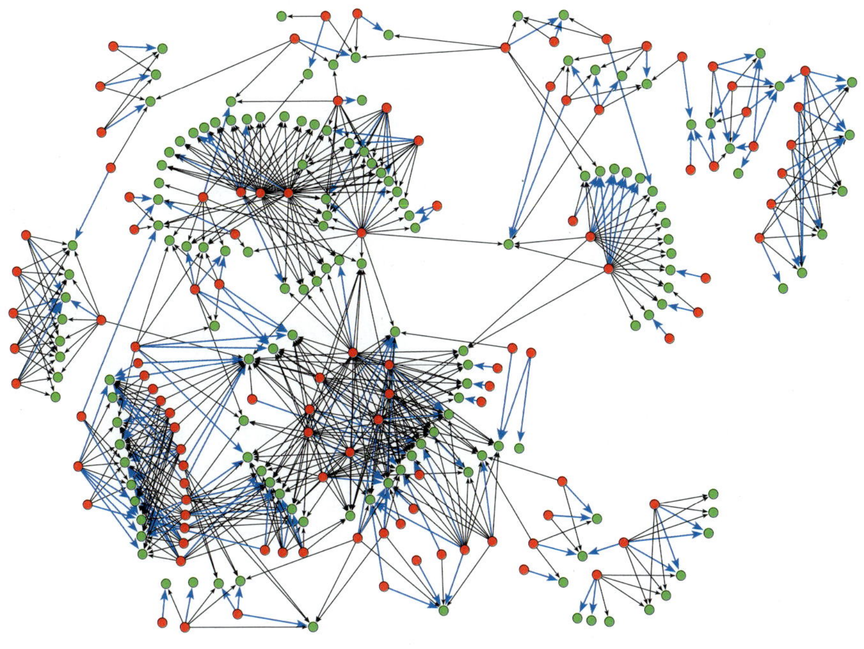

Figure 12.8 The largest component of the phage–bacteria infection network [666, 667] collected from seawater at positions in the Atlantic. Phages are red nodes and bacteria are green. Blue links refer to bacteria and phages collected at the same position and black links to cases where they are from positions that are separated by $> 100\,km$. On average, each phage infects 6.2 bacteria in the set, whereas each bacteria is infected by 4.7 phage types. The most promiscuous phages infect 31 bacteria species. Nearly all sampled bacteria that were sensitive to at least one phage were found to be in the *Vibrio* genus [668], a gram-negative proteobacteria that is in same superfamily as *E. coli*. The visualization here is done in collaboration with S. Maslov.

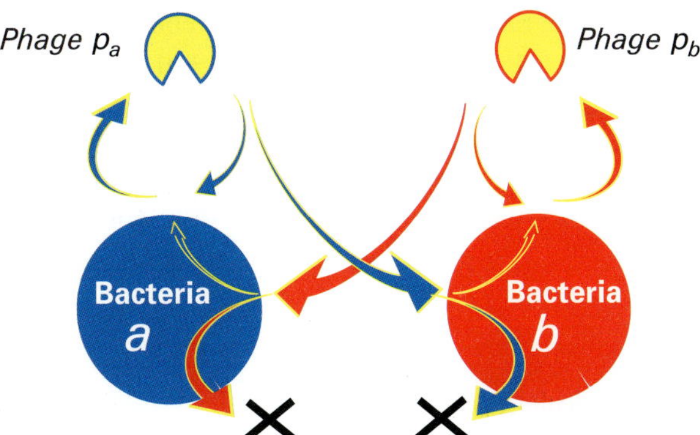

Figure 12.9 Figure for Question 12.1.3, illustrating the battle between two strains of bacteria, each with their restriction-modification system. When phages lyse one host, they infect hosts of the same strain without sensing any immunity. When these phages instead infect the opposing strain, then the chance of success is reduced by a factor ω_a and ω_b respectively.

12.1.5   Questions

12.1.1   Consider a restriction-modification system in an *E. coli* where the restriction enzyme cuts a certain sequence with efficiency $1\,\mathrm{min}^{-1}$, whereas the counteracting methylation happens $0.5\,\mathrm{min}^{-1}$. Consider a foreign phage that infects the bacteria, and which has four sites that can be cut/methylated. What is the probability δ that the phage DNA survives intact to be methylated at all four sites?

12.1.2   Repeat the above calculation with the invading phage expressing a protein that can prevent the restriction enzyme working within 1 minute.

12.1.3   Repeat Question 12.1.1 and 12.1.2 with the invading phage having eight DNA sites that can be cut/methylated.

12.1.4   Consider the idealized equations for two bacterial species a and b each with their restriction-modification system. The bacteria compete while exposed to one phage that acquires methylation from only the last bacteria it has successfully infected (modified from [671]):

$$\frac{\mathrm{d}a}{\mathrm{d}t} = a \cdot (1 - a - b) - a \cdot (p_a + \omega_a \cdot p_b) \tag{12.1}$$

$$\frac{\mathrm{d}p_a}{\mathrm{d}t} = a \cdot (p_a + p_b \cdot \omega_a) \cdot \beta - p_a \cdot (a + b + 1/\tau) \tag{12.2}$$

$$\frac{\mathrm{d}b}{\mathrm{d}t} = b \cdot (1 - a - b) - b \cdot (p_b + \omega_b \cdot p_a) \tag{12.3}$$

$$\frac{\mathrm{d}p_b}{\mathrm{d}t} = b \cdot (p_b + p_a \cdot \omega_b) \cdot \beta - p_b \cdot (a + b + 1/\tau) \tag{12.4}$$

where p_a and p_b are the population of phages that have been last methylated in a, and b, respectively β is the number of phage particles released per infection and τ is the lifetime of phages in the environment. Simulate the system for $\tau = 10$, $\beta = 100$ and relative deficiency of restriction-modification systems $\omega_a = 1$ and $\omega_b = 0.1$, respectively. Discuss the result in terms of bacteria that use phages to compete with each other. Hint: start with $a = 0.1$, $p_a = 1$ and fixed $b = p_b = 0$, and simulate until steady state using timesteps $\mathrm{d}t = 0.001$. Then introduce $b = 0.000001$ and simulate again until steady state.

12.2   Phage–bacteria ecology

12.2.1   Sustainability of a single virulent phage

Phages govern recycling of bacterial biomass in the global ecosystem, forming multiple predatory relationships with their prey, of the form outlined in Fig. 12.10. The equations in this figure are idealized and stripped of parameters. These parameters are experimentally accessible [673] and tabulated for some phages in Fig. 12.11. In terms of these

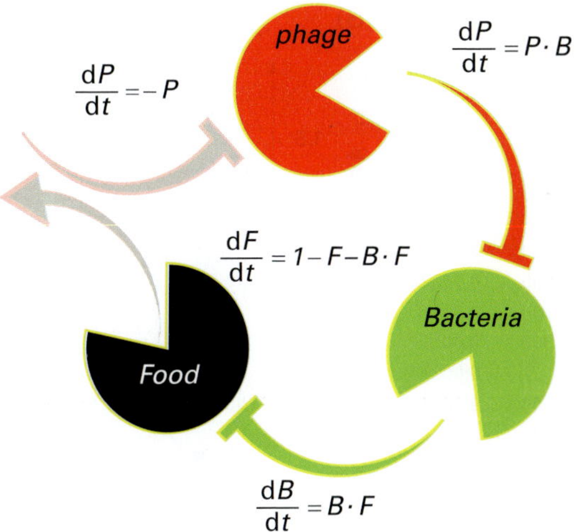

Figure 12.10 Idealized predator–prey model of phage–bacterial ecosystem, emphasizing an analogy to the rock–paper–scissors game suggested by Kerr *et al.* [598]. The analogy is not perfect, because food/available space does not demand the presence of other food for "growing." Similarly, a gray inhibition link marks that phages die spontaneously if only surrounded by food or free space. Notice that the food F can be eliminated by the steady-state assumption $\mathrm{d}F/\mathrm{d}t = 0 \Rightarrow F = 1/(1 + B)$, giving $\mathrm{d}B/\mathrm{d}t = B/(1 + B)$. This form of bacterial growth also holds when including the Monod growth equation [100] $\mathrm{d}B/\mathrm{d}t = B \cdot F/(1 + F) = B/(2 + B)$ with a limit "2" that depends on both nutrients and bacterial physiology. In the text we simplify further and use $\mathrm{d}B/\mathrm{d}t = B \cdot (1 - B)$, where B is measured in units of nutrient-dependent carrying capacity.

parameters and the maximum bacterial growth rate Λ_B, the dynamics of a bacterial population b and a virulent predator phage p can be described by a predator–prey model [619, 674]:

$$\frac{\mathrm{d}b}{\mathrm{d}t'} = \Lambda_B \cdot b(t) \cdot \left(1 - \frac{b(t)}{b_{\max}}\right) - \eta' \cdot b(t) \cdot p(t)$$

$$\frac{\mathrm{d}p}{\mathrm{d}t'} = \beta\eta \cdot b(t - \tau') \cdot p(t - \tau') - \delta' \cdot p(t) \tag{12.5}$$

where the bacterial growth itself is modeled by the Verhulst equation [675].[6] These equations express that the growth rate of the phage p depends on how often it encounters a bacterium b, the latency time τ' and the burst size β. In the absence of phages the bacteria grow exponentially until they reach the steady-state carrying capacity $b_{\max}$, a critical density that depends on the available nutrients. The last equation includes a time delay τ' that parameterizes the time it takes an infected bacterium to lyse. τ' does not influence population size in the steady state and therefore in the following we ignore it.

---

[6] The growth curve of an *E. coli* population of size b approximately follows the Verhulst equation $\mathrm{d}b/\mathrm{d}t = \Lambda \cdot b \cdot (1 - b/b_{\max})$ [675] giving $b(t) = b_{\max} \cdot b_0 e^{\Lambda t}/((b_{\max} - b_0) + b_0 e^{\Lambda t})$ where $b_{\max}$ is the carrying capacity and $b_0 \ll b_{\max}$ is the bacterial density at time $t = 0$ [675].

(A)

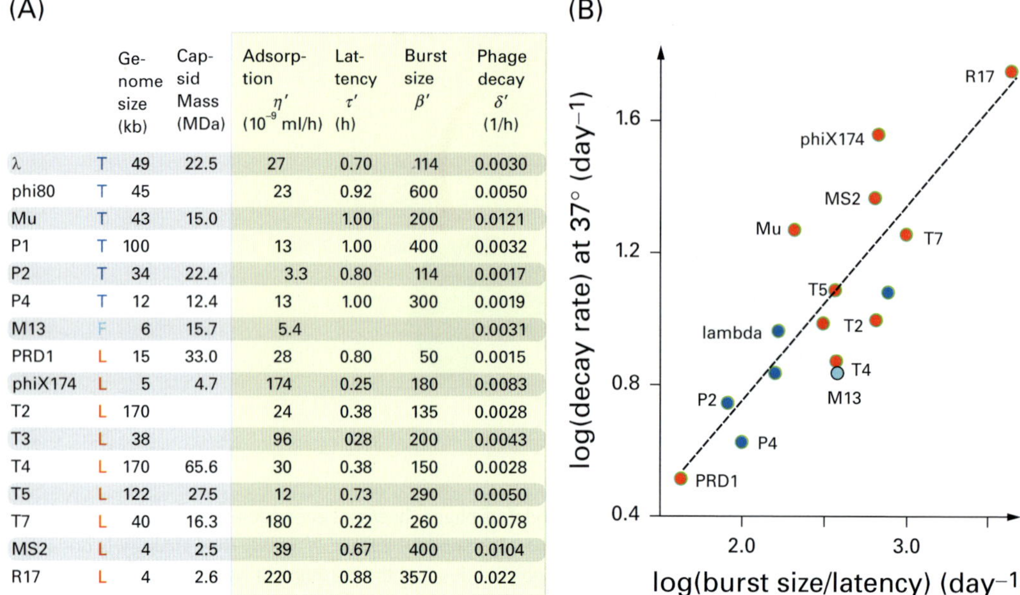

		Ge-nome size (kb)	Cap-sid Mass (MDa)	Adsorp-tion η' (10^{-9} ml/h)	Lat-tency τ' (h)	Burst size β'	Phage decay δ' (1/h)
λ	T	49	22.5	27	0.70	114	0.0030
phi80	T	45		23	0.92	600	0.0050
Mu	T	43	15.0		1.00	200	0.0121
P1	T	100		13	1.00	400	0.0032
P2	T	34	22.4	3.3	0.80	114	0.0017
P4	T	12	12.4	13	1.00	300	0.0019
M13	F	6	15.7	5.4			0.0031
PRD1	L	15	33.0	28	0.80	50	0.0015
phiX174	L	5	4.7	174	0.25	180	0.0083
T2	L	170		24	0.38	135	0.0028
T3	L	38		96	028	200	0.0043
T4	L	170	65.6	30	0.38	150	0.0028
T5	L	122	27.5	12	0.73	290	0.0050
T7	L	40	16.3	180	0.22	260	0.0078
MS2	L	4	2.5	39	0.67	400	0.0104
R17	L	4	2.6	220	0.88	3570	0.022

(B)

Figure 12.11 Measured properties of phages infecting *E. coli* [673] in rich medium (LB) at 37 °C. T refers to temperate phages, L to lytic (virulent) phages and F to filamentous phages that give chronic infection. M13 and R17 are RNA phages. The latency period is the time from infection to phage release. The capsid mass may be put in perspective by the 2.5 MDa mass of one ribosome, whereas 1 kb of DNA has a mass of 0.7 MDa. The right-hand panel [673] shows that the decay of free phages increases with their invasion rate raised to exponent 0.6. The four columns in the yellow box are parameters for predator–prey modeling, with η' being the rate that one phage binds to receptors at the bacterial surface. For comparison, the diffusion-limited on rate for the T4 phage with diffusion constant $D = 4\,\mu m^2\,s^{-1}$ [676] (from Eq. (5.18)) to a bacterium of size 1 μm is $\eta' = 4\pi \cdot D \cdot 1\,\mu m = 0.05 \cdot 10^{-9}\,\text{ml s}^{-1} = 180 \cdot 10^{-9}\,\text{ml h}^{-1}$.

The equations (12.5) can be rescaled by defining $B = b/b_{max}$, $P = p/b_{max}$, $\eta = \eta' \cdot b_{max}/\Lambda_B$, $\delta = \delta'/\Lambda_B$ and $t = t' \cdot \Lambda_B$. Here b_{max} is the maximum bacterial density in the given environment, and is in fact hugely dependent on the available food.[7] The rescaled equations read [619]:

$$\frac{dB}{dt} = B \cdot (1 - B) - \eta \cdot B \cdot P$$

$$\frac{dP}{dt} = \beta \cdot \eta \cdot B \cdot P - \delta \cdot P \tag{12.6}$$

where B and P now measure population sizes in units of bacterial carrying capacity b_{max} and t counts time in units of $1/\Lambda_B$. These equations can also be approximated from the phage–bacteria–food interplay outlined in Fig. 12.10, with the main lesson being that

[7] The nature of b_{max} may be more easily understood from a near equivalent formulation of the above growth equation: $db/dt = b/(b_{max} + b) - 0.5 \cdot (b/b_{max}) - \eta' \cdot p \cdot b$, an equation that follows from limited flow of food, as shown in Fig. 12.10, supplemented with external dilution. In practice $b_{max} \sim 3 \cdot 10^9\,\text{ml}^{-1}$ for *E. coli* when grown in an initially nutrient-rich growth medium. b_{max} will be much smaller in the nutrient-poor oceans.

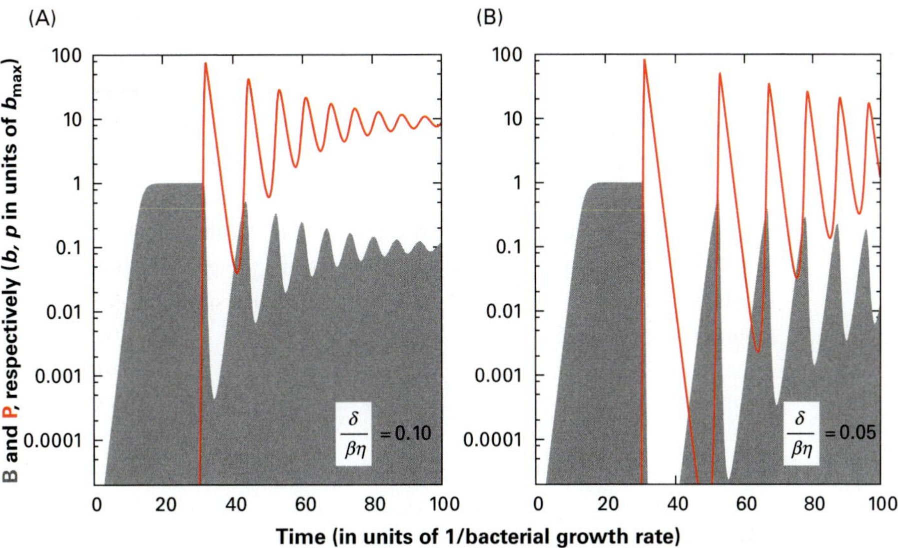

Figure 12.12 Simulation of a phage–bacteria ecosystem with one susceptible bacterial species, using Eqs. (12.6). The simulation is initialized with $B = 10^{-6}$, $P = 0$, and the phage is introduced at $t = 30$ units with $P = 10^{-6}$. (A) Infection rate $\eta = 0.1$, phage death $\delta = 1$ and burst size $\beta = 100$. Notice that system approaches a steady-state ratio between phage and bacteria $P/B = \beta/\delta = 100$. (B) Larger infection rate $\eta = 0.2$, $\delta = 1$ and $\beta = 100$, giving an identical steady-state ratio (P/B) as (A) but with a factor of two smaller populations. Population variations become huge as more efficient phages are introduced, implying that extinctions may occur in real samples. Notice that the simulations use $\eta << 1$ whereas the estimated $\eta > 10$ (in which case the growth speed cannot exceed the limit set by $\eta = 1$).

b_{max} in fact represents a limit set by both bacterial physiology and by the amount of food in the environment. Eqs. (12.6) are simulated towards the steady-state solution in Fig. 12.12, whereas the more extended equations are analysed in Question 12.2.4.

Apart from the saturation term, the above equations are identical to the famous Lotka–Volterra equations for predator–prey systems [672, 677]. The equations are simplified, as they, for example, ignore that the phage burst from bacteria close to starvation is small [678, 679, 680], a feature that has been explored theoretically in [681, 683].[8] Also, the equations ignore that cell-to-cell variation in surface receptors might be high (variation in η), thereby allowing some bacteria to evade predation by having low expression of particular receptors.

Equations (12.6) predict steady-state populations:

$$B = \frac{\delta}{\beta\eta}; \quad P = \frac{1}{\eta} \cdot \left(1 - \frac{\delta}{\beta\eta}\right) \tag{12.7}$$

---

[8] Saturation of phage growth can be implemented by assuming that the phage growth is proportional to bacterial growth when this is slowed down as $B \to 1$ [681]. More realistically, one may also include that the phage burst size is reduced [683], $\frac{dB}{dt} = B \cdot (1-B) - \eta \cdot B \cdot P \cdot (1-a_b B)$ and $\frac{dP}{dt} = \beta \cdot \eta \cdot B \cdot P - \delta \cdot P \cdot (1-a_p B)$. These modified equations still allow for extended co-existence [683], provided that a_p is not to high.

measured in units of the carrying capacity for the bacteria. Thus, the number $\frac{\delta}{\beta\eta}$ is a fitness measure of the phage, a small value implying a phage that predates hard on its prey. Noticeably when two phage types rely on the same bacterial host, the phage with the smallest value of $\frac{\delta}{\eta\beta}$ eliminates the opponent.

To translate these results into real ecosystem numbers, consider the case of phage T4 living on one bacterial species in the ocean. The rates in Eq. (12.7) are then measured in units of the typical prokaryotic recycling rate in the oceans $\Lambda_B \sim 1/(10\,\text{days}) = 0.004\,\text{h}^{-1}$ [129]. Consider δ', η' and β for the T4 phage as tabulated in Fig. 12.11. The tabulated $\delta' = 0.003\,\text{h}^{-1}$ for T4 from Fig 12.11 corresponds to $\delta = \delta'/\Lambda_B = 0.75$. With a burst size of $\beta = 150$:[9]

$$\frac{B}{P} \sim \frac{\delta}{\beta} = \frac{\delta'}{\Lambda_B \cdot \beta} \sim 0.005 \tag{12.8}$$

which is smaller than the $B{:}P$ ratio of 0.05 to 0.1 that is observed in oceans [133, 134]. However, considering a phage decay in oceans of 0.04–0.11 per hour [682], the estimated $B{:}P$ ratio in fact becomes close to the measured factor of 0.1.

Noticeably, the numbers of both bacteria and phages decline with phage predation efficiency η, according to:[10]

$$b = B \cdot b_{\max} = \frac{\delta}{\beta\eta} \cdot b_{\max} = \frac{\delta'}{\beta \cdot \eta'} = \frac{0.1\,\text{h}^{-1}}{150 \cdot 30 \cdot 10^{-9}\,\text{ml h}^{-1}} = 20\,000\,\text{ml}^{-1} \tag{12.9}$$

which is smaller than the $\sim 500\,000$ prokaryotes ml^{-1} that typically is measured in the upper layers of the oceans [129]. This discrepancy might in part be due to poorer nutrients in the oceans causing the bacteria to have fewer receptors and thereby a lower η' than in the rich growth medium used in Fig. 12.11. Also, different bacteria can each be predated by each their phage type, thereby adding to B total without changing B/p. Bacteria can, in addition, be supported by heterogeneous growth conditions, including biofilm [684], that may well increase the chances that bacteria survive dramatic transients [172] such as the ones illustrated in Fig. 12.12.

The main message is that a single bacterial strain can co-exist with its virulent predator, adjusting itself to such a low level that the phage can barely propagate itself (resembling a self-organized near-critical threshold of existence). The predicted bacterial population will be much smaller than the phage population by a factor:

$$\frac{B}{P} \sim \frac{\delta}{\beta} \sim \frac{\text{(generation time of one host cell)}}{\text{(burst size)} \cdot \text{(life time of phage particle)}} \tag{12.10}$$

which is a relation that states the counter-intuitive result that the bacteria gains by growing more slowly. At the same time the bacterial biomass is independent of bacterial

---

[9] $B/P = \delta\eta/(\beta\eta - \delta)$, where δ is typically small compared to the rate that one phage from a lysis burst finds a bacterium at carrying capacity. For simplicity we ignore this small correction in the text.

[10] The data from Fig. 12.12 will give $\eta \gg 1$ in most environments: for bacteria in oceans with $b_{\max} \sim 10^7\,\text{ml}^{-1}$, $\Lambda_B = (1/10)\,\text{day}^{-1}$ the T4 phage have rescaled $\eta = \eta' \cdot b_{\max}/\Lambda_B = 7.2 \cdot 10^{-6}\,\text{ml} \cdot b_{\max} > 1$ for $b_{\max} > 10^5$. In LB medium with a maximum bacterial density of $b_{\max} = 3 \cdot 10^9\,\text{ml}^{-1}$ and growth rate $\Lambda_B = 1.5\,\text{h}^{-1}$, then $\eta = (30 \cdot 10^{-9}\,\text{ml h}^{-1}) \cdot (3 \cdot 10^9\,\text{ml}^{-1})/(1.5\,\text{h}^{-1}) = 60$.

growth rate, and a factor $B = \delta/(\beta\eta)$ smaller than the carrying capacity of the bacteria in contemporary environments.

12.2.2 Sustainablity of phages with resistant hosts

Importantly, in real ecosystems, one bacterial species exists together with many others. In fact, if the bacterial host B is with another bacterial species/strain R that is resistant to the phage [619], the total bacterial biomass changes dramatically. The dynamics can again be modeled in terms of simplified equations:

$$\frac{dB}{dt} = B \cdot (1 - (B + R)) - \alpha \cdot B - B \cdot \eta P$$
$$\frac{dR}{dt} = g \cdot R \cdot (1 - (B + R)) - \alpha \cdot R$$
$$\frac{dP}{dt} = \beta \cdot B \cdot \eta P - \delta \cdot P \tag{12.11}$$

Here g quantifies the growth rate of R in units of the growth rate of B, whereas α is the dilution of bacteria due to external factors. If grown in a chemostat, as in Fig. 12.13, α would be the fractional dilution rate, and δ in this case would be the sum of this dilution rate and the natural decay from Fig. 12.11. α could also include dilution by external eukaryotic predators.

Without phages, the fast-growing bacteria eliminates the slow-growing one [619]. Therefore the phage can only exist if it shifts this balance, i.e. it has to prey on the faster-growing bacteria [619].[11] For $\frac{\alpha}{1-\delta/\beta\eta} < g < 1$, there is a stable steady-state co-existence with:

$$B = \frac{\delta}{\beta \cdot \eta}, \quad R = 1 - \frac{\alpha}{g} - \frac{\delta}{\beta \cdot \eta}, \quad \eta P = \alpha \cdot \left(\frac{1}{g} - 1\right) \tag{12.12}$$

Here the faster-growing bacteria B is insensitive to the slower-growing one, and is in fact fixed at exactly the same density as it would have without the slow but resistant strain. Remarkably, the phage population declines as g increases, reflecting the implicit antagonism of R towards P, thereby again forming a cyclic "rock–paper–scissors like" relation $P \to B \to R \to P$ (see Fig. 12.14). Using B:P ratio from Fig. 12.13,[12] suggest that η' for the conditions of the Levin experiment may be a factor of 10 smaller than η' from Fig. 12.11 or that $g \approx 0.9$.

---

[11] This statement relies on the assumption that both bacteria are exposed to the same dilution α. If dilution/external killing of the slower-growing bacteria is much slower, it may win. Furthermore, if the bacteria, in part, also grow on different food sources, then they may constrain each other in asymmetric ways. This may allow a slow-growing species to grow and co-exist at nutrient conditions where the fast-growing species has already reached saturation [688]. In the text we simplify our discussion to co-existence of metabolically equivalent bacteria.

[12] Consider the experiment shown in Fig. 12.13, with a T2 phage infecting a host with carrying capacity $2 \cdot 10^8$. Dilution in the chemostat sets a decay rate $\delta = \alpha = 0.2$ in units of bacterial growth rate. Thus $B/P \sim \delta/(\beta \cdot \alpha) \cdot g/(1 - g) \sim g/(\beta \cdot (1 - g)) \sim (1/135) \sim 0.01$, apart from a growth penalty factor $g/(1 - g)$. Thus P:B should be about $100 \cdot (\frac{1}{g} - 1)$ whereas the final P:B ~ 10 in Fig. 12.13, consistent with $g \simeq 0.9$, or a growth penalty of $\sim 10\%$.

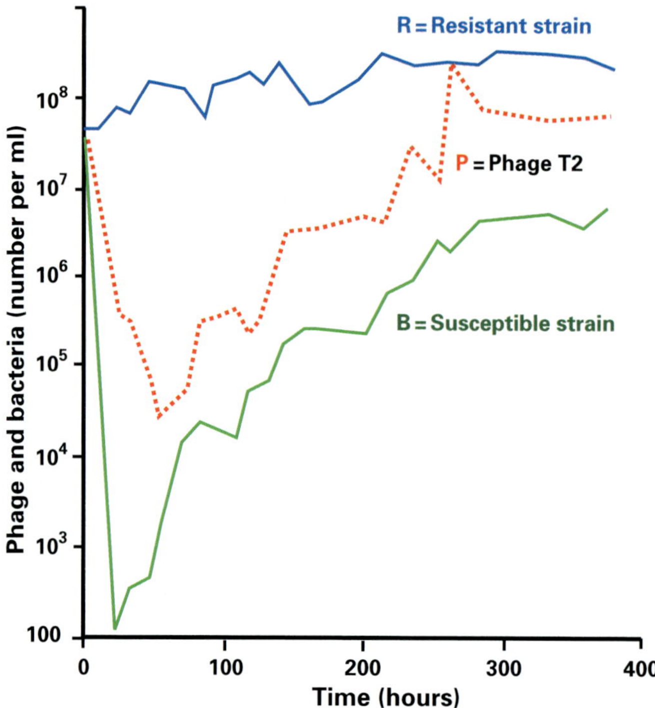

Figure 12.13 Time development in an experiment by Levin *et al.* [674], containing T2 phages and two strains of bacteria, one that is susceptible (B) and another that is resistant (R) to the phage (P). The experiment is conducted in a solution that is continuously diluted with a fractional rate of $0.27\,h^{-1}$. Co-existence is obtained, as predicted, when the resistant strain is the slow grower [619, 674].

Ecosystems with one phage and one or two bacterial strains are examples of sustainable microbial relationships. In fact, one may consider ecosystems with many phage strains M and many bacterial strains N, generalizing Eqs. (12.11) into $M + N$ equations that are then coupled through phages that prey on one or more of the bacterial strains [685, 686, 687]. Because the phage predators do not interact directly, these systems fulfill the comparative exclusion principle, stating that the predator diversity M has to be less than or equal to the prey N [685], a statement that can be proven by considering the steady-state solution in terms of an $N + M$ dimensional matrix equation, of which the lower right M product $\times$ M matrix is all zero (phages do not interact) [687]. Furthermore, for bacterial strains that are limited by the same resource, one can prove that a stable steady state requires that $M = N - 1$ or $M = N$ [687].

The coupling between N and M reflects that phages prevent the fastest-growing bacteria completely taking over food sources. Instead, the presence of phages transforms the ecosystem into an alternating arms race between bacteria aiming at high growth rates and phages sustained by large values of $\frac{\beta\eta}{\delta}$. This competition also exhibits a very narrow window of sustainability for large diversity, a window that, however, can be stabilized to large values of M and N if one considers spatially separated niches.

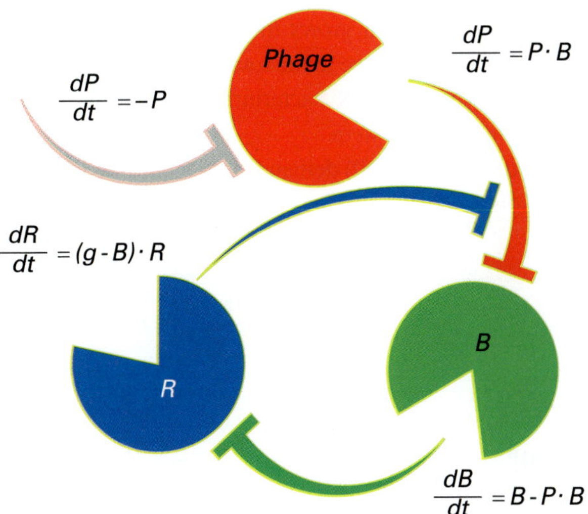

Figure 12.14 A cyclic relation between phages, their susceptible host and the resistant bacteria. In the text we show that the phage population is reduced as the growth of R is increased, although the population of B is insensitive to this protection. However, R may be immune instead of resistant, in which case the phage population may suffer a loss due to non-production adsorption in to R. The resulting B population then increases to $B = \delta/\beta\eta + \mathrm{const}/\beta$ where the constant is of the order of const ~ 1.

12.2.3 Partial resistance in a virulent world

An important aspect of the two-bacteria system above is that there often remains a tiny subpopulation of one host that has immunity to the infecting phage [689]. Such immunity comes at a cost in the bacterial growth rate, as quantified in Fig. 12.7. Without phages, these mutant bacteria only exist transiently and their existence therefore depends on an ongoing supply, by mutations from the faster-growing bacterial strain. With phages, however, the two bacterial strains do indeed co-exist, with the slow grower occupying the major part of the biomass in the system (see Fig. 12.13).

In ecosystems there are many phage and bacterial species [637, 666]. Immunity or resistance[13] to any particular phage therefore only provides a small contribution to the overall immunity, when viewed at the scale of all potential phages. We therefore reconsider Eqs. (12.11) in terms of partially resistant strain R which is resistant to a fraction $r \in [0,1]$ of the phages and therefore has an overall reduced adsorption $\eta^1 = (1 - r) \cdot \eta$ [640, 690]:

$$\frac{dB}{dt} = B \cdot (1 - (B + R)) - \alpha \cdot B - B \cdot \eta P$$

$$\frac{dR}{dt} = g \cdot R \cdot (1 - (B + R)) - \alpha \cdot R - (1 - r) \cdot R \cdot \eta P$$

$$\frac{dP}{dt} = \beta \cdot B \cdot \eta P + \beta \cdot (1 - r) \cdot R \cdot \eta P - \delta \cdot P \tag{12.13}$$

[13] Commonly, resistance is associated with lack of a surface receptor for the phage, whereas immmunity is associated with prevention of phage development once it has entered the bacterium.

R may co-exist with B provided that it grows more slowly than the fully susceptible strain B. $r = 1$ implies complete resistance to the total population of all phages, whereas $r = 1/2$ reflects a situation where each R bacterium has resistance against 1/2 of the phage strains in this well-mixed ecosystem.

Coexistence of B, R and P ($R, B, P > 0$) demands that the following three steady-state equations have positive solutions:[14]

$$B + R = 1 - \alpha - \eta P$$

$$B + R = 1 - \frac{\alpha}{g} - \frac{(1 - r)}{g} \cdot \eta P \tag{12.14}$$

$$B + (1 - r) \cdot R = \frac{\delta}{\beta \eta}$$

giving the solutions [640]:

$$\eta P = \frac{\alpha \cdot (1 - g)}{g + r - 1} \tag{12.15}$$

$$R = \frac{1}{r} \cdot \left(1 - \alpha - \frac{\delta}{\beta \eta} - \frac{\alpha \cdot (1 - g)}{g + r - 1} \right)$$

and $B = 1 - \alpha - R - \eta P$. In particular this means that $g - \alpha > (1 - \alpha) \cdot (1 - r)$, a lower growth speed limit, that is explicitly drawn as a black line in Fig. 12.15.

The co-existence region is shown in Fig. 12.15, where the colors highlight which of the three species is closest to extinction: adjacent to the minimal growth of the partially resistant strain, one may notice R is the least populated species. In contrast, for high resistance and growth of R, the bacterial strain B is under threat.

Figure 12.15 teaches us that a slower-growing bacteria may out-compete a fast grower provided that it has $\eta^1 > 0$ and $\eta^1 < \eta$. Thereby it sacrifices part of its population to eliminate its competitor. Figure 12.15 contains another noteworthy lesson. Imagine a bacterial species B that can obtain resistance against a phage. Given that there presumably are multiple phage types in any ecological system, the obtained r for the mutant that we coin R, would then be slightly larger than $r \sim 0$. At such small differences in resistance beween R and B, there is essentially no cost value that allows for co-existence of the ancestor and the mutant phage. The predation of phages therefore favor increasing immunity of bacteria accompanied by extinction of any less-resistant ancestor (moving nearly horizontally away from the $r = 0$ ($\eta^1 = \eta$) and $g = 1$ strategy in Fig. 12.15).

Gradual Speciation that also maintains the parental $r = 0$ strain is almost impossible, a problem reminiscent of what Darwin faced when trying to explain the divergence of closely related species in his theory of evolution. As with Darwin and Wallace [691], we also turn to a central feature that we have ignored until now: that development of diversity of immunization is driven by spatially localized adaptation [639].

---

[14] Solving $dB/dt = 0 \Rightarrow B(1 - B) - \alpha B - B \eta P = 0$, which can be divided by B, provided that we assume that $B > 0$. A similar approach is used for the other equations.

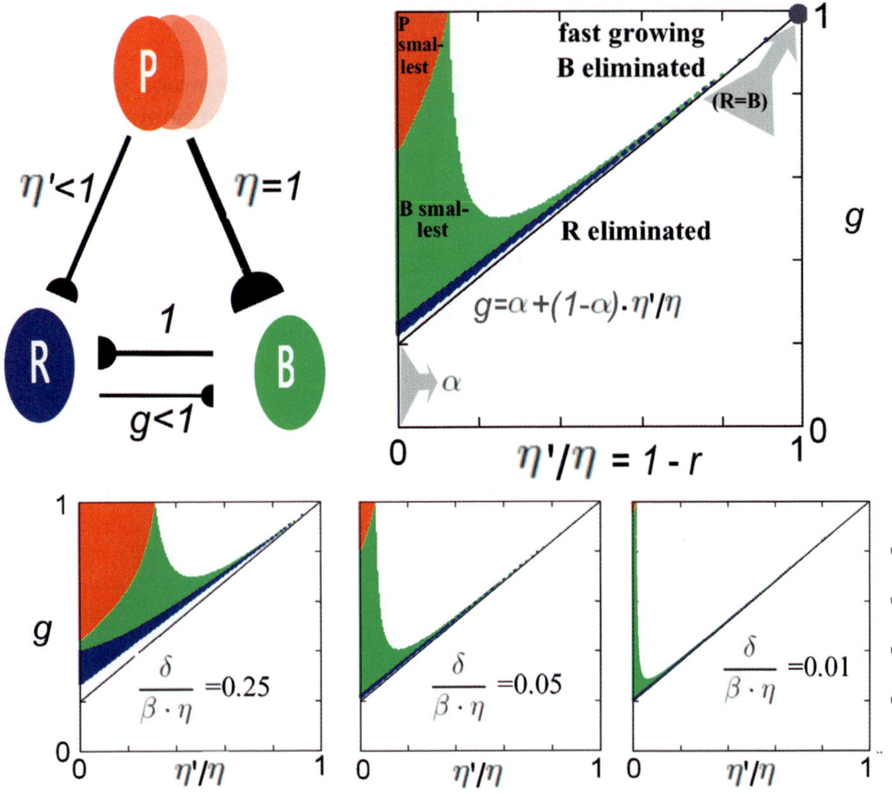

Figure 12.15 Co-existence plot [640] between a fully susceptible bacterial strain B with growth rate 1, and a partially resistant strain R with reduced growth rate. The upper left-hand panel shows the phage as a series of circles, representing the many species of which a fraction $1 - r\,(= \eta^1/\eta)$ may infect the partially resistant strain R. The upper right-hand panel shows the steady-state co-existence region, with $\alpha = 0.2$ and $\delta/\beta\eta = 0.1$. The blue parameter region is where R is close to extinction, and the green region is where B is close to extinction. The lower panels examine more efficient phages, $\delta/\beta\eta \to 0$, emphasizing that the co-existence region shrinks to a narrow band close to $r \sim 1$ or equivalent by $\eta^1 = 0$. This then resembles the B–R competition from Eq. (12.11).

12.2.4　CRISPR and spatially distributed immunity

Consider a bacterial species B that can acquire immunity to phages through incorporation of spacers using a CRISPR-like system. In a well-mixed system there will be numerous phage species, and presumably any acquired immunity will come at some cost to growth rate. Thus, the growth rate g will decrease with increasing number of immunities against phage species (r).

From Fig. 12.15 we conclude that in order that an $r > 0$ mutant can co-exist with its susceptible ancestor ($g = 1, r = 0$), it requires that the mutant acquires immunity to a fair fraction of all phages ($r \sim 1$). Co-existence is difficult when the immune strain is only partially immune. In real ecosystems there are both bacteria without CRISPR, and

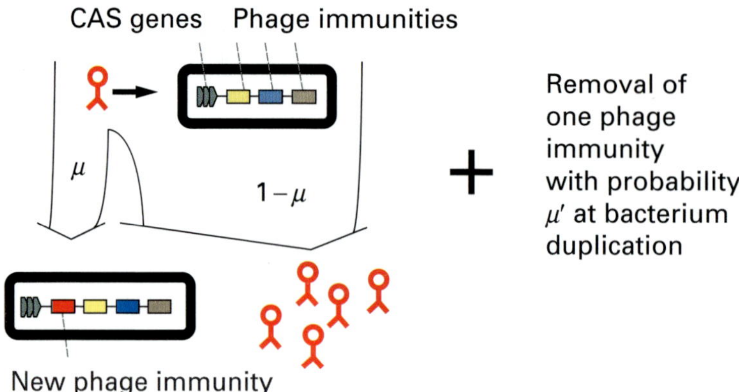

Figure 12.16 A model for the evolution of a system of phages and bacteria that may modify their immunities using a CRISPR-like system. Phages are allowed to mutate at a low rate (say 0.0001), whereas a bacterium that is infected can gain instant immunity with rate $\mu = 0.05$. If they have no immunity, and did not evolve it during infection, the bacterial colony is eliminated and replaced by P, which subsequently can infect neighboring sites. Bacteria can also lose immunity at a self-organized rate that is optimized by evolution, μ', here assuming that it costs 5% in growth for each immunity a bacterium has [640]. Notice that this model relies entirely on growth detriment, whereas another mechanism may be genetic instability, where occasionally large segments of especially repeated DNA are lost.

bacteria with a number of spacers that is much smaller than the estimated diversity of phages [692, 693].

In contrast, it is feasible for a bacterium to obtain near total immunity against a *local* phage population while at the same time maintaining co-existence with its susceptible ancestor. This can be explored by use of an agent-based model where small bacterial colonies are allowed to proliferate and spread on a square lattice. Bacteria are allowed to acquire or lose immunity to phages, and different bacterial strains will therefore have different sets of immunities, and grow with rates that decrease with the number of phage types they are immune to. Phages are taken into account by allowing lattice sites to be occupied by either a healthy bacterial colony (B), or by a colony that is infected by a phage of one particular type (P_i).

The system is simulated in discrete stochastic steps, each consisting of either a bacterial duplication event or a phage replication event, with rates set by the bacterial growth rates and the phage infection rate, respectively. For bacterial growth, a duplicate of a bacterial colony at a given site invades and replaces whatever was occupying one of its four neighboring sites.[15]

For phage spreading, a phage-infected colony replicates and attempts infection of all of the four nearest sites on the square lattice. The fate of this infection attempt is set by the immune status of the corresponding bacterial colonies, and whether they

---

[15] Thus bacteria can replace and out-compete another bacterial strain. Without phage predation, the fastest-growing but less-immune bacteria will recapture the dominant biomass in the system.

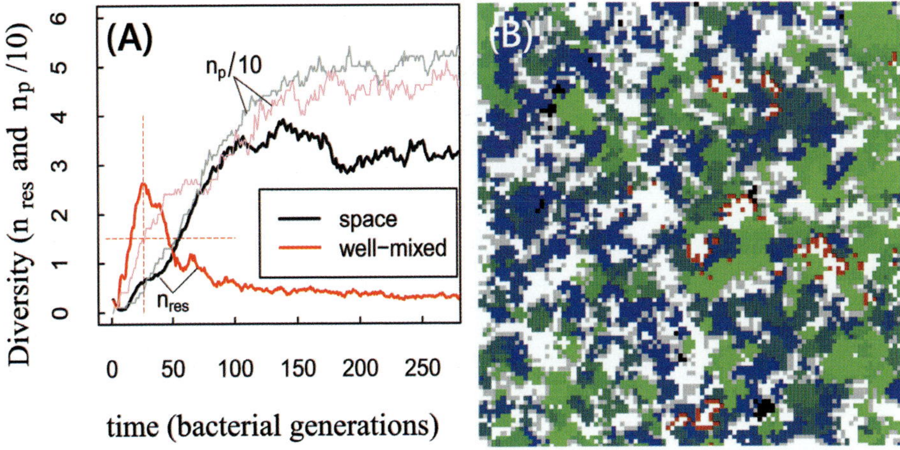

Figure 12.17 (A) Evolving system of phages and bacteria with parameters from Fig. 12.16. The time dependence has a unit where, on average, each site has been updated once. The curves in (A) refer to the number of different phages (n_p) and the average number of immunities per bacterium (n_{res}), respectively. (B) Phage–bacteria ecology emphasizing the localization of one particular "red" phage, which in fact has to move in order to stay alive, mimicking a red queen race [694] against the acquired immunity of its bacterial hosts [640]. Blue or black sites represent bacteria with low n_{res}, whereas light green shows colonies with high n_{res}.

acquire immunity by CRISPR during the infection,[16] as described in Fig. 12.16. After the infection attempt, the previously infected site is left vacant, and any phage invading an empty site is lost. The random death of phages is simulated by selecting a site randomly and, if it contains a phage-infected colony, eliminating it with probability $\propto \delta$.

In addition to the ongoing invasions, the bacteria are allowed to mutate at some low rate to gain or lose immunity against specific phages. Similarly, the phages are allowed to mutate into new strains that are not yet known by any bacteria in the system.

Figure 12.17A from [640] shows the development of the number of spacers carried by the bacteria, n_{res}, and the number of distinct phage species in the system, n_p. One notices that n_{res} initially increases, but eventually saturates at a value between 40 and 50, for both the spatial and the well-mixed system. However, the bacterial response to the growing n_p number is different: While the spatially structured bacteria acquire more and more spacers until settling at about 3–4, the well-mixed bacteria only acquire more spacers until n_p reaches 15 and then decay to a value close to zero. This is explained by the pay-off of incorporating a spacer: in a well-mixed system the gain is proportional to $1/n_p$, as the probability of encountering the same phage species again decreases with the number of species. In contrast, in the spatially structured system, the chances are much higher and nearly independent of n_p, as any phage species at any given instant remains fairly localized. Noticeably, with time, phage species are constantly moving, always predating on new bacteria that have not yet gained resistance to them.

---

[16] As a bacterial colony is assumed to be of size 10^5 and CRISPR is spontaneously integrated with probability 10^{-6} bacteria, one may assume that between 0.05 to 0.1 of the infection attempts lead to incorporation of a spacer.

12.2.5   Questions

12.2.1   A more complete set of equations for a bacteria–phage ecosystem, where bacteria and phages are also removed by external sources, is:

$$\frac{\mathrm{d}b}{\mathrm{d}t} = \Lambda_B \cdot \frac{b}{1 + b/b_{\max}} - \eta' \cdot b \cdot p - \delta'_b \cdot b \qquad (12.16)$$

$$\frac{\mathrm{d}p}{\mathrm{d}t} = \beta \cdot \eta' \cdot b \cdot p - \delta' \cdot p$$

Use a saturation concentration of bacteria $b_{\max} \sim 10^8\,\mathrm{ml}^{-1}$, and a maximum growth rate $\Lambda_B \sim 0.0004\,\mathrm{s}^{-1}$. Estimate η' from diffusion-limited localization of a phage with $D \sim 4\,\mu\mathrm{m}^2\,\mathrm{s}^{-1}$ to a bacterium with radius $\sim 1\,\mu\mathrm{m}$. Reduce the number of parameters by rescaling $t \to t \cdot \Lambda_B$ etc.

12.2.2   Simulate the Lotka–Volterra equations: $\mathrm{d}b/\mathrm{d}t = b - b \cdot p$ and $\mathrm{d}p/\mathrm{d}t = \beta \cdot b \cdot p - p$, starting with different initial conditions, $p = 3, b = 1$ respectively $p = 2, b = 1$.

12.2.3   Compare the standard Lotka–Volterra equation with two versions that include saturation terms on the prey population: (1) $\mathrm{d}b/\mathrm{d}t = b/(1 + b) - b \cdot p$, $\mathrm{d}p/\mathrm{d}t = \beta \cdot b \cdot p - p$ and (2) $\mathrm{d}b/\mathrm{d}t = b \cdot (1 - b) - b \cdot p$, $\mathrm{d}p/\mathrm{d}t = \beta \cdot b \cdot p - p$. In all cases start the simulation with $p = 3, b = 1$.

12.2.4   The time delay due to latency can, in principle, be implemented as a third state in Eqs. 12.17, which in rescaled units then translates to

$$\frac{\mathrm{d}B}{\mathrm{d}t} = \frac{B}{1 + B} - \frac{1}{2}B - \eta \cdot B \cdot P$$

$$\frac{\mathrm{d}I}{\mathrm{d}t} = \eta \cdot B \cdot P - I \qquad (12.17)$$

$$\frac{\mathrm{d}P}{\mathrm{d}t} = \beta \cdot I - \delta \cdot P - \eta \cdot I \cdot P$$

where we implicitly parameterize the latency time with an I state with lifetime $\tau = 1$. Simulate these equations with $\eta = 0.1$, $\beta = 100$ and death rate $\delta = 1$. Start the simulation with $B = 10^{-6}$ and $P = 0$ at time $t = 0$, and introduce phage $P = 10^{-6}$ at time $t = 30$.

12.2.5   Bacterial species may limit their own growth more than others, because they will partly depend on different food resources. Such inhibition may be modeled through:

$$\frac{\mathrm{d}B}{\mathrm{d}t} = \frac{B}{0.5 + B} \cdot (1 - B - R) - 0.8 \cdot B$$

$$\frac{\mathrm{d}R}{\mathrm{d}t} = \frac{0.2 \cdot R}{0.5 + R} \cdot (1 - B - R) - 0.1 \cdot R \qquad (12.18)$$

Solve these equations numerically by starting both populations at size 0.000 001 and follow them until time $t = 200$. Notice that stable co-existence is indeed possible, and notice that the slow grower in the end dominates the population.

12.3 Phage games: lysogeny or lysis

12.3.1 Phages in varying environments

Phages live in high-risk environments [619, 622, 695], and local populations of individual phage species should routinely experience extreme fluctuations caused by changes in availability of hosts that can support lytic development [696, 697]. In real ecosystems these fluctuations may be caused by the lack of nutrients for hosts, development of host resistance, and interference from competing phages or other bacterial predators, for example [598, 619, 674, 696, 698, 699]. As an alternative, the temperate phages can opt for transition to the lysogenic state where the future fate of the incorporated prophage is aligned with its host bacterium [135, 700]; a strategy that is always accompanied by an exit possibility, either by being sensitive to DNA damage, like phage λ, or by a sizeable spontaneous lysis frequency (of the order of 10^{-3} per generation for P2 or VHSI [701]).[17] Noticeably, bacteria can also sometimes be cured of a lysogen, an event that for some phage–host systems happens relatively frequently [701, 702].

The lysis–lysogeny decision of temperate phages can be studied using game theory [703]. Imagine one single phage entering one host cell. It is faced with a classic decision: (a) should it go lytic, and thereby make 100 new copies of itself, or (b) should it enter lysogeny and couple its fate to the future fate of the bacteria. We approach this decision from two angles, first arguing that the phage has to behave stochastically to minimize the chance of extinction in the long run. Second we will use optimization on a population level to calculate the optimal frequency of lysogeny.[18] The first approach serves to emphasize the power of collective optimization and the altruistic sacrifice that individual phage particles typically show when becoming lysogenic. The second approach in fact allows a quantitative calculation, as well as developing an understanding of what abilities a phage needs in order to be persistently virulent.

12.3.2 Why phages should play dice

Most simply, one may score the success of the single phage taking a decision in its host as the probability that one copy of it survives to the next infection round. This

---

[17] Phage λ also has a spontaneous exit from lysis of about 10^{-5} per generation associated with random activation of the RecA pathway.

[18] The presented analysis finds parallels in more general considerations which emphasize mixed strategies as a natural evolutionary response to stochastic/unpredictable environments [704, 705].

probability may be written:

$$P_{\text{single}} = (1 - x) \cdot p_{\text{lysis}} + x \cdot p_{\text{lysogeny}} \tag{12.19}$$

where p_{lysis} is the probability that the phage survives and allows at least one of its descendants to enter another host if it chooses the lytic pathway. This probability will, in practice, be determined by the probability that the density of hosts is below some critical value. For the T4 phage this critical density of host bacteria is estimated to be $7000\,\text{ml}^{-1}$ [696].

In the above equation, p_{lysogeny} is the probability that an offspring of the phage survives to enter a new bacterium sometime in the far future. x parameterizes the strategy space for the phage in terms of its encoded probability for choosing lysogeny. This strategy is under the control of the phage, by regulating the activity of deciding genes (for phage λ this regulation in particular goes through the protein CII).

The above strategy space is easy to optimize: For any value of p_{lysis} and p_{lysogeny} one simply chooses the strategy that has the largest overall P_{single}. For $p_{\text{lysogeny}} > p_{\text{lysis}}$, $x = 1$ (never lysis), whereas $p_{\text{lysis}} > p_{\text{lysogeny}}$ set to $x = 0$ is the optimal strategy. So why do phages not always decide deterministically for one or the other pathway, or at least always decide deterministically a path for a given environmental input?

Importantly, a phage is not alone. It is produced in a burst with many identical siblings, quantified by a burst size $\beta \gg 1$, as tabulated in Fig. 12.11. The long-time average of phage growth in a real ecosystem only allows an average of one of these to be sufficiently successful to complete the cycle to produce a new burst. Consider the situation where a phage has to take the lysis–lysogeny decision, based on the information that there is another copy of it that takes a parallel decision somewhere else. Success for the phage is now given by the probability that at least one phage copy survives. If the two events are uncorrelated, then this can be calculated as a simple product:

$$P = 1 - (1 - P_{\text{single}})^2 \tag{12.20}$$

and the optimal strategy still remains deterministic.

If the two survival probabilities in the two phage–host situations are not independent, then the calculation changes fundamentally. In particular, since the lysis decision for two sibling phages most likely occurs in a common environment, the two phages share the event that there are no hosts left in this environment. In contrast, the fate of the lysogeny choice relies on the fate of a bacterium and its encounters with other bacterial predators. Therefore, the probability that both phages become extinct is estimated as:

$$
\begin{aligned}
1 - P = {} & (1 - x)^2 \cdot \left(1 - p_{\text{lysis}}\right) \\
& + 2(1 - x) \cdot x \cdot \left(1 - p_{\text{lysis}}\right) \cdot \left(1 - p_{\text{lysogeny}}\right) \\
& + x^2 \cdot \left(1 - p_{\text{lysogeny}}\right)^2
\end{aligned}
\tag{12.21}
$$

Here it is explicitly assumed that the fate of the lysis of one phage is equal to the fate of the lysis of the other phage in the other bacterium [703]. Accordingly $1 - p_{\text{lysis}}$ is not squared in any term of the equation.

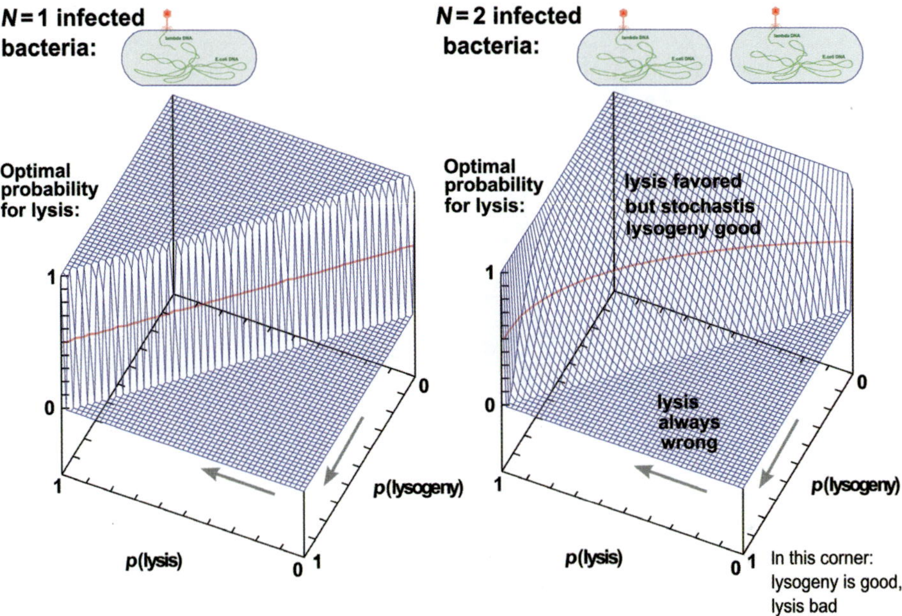

Figure 12.18 Optimal strategy for survival in terms of probability for lysogeny, as a function of probability of success for lysis, and probability of success as a lysogen. If only one game is played, the phage should deterministically optimize the choice. If there are two phages, there will be better survival if both strategies are played, a choice that has a high likelihood of happening if each phage chooses its path randomly.

For each value of p_{lysis} and p_{lyosgeny} one may calculate $x \in [0, 1]$ that minimizes $1 - P$. Figure 12.18 shows this optimal strategy, which is indeed stochastic for all conditions where lysis is, in principle, the best deterministic choice. This stochastic bet-hedging, based on collective optimization, is illustrated in Fig. 12.19. In the following section we extend this analysis by taking into account the exponential growth associated with successful lysis events.

12.3.3   The well-tempered phage

We concluded that the lysogenic strategy could be understood in light of survivablity on a population scale, which as a whole hedges its bets against an uncertain future by a small "investment" in lysogens. The size of this "investment" is a theoretical problem, presumably reflecting optimization of the long-term growth rate in much the same way as gamblers on horse races may optmize their portfolio [706]. In the context of biological evolution, game theory for diversifying phenotypes in varying environments has been discussed by Bergstrøm and Lachman [705], whereas bacterial hedging in persistently changing environments has been analyzed by Kussel and Leibler [707, 708].

It is plausible that naturally occurring temperate phages have evolved a probability of entering lysogeny that optimizes their long-term growth. Reference [709] considered a population of temperate phages exposed to a sequence of risky events. The population

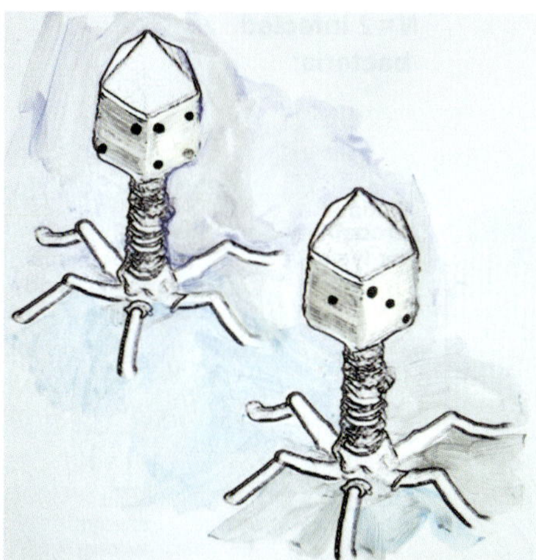

Figure 12.19 Artistic interpretation (courtesy of Mette Høst) of phages playing dice, and the advantages of doing this when more than one phage is present. In fact, under good growth conditions a λ-phage behaves deterministically when alone and stochastically when together with other copies of the same phage in an infected bacterium [150, 153, 700].

is assumed to grow or decline, depending on an environment that randomly switches between "good" conditions that favor lytic growth and bad conditions where the lytic population collapses. Rapid growth during "good" conditions is quantified by the amplification factor $\Omega > 1$. Bad conditions are treated as transient events of indefinite length that occur with probability $p \ll 1$. The temperate phages have no control over their environment, but are free to choose what fraction x of their population is preserved as lysogens.[19]

Assume for simplicity that the lysogens are fully protected from changes in the environment and therefore do not change with time. In contrast the lytic subpopulation is invested in a "risky asset," subject to substantial and unpredictable multiplicative fluctuations. The expected value of the (logarithmic) growth rate Λ of the entire phage population is then given by the mean trajectory with an average of $1 - p$ good events and p bad events per time unit.[20] Therefore the average long-term growth rate is:

$$\Lambda(x) = (1 - p) \cdot \log(\Omega(1 - x) + x) + p \cdot \log(x) \qquad (12.22)$$

The first term is the logarithmic growth rate under good conditions, where the lytic fraction $1 - x$ is multiplied by Ω, while the lysogenic subpopulation x remains unchanged.

---

[19] Phages only seem to be free to choose between lysis and lysogeny just after infection. In good periods the phage population is assumed to grow, $\Omega \gg 1$, and accordingly the major faction are active and our assumption of correct partition at infection will be close to the history integrated partitioning.

[20] The optimization of growth over all possible histories of duration t would be obtained by maximizing $N(t) = N(0) \cdot \sum_{b=0}^{t} C(t,b) \cdot p^b \cdot (1-p)^{t-b}((1-x)\Omega + x)^{t-b}x^b$, with $C(t,b)$ being the number of ways b bad events can be distributed among t total events. Optimization of all historical possibilities favor a purely lytic strategy, since a singularly rare lucky strike of t "good" times contributes enormously. In contrast, the maximization of a typical trajectory has an expected number of bad events $b = p \cdot t$. This is associated with optimization of $N(t) \sim N(0) \cdot \left(p^p \cdot (1-p)^{1-p}((1-x)\Omega + x)^{1-p}x^p \right)^t \propto e^{t\Lambda(x)}$ with respect to x.

The second term is the logarithmic growth rate under bad conditions, when only lysogens survive.

We emphasize that the growth rate weights the *logarithms* of multiplicative growth factors of the entire phage population, under two conditions, with their respective probabilities of occurrence. Maximization of Λ with respect to x secures the long-term optimal growth rate [706]. This should not be confused with optimization of the expected population growth after only one growth cycle. Such a short-term average would always favor a purely lytic strategy with $x = 0$ provided that $(1 - p)\Omega > 1$. The last condition is almost always fulfilled during good times, since phages have big lytic bursts ($\beta >> 1$). On the other hand, following a purely lytic strategy for a long time would almost certainly bring the population to total collapse.

In contrast to its short-term counterpart, the long-term logarithmic growth rate $\Lambda(x)$ usually reaches its maximum at some x^* between 0 and 1. In the economics literature this is denoted the Kelly optimal-investment ratio [706]. It describes the optimal fraction of capital that a prudent long-term investor should keep in relatively safe financial assets such as bonds, while investing the rest in more risky assets such as stocks [710]. Here x^* corresponds to the optimal fraction of the phage population in the lysogenic state. At the Kelly optimum the derivative should be zero:

$$\frac{d\Lambda(x)}{dx}\Big|_{x^*} = -(1 - p) \cdot \frac{\Omega - 1}{\Omega(1 - x^*) + x^*} + p \cdot \frac{1}{x^*} = 0 \Rightarrow$$

$$x^* = p \cdot \frac{\Omega}{\Omega - 1} \tag{12.23}$$

This equation assumes that the lytic phage population is completely eliminated during one bad event. This can be relaxed, instead assuming that a small but finite multiplicative ratio $\omega < 1$ quantifies the total decay of the lytic population during an entire bad event. The logarithmic growth rate of the phage population is then:

$$\Lambda(x) = (1 - p) \cdot \log(\Omega(1 - x) + x) + p \cdot \log(\omega(1 - x) + x) \tag{12.24}$$

and the Kelly optimal lysogenic fraction is:

$$x^* = p \cdot \frac{\Omega}{\Omega - 1} - (1 - p) \cdot \frac{\omega}{1 - \omega} \tag{12.25}$$

an optimum that can be examined in Fig. 12.20.

The important new result is the finite threshold for transition between purely lytic (virulent) and temperate strategies of phages. Assuming that $\Omega \gg 1, p \ll 1$ and $\omega \ll 1$, then:

$$x^* = p - \omega \tag{12.26}$$

implying that virulence "wins" when the likelihood of disaster p is smaller than the probability of surviving the disaster ω:

$$p < \omega \tag{12.27}$$

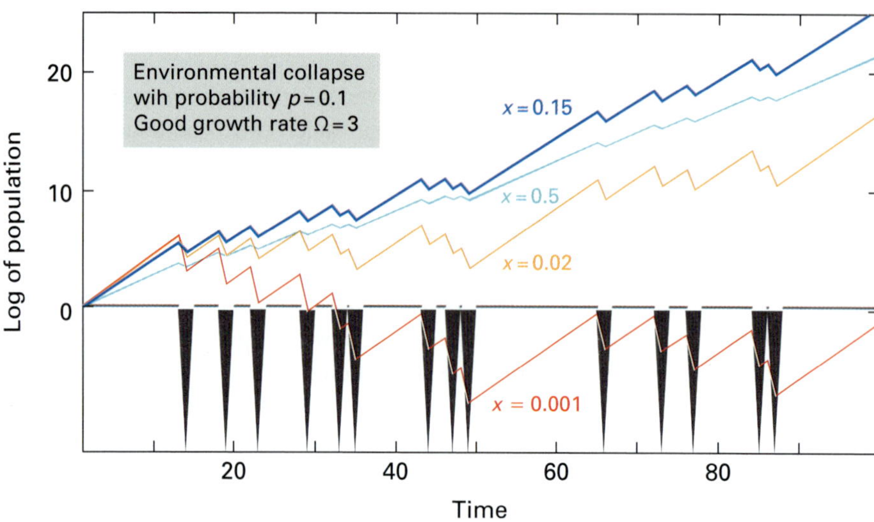

Figure 12.20 Dynamics of a phage population when growing under long periods of good conditions, interrupted by events of probability $p = 0.1$, where the lytic sub-population drops by a factor $\omega = 10^{-12}$. Under good growth conditions, the lytic population is amplified by a factor $\Omega = 3$ per timestep. The blue curve is the growth of a well-tempered phage following the Kelly optimal strategy with lysogenic fraction $x^* = p \cdot \Omega/(\Omega - 1) = 0.15$, whereas the orange and red curves show suboptimal strategies with $x = 0.01$ and $x = 0.001$. Conversely, the cyan trajectory simulates an over-cautious phage with $x = 0.5$.

If, on the other hand, the growth advantage of lytic over lysogenic states under good conditions, Ω, approaches the growth rate of lysogens (here set to 1) then it is best to allocate a progressively larger portion of the population to the safety of lysogens. For $p \cdot \Omega/(\Omega - 1) \geq 1$ phages should abandon the possibility of lytic growth and instead transfer their genomes to plasmids or to the bacterial genome.

In between these extremes, for a moderate likelihood of bad times p, the temperate strategy wins. The well-tempered phage is the one where lysogenization frequency during good times is close to the likelihood of occurrence of bad times within the duration of one lytic burst cycle. Figure 12.21 shows the Kelly optimal lysogenic fraction x^* as a function of probability p and severity ω of population collapse during bad times.

The choice between virulent and temperate strategies also depends on the ability of phages to explore independent, spatially separated environments by diffusion, reminiscent of portfolio management through independently fluctuating assets [711]. When the number of uncorrelated spatial locations that phages can access becomes large, the minimum growth rate becomes bounded from below. This limits the effect of local disasters, which in turn favors the lytic strategy. Distinct boundaries between different environments may also support continued supply of susceptible hosts, and thereby support a sustainable virulent strategy [684].

Overall, it is expected that virulent phages should have high diversity in host types, or be able to diffuse between many independent environments as a phage particle before

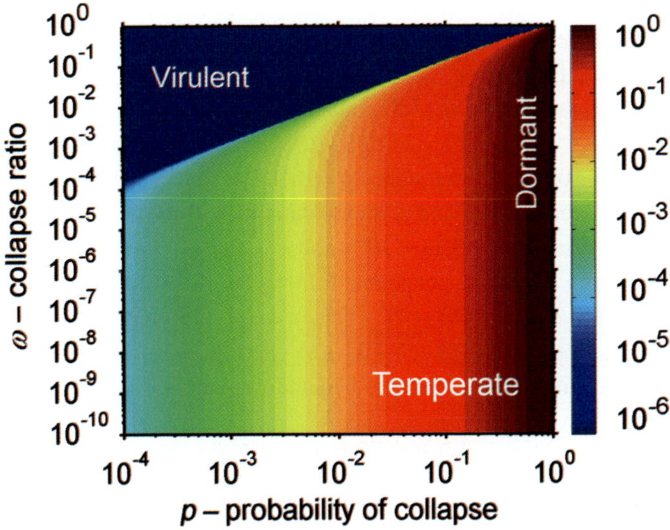

Figure 12.21 Kelly optimal lysogenic ratio (x^* from Eq. (12.25)) as a function of probability p and severity (ratio ω) of population collapse during bad times. Note the sharp boundary separating purely virulent (the blue region in the upper left-hand corner) and temperate states at $p\Omega/(\Omega - 1) = (1 - p)\omega/(1 - \omega)$. Another sharp boundary separating temperate ($0 < x^* < 1$) and dormant ($x^* = 1$) states is less visible in this plot. Here we used $\Omega = 3$.

they lose their ability to infect. Said in another way, given that a phage is lysogenic, its other abilities would constrain a virulent mutant of the phage to be less fit. The predicted fitness change of a virulent mutant will be given by the likelihood of disasters p and their severity ω:

$$\Delta\Lambda \sim p \cdot \log(\omega/p) \tag{12.28}$$

relative to its temperate ancestor (for an explanation see Question 12.3.4). The prediction of non-viabilty of virulent mutants of temperate phages is in agreement with the observation that proteins in virulent phages have essentially no functional overlap with proteins in temperate phages [615] (see Fig. 12.2).

12.3.4 Questions

12.3.1 Compare the two phage, two host scenario explored in Eq. (12.21) for the case where there are three phages, each in their host. Plot the optimal strategy as a function of p_{lysis} and p_{lysogeny} when these two vary between 0.02 and 0.98. Comment on the limitation of the used life/death approximation from the perspective of the prediction that the probability of choosing lysis decreases with increasing phage number.

12.3.2 Estimate the phage decay rate, given that a burst of $\beta = 100$ phages on average allows just one phage to find a host when the density of hosts is $7000\,\text{ml}^{-1}$ [696]. Assume a diffusion rate of phages of $4\,\mu\text{m}^2\,\text{s}^{-1}$ [676]

and assume that a phage needs to come within 1 μm of the center of a bacterium to infect it.

12.3.3 Simulate the long-term (500 updates) development of a phage population that grows with rate $\Omega = 2$ during good times, but is exposed to events of size $\omega = 10^{-12}$ with frequency $p = 0.1$. Use the Kelly optimum value of x, $x = 0.01$ and $x = 0.9$, and compare outcomes. Repeat the simulation for smaller disasters, with $\omega = 10^{-2}$ and $\omega = 0.5$.

12.3.4 Consider a temperate phage that follows the Kelly optimal lysogeny frequency $x^* = p\Omega/(\Omega - 1) - (1 - p)\omega/(1 - \omega)$ with its associated fitness $\Lambda(x) = (1 - p) \cdot \log(\Omega \cdot (1 - x) + x) + p \cdot \log(\omega(1 - x) + x)$. Show that its fitness changes by mutating to become virulent is:

$$\Delta\Lambda = \Lambda(0) - \Lambda(x^*) \sim p \cdot log(\omega/p) \tag{12.29}$$

in the limit where $\Omega \gg 1 \gg \omega$. Interpretation: the fitness loss at relatively frequent disasters $p > \omega$ comes from treating the probability of disasters as if it was as small as the severity of disasters.

12.4　Speciation and biodiversity of sessile species

Biological organisms co-operate or compete with each other and thereby form complicated ecological systems with an intriguing ability to sustain themselves over an incredibly long timespan. This stability is not easily understandable, as exponential growth with competition is unstable. For example, from Eqs. (12.11), we have already seen that two bacterial species that compete for the same resource lead to extinction of the slower grower (provided that they depend on the same resources). Ecosystems with even more species are similarly, in theory, expected to collapse into a state with low diversity D. In a seminal paper R. May pinpointed that this instability is reduced by reducing the numbers of interactions in the ecosystem [712]. Such reduced competition might be obtained by relatively weak[21] interactions between species [713] organized in a constrained and nested ecological network [714, 715]. Another way to reduce both interactions and exponential growth is spatial segregation [596, 598, 609, 690, 716, 717].

Many ecosystems not only have many competing species, but these species may form niches for each other, exemplified by the so-called keystone species [718, 719, 720]. A keystone species may, for example, be a predator that indirectly secures the existence of several other species in the system. A canonical example is the sea otter along the coast of the western USA, which eats sea-urchins and thereby indirectly prevents over-predation of sea vegetation [721].

---

[21] In population equations like $db_1/dt = b_1(1 - c_{11}b_1 - c_{12}b_2)$ a self-interaction c_{11} that is much larger than the interaction term c_{12} effectively allows each species to exist in spite of big populations of other species.

Figure 12.22 Lichen, as observed on a stone in the Italian Alps. The scale across the picture is about 30cm. The different lichen species are separated by sharp boundaries, defining a patchwork ecology of these non-motile species.

As a scenario for self-organized niche formation, consider competing lichen species on the two-dimensional surface of a rock (Fig. 12.22). One can observe a pattern of mutually exclusive zones of different lichen species. The photographed stone is entirely covered by these species, leaving a system that only develops further if some lichens have the ability to invade others. In fact, such invasions can be identified in the picture, in terms of fronts, where the species behind one front "bulges" into the area of another species. The ecosystem of such a surface of rock may serve as a model system for competing sessile (non-motile) species with complex invasion patterns, forming a directed network of opportunities that does not leave any particular species as the most fit or superior. Instead the species form cyclic relationships, as also observed in, for example, coral reef ecosystems [608].

The dynamics are modeled in terms of an agent-based model [717, 722], where each site i can, at most, be occupied by one individual. The model further investigates an open system, in the sense that new species can arrive or evolve with a rate α. The limit $\alpha \to 0$ is interesting, as it represents a clear timescale separation between population dynamics and evolutionary dynamics.

The central parameter is the likelihood that one species will interact with another specific species. This probability is given by the parameter γ. If $\gamma = 0.10$ then a species i will be assigned the property that it can invade species j with probability $\gamma = 0.10$. However, the invasion can only take place if they are neighbors on the two-dimensional lattice.

The model is defined in terms of discrete updates. At each timestep:

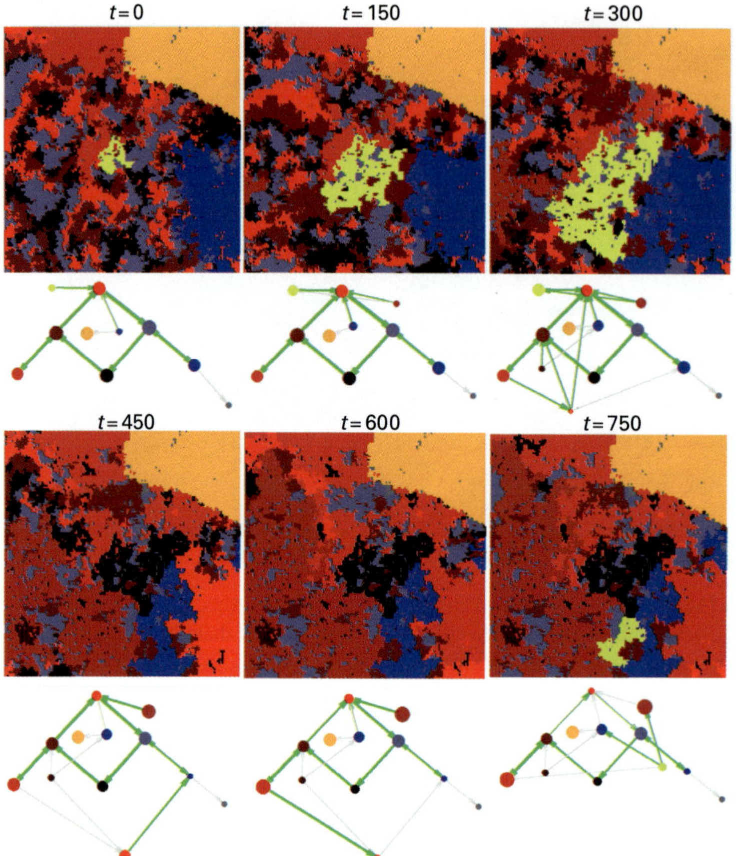

Figure 12.23 Results of a model simulation with $\gamma = 0.1$ and $\alpha = 0.1$ for an $L = 200$ system [717]. The networks below the ecosystems show corresponding interaction networks with the thickness of green arrows measuring the current predation activity. The gray arrows show potential interactions that are not active because the species are not in contact with each other. Remarkably, when taking the last configuration, and then changing the rules such that invasions can take place between any pairs of points across the lattice, one observes a fast collapse to a state with diversity $D \sim 1$ (not shown).

- Select a random site i and one of its nearest neighbors j. If species $s(i)$ at site i can invade species $s(j)$ on site j, i.e. $\Gamma(s(i), s(j)) = 1$, then the occupation of site j is updated: $s(j) = s(i)$.
- With probability $\alpha \times \gamma/N$, a new random species s is introduced at a random point j and assigned random interactions $\Gamma(s, u)$ and $\Gamma(u, s)$ with all existing species u in the system. Each of these interactions are assigned the value 1, with probability γ, or otherwise set $= 0$. The introduced species s is assumed to be able to invade the previous species at the site j, $s(j)$: $\Gamma(s, s(j)) = 1$.

Notice that the interaction between any pair of species is fixed for their entire existence. New interactions only appear with new species. The model is implemented on a two-dimensional square lattice with four neighbors per lattice location.

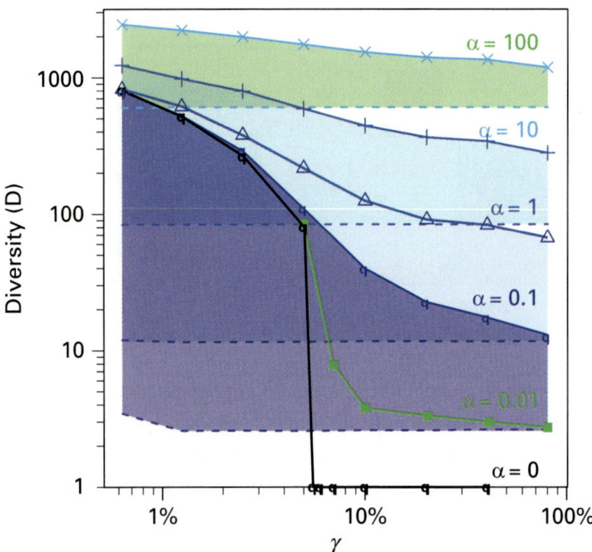

Figure 12.24 Diversity D as function of γ, for different α values. The horizontal lines refer to results obtained in a corresponding random-neighbor model, a scenario where every site can invade any other site irrespective of location. One can see that diversity is sustained by the geometry of two. The thick black curve reflects behavior for $\alpha \to 0$, a limit that can be obtained in a quasi-static simulation where species are only added when all dynamics are frozen. Figure from [717].

Figure 12.23 shows snapshots for a system of size $N = L \times L = 200 \times 200$. The sequence of pictures shows the steady-state dynamics at moderate γ and α values. The different colors indicate different species, of which some can invade others while some cannot. As a result, some parts of the lattice are inactive (e.g. the orange color in the upper right-hand corner), whereas other parts are exposed to an active cycle of four mutually invading species: *black* $\to$ *brown* $\to$ *red* $\to$ *gray* $\to$ *black*. This cycle is also illustrated in the lower panels, where the width of the green arrows measures the amount of active invasion. The sequence of figures also illustrates the advance of the hostile dark-red species as it displaces the red species. In the end, this ongoing depletion of the emerging red top predator results in a breakdown of the four-member cycle. After its collapse, the surviving species are fragmented into disconnected patches.

As the interaction density γ is decreased below a critical level $\gamma = \gamma_c \approx 0.06$, the system is found to exhibit a sharp transition from a one-species state ($D \sim 1$) to a state of high diversity D ($D > 20$) (see Fig. 12.24).

To emphasize the importance of space, the model can also be compared to its non-spatial counterpart, a model where an individual on any site can attempt invasion of any other site. This non-spatial model sustains much lower species diversity than the spatial model, and in fact collapses to diversity $D = 1$ for $\alpha \to 0$, independently of γ.

In the model, species diversity can only increase if an invading species cannot access the full population of its prey. This will occur when the prey is subdivided into spatially separated regions. Accordingly, habitat fragmentation into isolated patches is a

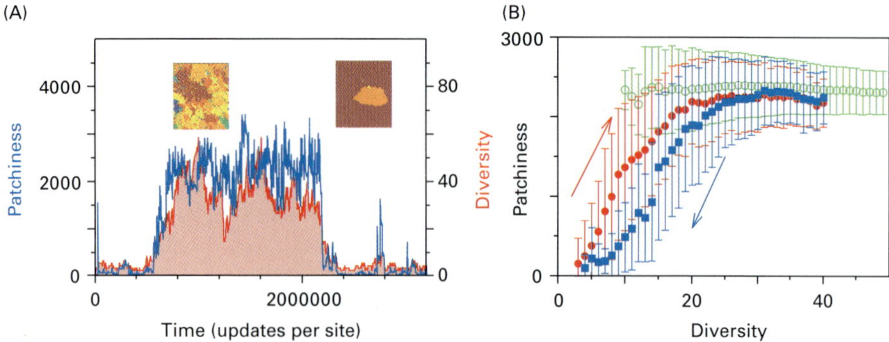

Figure 12.25 Patchiness → Diversity: (A) Time evolution of the patchiness (line) and diversity (line with shaded area) for $\gamma = 0.07$ for a system of size $L = 200$. Typical snapshots are shown for low-and high-diversity states. (B) The average trajectory in patchiness and diversity when the system switches between low-and high-diversity states. Figure from [722].

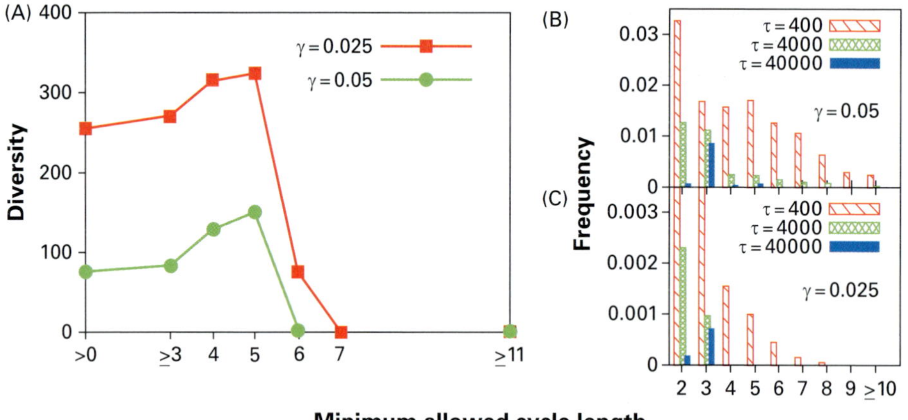

Figure 12.26 Cycles → Patchiness (A) Quasi-static simulation ($\alpha \to 0$) of an $L = 200$ system where species are only introduced when they do not form cycles shorter than specified on the x-axis. (B,C) Shows that the frequencies of eventual cycles are low, and that especially the long cycles only exist shortly. However, diversity can be obtained from cycles of length five, in spite of their fast decay. What is important is that they fragment space into many patches. Notice that the probabilities of various cycle lengths do not add up to one, because the vast majority of species additions do not cause cycles. Figure from [722].

prerequisite for speciation. Figure 12.25 examines typical switching between low- and high-diversity states, and shows that the number of patches increases before diversity increases.

Spatial separation and the creation of patches require spatial reshuffling events that occur when cycles are formed and subsequently break down. Thus, cycles are needed to obtain a sufficient amount of patches (see Fig 12.26).

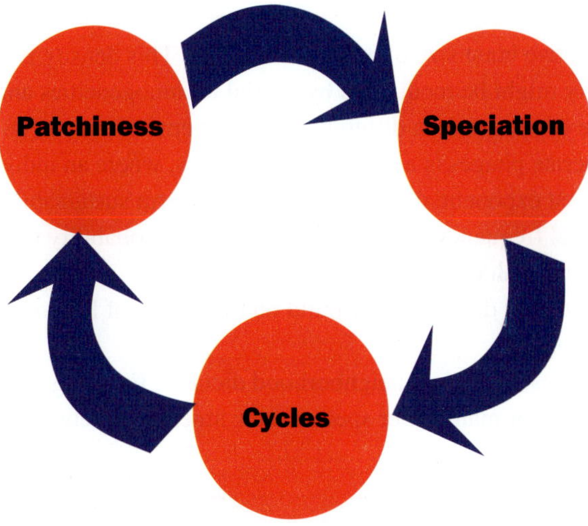

Figure 12.27 Positive feedback for sustainable diversity in our model ecology: patches of species create the heterogeneous environment that is needed for speciation, where both a new mutant and its ancestor remain in the system. If sufficient amounts of different species are present in the system, cyclic relationships may form. When these collapse due to stochastic fluctuations, it generates a quasi-static pattern of species that each are distributed among many segregated patches.

Overall, the model provides for a fairly robust co-existence of multiple species based on a dynamic maintenance of the feedback outlined in Fig. 12.27. The obtained co-existence of multiple species depends on both the limitations imposed by the interaction network (γ), and on spatial separation. If either are absent, the diversity of the species collapses. Fitness in this model is truly context dependent, and may collapse, just as a species expands and connects to a predator.

12.5 Summary

- Phages have elaborate strategies to invade bacteria, whereas bacteria in particular use two complementary strategies; restriction-modification systems deal with unknown phages, whereas CRISPR protects against known phages. In addition, the bacterium often loses a surface receptor, a loss that often will cost substantially in growth rate.
- Virulence is also sustainable, in well-mixed environments, because the host population adjusts itself to so low a level that the phage is only marginally propagated [619, 674]:

$$\text{Host population} \propto \frac{1}{\beta} \tag{12.30}$$

where β is the effective burst size of a phage infection. This density is remarkably small, and in reality might be modulated by spatial inhomogeneities and possible co-existence at the edges between different microbial environments [684].

- Lysogeny of temperate phages is a bet-hedging strategy, where an individual phage sacrifices its own fast exponential growth for the long-term sustainability of its "germ" line. Optimal frequency for lysogeny equals the probability that good conditions for lytic growth collapse.
- Speciation and biological diversity may build on the ability of life to provide protection from other life.
- A species is as fit as its ecological neighborhood dictates. For phages, this means the presence of hosts, for many other species their fitness also relies on not being exposed to antagonists.

13 Evolution and extinction

13.1 History

Evolution [1, 4, 689, 723, 724, 725, 726] and the ability to evolve is basic to life. It is perhaps the most unique ability that separates life from non-life. The process of evolution has brought us from inorganic material in $\sim$ 3.8 billion years, from our bacterial ancestors in $\sim$ 2 billion years, from "primitive" sea-living chordates in $\sim$ 500 million years and from common ancestors to mice in only $\sim$ 100 million years. Evolution has inspired our way of looking at life processes throughout modern biology, including the idea that evolvability is an evolving property in itself [727].

Evolution is a costly process where most attempts are either without consequence or aborted by the error-correcting process of selection. That evolution has led to the present-day diversity of species, reflects the capacity of life to copy itself so abundantly that it can sustain the costly evolutionary attempts.

Evolution is the logical consequence of:

(1)   Heredity (memory)
(2)   Variability (mutations)
(3)   Each generation providing more individuals than can survive (copying).

These points imply selection, based on the principle of "survival of the fittest" formulated by Darwin and Spencer.

A noticeable facet of evolution is that it works with a number of individuals N, that is much smaller than the combinatorial possibilities of evolving their genome. There are $5 \cdot 10^{30}$ prokaryotes on Earth [129], but a bacterial genome provides a number of codes of the order of 4^L with $L \sim 10^6 \rightarrow 10^7$. The implication is that evolution is a historical process [1, 4, 19], where new variants arise by rather small variations of what is already present. Or stated in another way, if one imagined starting life again, the outcome would with all likelihood be quite different.

An example of divergent solutions to the design of life are found in the rich variety of body forms that had already emerged by Cambrian times (530 million years ago) [4, 731, 732]; a variety that originated within a remarkably short time interval. Some of the designs did not survive, whereas others proliferated. The subsequently evolved animals were constrained to the taxonomic phyla defined at this relatively early event [4, 731, 732]. Figure 13.1 illustrates life history since the Cambrian. Notice the coherent

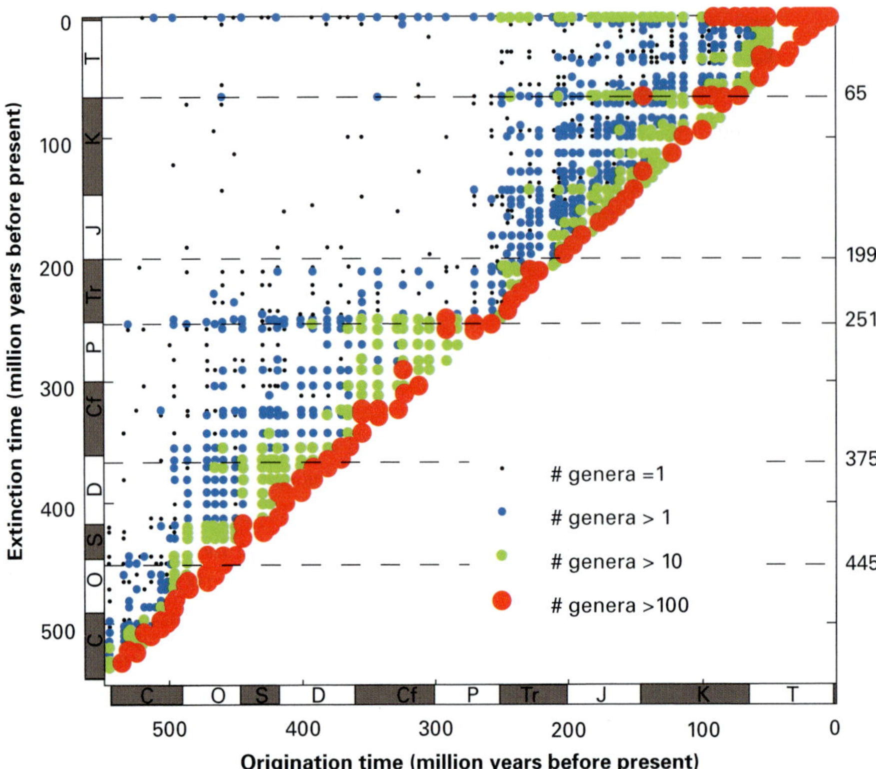

Figure 13.1 Origination and extinction of 35 000 genera in the Phanerozoic [728], as visualized by [729]. Every event is quantified by number of genera = group of related species, for example all types of humans (neanderthals, homo sapiens, cro-magnon) are one genera. The vertical distance from a point to the diagonal measures the lifetime. Notice the many points close to the diagonal, reflecting the fact that most genera exist for less than the overall genera average of about 30 Myr (million years). Notice also the division of life before and after the Permian extinction 250 Mybp (million years before present).

emergence and collapse of whole ecosystems, as is evident by the close to block diagonal form of the origination–extinction "matrix." Basic statistical characteristics can be examined in the log–log plot in Fig. 13.2B. In particular, notice that the genera existence times have a broad distribution, and also that the distribution of genera diversity within higher-level taxonomic orders is remarkably scale free [592, 730, 733].

Figure 13.3 examines another interesting feature of the fossil record: that survival of genera is random in the sense that within each order, extinction can be characterized by a probability that is constant with time:

$$P(t_{\text{life}}) \propto e^{-t_{\text{life}}/\tau} \Rightarrow dP = -\frac{P}{\tau}\, dt \tag{13.1}$$

This is an exponential distribution, with a coefficient of variation CV that is equal to 1. The characteristic extinction rate $1/\tau$ varies between taxonomic orders, reflecting the

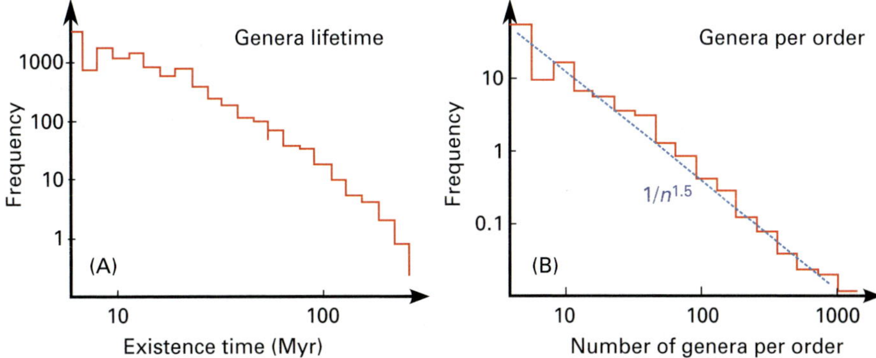

Figure 13.2 Left: "Lifetime" distribution of the 31 071 extinct genera in the Sepkoski database [728]. The average lifetime of genera is 27 ± 36 million years (Myr). Right: Number of genera in "orders" counted in same database. An order is a taxonomic group of animals that share certain characteristic features. There are 522 orders in the dataset. The size distribution of these orders in terms of genera diversity is scale-free, a classical observation reported by [592, 730].

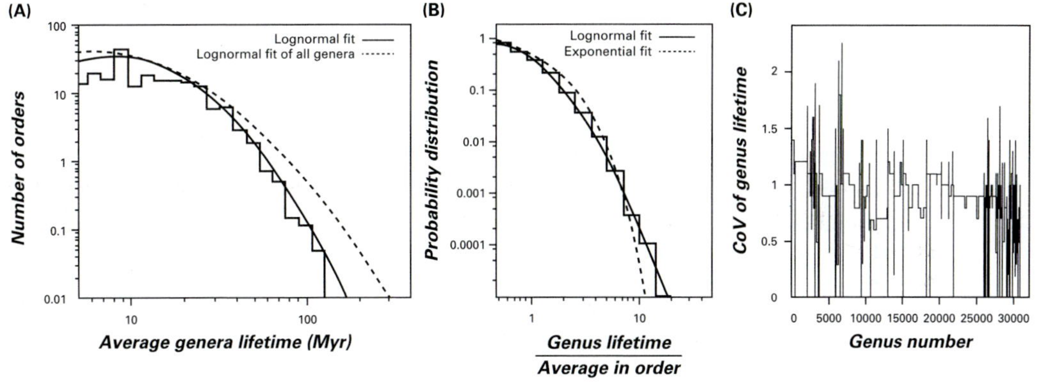

Figure 13.3 Genera survival time statistics [729] using data from [728]. (A) Log-normal fit to distribution of genera existence times is indicated by dashed curve. The histogram shows the average genera existence times within taxonomic orders, demonstrating that a few taxonomic orders have genera that typically exist for $\tau_O \sim 100$ Myr, although many orders consist of genera with much shorter existence times $\tau_O \sim 10$ Myr. (B) Genera lifetime t_{life} rescaled by τ_O for its taxonomic order. Comparison with the exponential fit (dashed line) shows that genera survival is near exponential within orders. (C) CV = standard deviation divided by the mean of genera lifetimes within orders. CV = 1 corresponds to an exponential distribution, consistent with an extinction risk that is independent of time. Calculation was done at mutation rate $\mu = 0.005$

fact that species within some orders are more robust than species within others. Investigating how existence times τ vary between early and late in the history of the orders, Van Valen [694] reported that τ did not increase as evolution "progressed." This led Van Valen to propose the "Red Queen" hypothesis: a species has to "run" just to keep its ability to survive constant.

13.1.1   Questions

13.1.1   How many bacterial generations have passed since life started on Earth? Assume near optimal growth, with 1 hour doubling time.

13.1.2   Assume that there is on average one base pair mutation per bacterium per generation. Assume that there has always been about $5 \cdot 10^{30}$ bacteria on Earth. Give an estimate of how many of all the possible bacterial mutants have been tested?

13.1.3   With 5 000 000 million base pairs in the genome, what fraction of all mutants have been tested until now?

13.2   Fitness landscapes

13.2.1   Fitness

The central theme in this section is the fitness Λ, a scalar quantity that measures the typical growth rate for the number of individuals N with a particular gene sequence:

$$\frac{dN}{dt} = \Lambda \cdot N \tag{13.2}$$

Here the fitness $\Lambda = \Lambda(x)$ depends on a genomic co-ordinate x, which in principle has a very high dimension, thus defining a landscape [724] on which a population may spread [723, 734, 735, 736] and move by mutations [723, 737, 738]. There are some concepts in evolution that are often discussed: the genetic drift on neutral landscapes by Kimura [734], and the adaptive trait concept (Fisher [723] and Wright [724], see Fig. 13.4):

(1)   The neutral scenario where most mutants are equal, thereby allowing the population to diffusively traverse large distances in genome space [734, 739, 740].

(2)   Fitness climbers [723, 724]. Evolution is modeled as an ongoing optimization process, where persistently there are individuals that are better than the best of the earlier generations.

13.2.2   Neutral walks

The neutral theory has experimental support in molecular evolution, where the sequences of most proteins/RNAs vary substantially between species. Probably, many variants of a particular protein do their job well, and large-scale evolution is not only associated with developing better molecules. Instead evolution of new species designs could be associated with reallocation of functions between cells. Such a reallocation can be done through rewiring of genetic regulatory networks for cellular differentiation.

The neutral evolution model for regulatory networks presented in [742] explores the fact that every existing network must have an unbroken line of viable mutations that

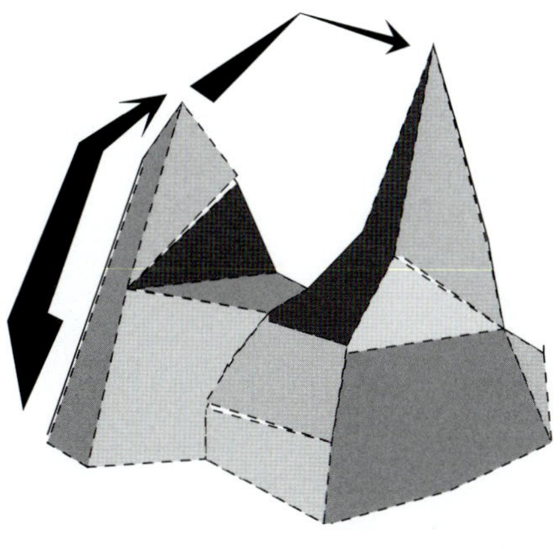
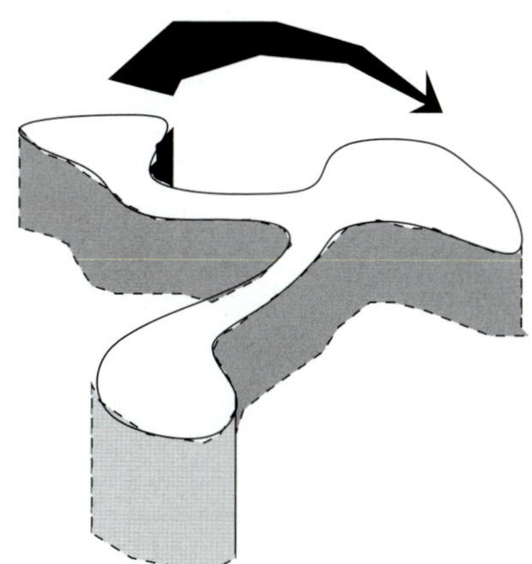

Figure 13.4 Fitness landscapes: evolution may take place under conditions where the fitness may decline or increase as a function of the genome composition, giving dynamics dominated by hill climbing, interrupted by barriers, as D. Fisher and S. Wright respectively envisioned [723, 724]. Alternatively, evolution may mainly be characterized by diffusion on large fitness plateaus as suggested in the neutral evolution scenario of Kimura [734]. The latter is explored for folding of RNA by [741].

connects it to very simple regulatory designs. Thereby one may explore an interplay between neutral evolution in each environment, and possible evolution associated with environments that may change and open new "platforms" of neutrality between each evolutionary step. Noticeably, the requirement for backwards compatibility in itself predicts networks where genes are either persistently on or persistently off, a design that is compatible with multiple genetic switches.

13.2.3 Adaptive walks

Conceptually, the simplest model of evolution is based on the fitness climbing in Fig. 13.5. In this, one follows a population with a variation in genotype and characterizes the population in terms of a Gaussian distribution with width σ in a one-dimensional genome space [723] (see Fig. 13.5). This width will be proportional to the mutation rate. The population will have a substantial fraction that is a genomic distance σ ahead of the mean fitness. Further, this subpopulation will increase its fraction with a rate proportional to σ times the gradient in the fitness. Thus, the adaptive walk predicts a rate of fitness increase that is proportional to σ^2 [723].

Fisher's description of adaptive walks along an ever-increasing fitness slope is idealized in many ways. One of the simplest limitations is associated with the possibility that the walk experiences a local fitness maxima. If the evolving population reaches an impasse, where no small mutations can bring it to better fitness, evolution would be arrested and the species characteristic frozen for a considerable time. If nothing changes in the surroundings of the species, it can only change when a mutant appears

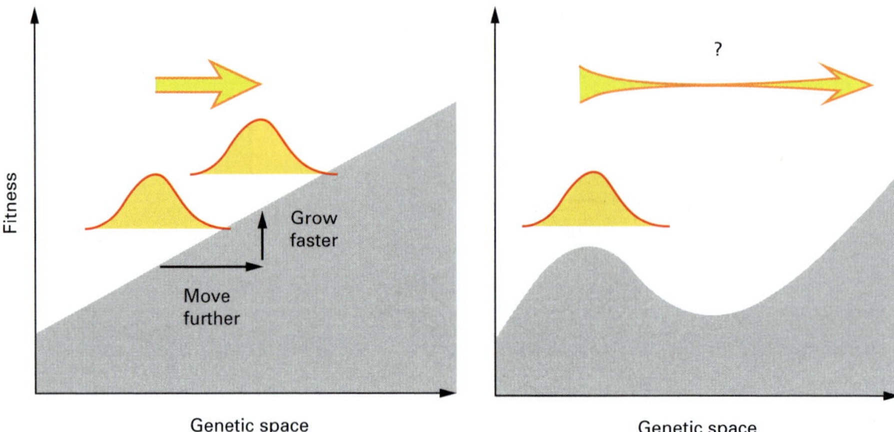

Figure 13.5 Walks in adaptive landscapes. The left figure illustrates the Fisher theory of evolution [723], where the population indicated by the Gaussian moves steadily to higher fitness by selecting the individuals with highest fitness to dominate subsequent generations. Evolutionary drift predicted to be proportional to the variance of the genetic variation, σ^2, which in turn depend on the mutation rate μ. The right figure illustrates the trapping of a population on a local fitness peak, in which case the evolution does not proceed steadily. It has to overcome a barrier and may therefore proceed in discrete steps, coined punctuated equilibrium.

with a higher fitness. In the framework of fitness landscapes, the chance of such a fluctuation depends exponentially on both the size of the fitness barrier, and the effective size of the population N_e[1] [724, 743]. Thus on long timescales one expects a punctuated evolution, with large waiting times on peaks, and fast adaptations when the population shifts [743]. Such step-like evolutionary dynamics have been emphasized by Simpson's "quantum evolution" [725, 726], as well as by the punctuated equilibrium concept of Gould and Eldredge [744, 745, 746].

Overall, the adaptive walks are promising for local exploration of nearby mutants or for selective breeding; but such walks are hopeless when large-scale jumps are needed. Neutral landscapes, on the other hand, allow large-scale meanderings and are in this way a more promising scenario for larger-scale evolution.

13.2.4   Limits of the fitness concept

The relevance of fitness defined in terms of growth rates was already challenged when comparing two bacterial species exposed to the same predatory phage. Furthermore, growth is challenged when considering the history of life on larger scales. Exponential growth is inherently so fast that, on an essentially instant geological timescale, it reaches a level set by environmental constraints. Therefore growth rates should mainly serve to modulate the already very short time that is needed to adjust to new environmental options.

---

[1] Because a big population tends to reduce the fluctuations, unless the lucky mutant species appears directly from the peak population.

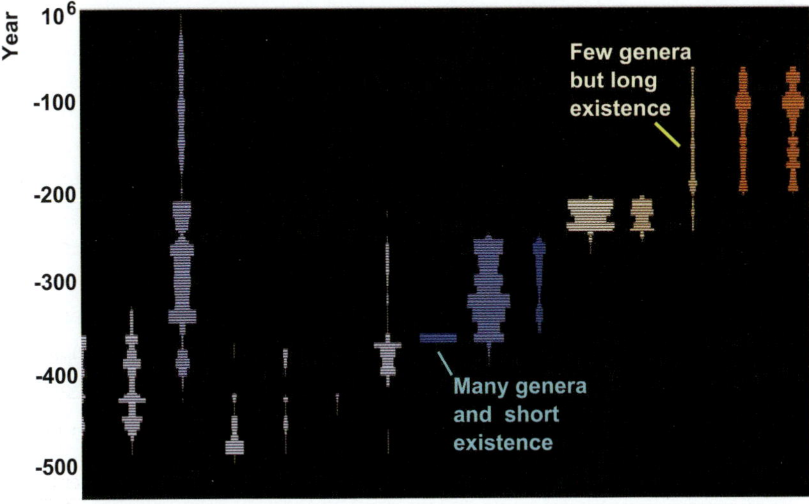

Figure 13.6 Diversity (shown as the width of the objects) of some taxonomic orders in terms of genera abundance, as a function of time ([729] using data from [728]). Orders are colored according to their origination time. Growth and decline are intermittent, with long periods of stasis interrupted by sudden growth or collapse. Notice very different "trajectories": sometimes diversity is largest at the birth of an order, whereas for other orders, it peaks late in its history. Also notice that the survival of an order does not increase just because it has higher diversity.

As suggested in the previous chapter on microbial and lichen competition, evolution on the scale of species may often be associated with fragmentation of subpopulations into independently changing ecological niches [1]. Thereby the successfully evolving subpopulations can take advantage of local fitness opportunities, and bypass the eventual bottlenecks that would restrict evolution in a homogeneous and static environment. One way to start thinking about a selection process exposed to dynamic changes is to consider the surrounding ecosystem in terms of other species. Because other species also evolve, they may open up new opportunities for each other: co-evolution may speed up evolution.

On the scale of origination and extinction of taxonomic groups, shown in Fig. 13.1, a million-year success with high diversity of species does not increase long-term survival of the taxonomic order [729]. This is illustrated in Fig. 13.6,[2] where one can indeed see that taxonomic groups with many species may well collapse as a whole when exposed to the new environment of a new geological/ecological period in the history of the Earth. The intrinsic design properties of related species makes them vulnerable to the same types of stress, suggesting that stability against extinction may be inherited, but is not a property that seems to be selected for when one considers timescales that exceed a million years.

---

[2] Species are hierarchically organized into related taxonomic groups. Genera consist of several species, for humans, the genera would include, e.g. neanderthal and cro-magnon [747]. On larger scales the genera are organized into families, which for humans include chimpanzee, gorilla and orangutan. On an even higher level there are "orders," as examined in Fig. 13.6.

13.2.5 Questions

13.2.1 Draw random numbers from a continuous distribution, and remember the largest you have had until now. Simulate the times at which this largest number increases. Show that the times for such changes get subsequently larger as time passes. What is the distribution of these times? The numbers may be fitness values, and each new number a mutation attempt (only a fitter species will "outrun" the current species).

13.2.2 Study numerically the space–time trajectory of a particle in a double well potential $V(x) = -2 \cdot x^2 + x^4$ using first-order Langevin dynamics and a noise term that sustains bistable behavior. Discuss the results in terms of evolution in the fitness landscape $F = -V$.

13.3   Punctuated equilibrium and co-evolution

13.3.1   Scale-free punctuations in the fossil record

The evolution of species during the last 540 million years shows signs of large-scale co-operative behavior: often during the history of life there have been major

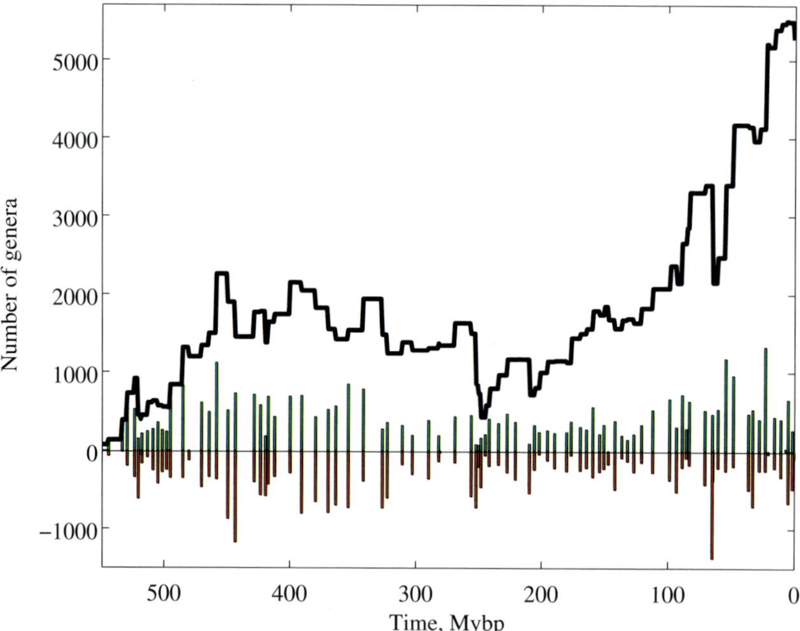

Figure 13.7 Total number of genera plotted as a function of time since the Cambrian, using data from [728]. This diversity increases or decreases in steps at geological boundaries. The size of extinction events in terms of number of genera that went extinct are shown as red "bars" below zero, emphasizing that they lead to a decline in diversity. Origination events are marked with "bars" above the x-axis. Extinction and origination are the horizontal and vertical projection of activity from Fig. 13.1.

"revolutions," where many species have been replaced "almost" simultaneously [725, 726, 733]. This is illustrated in Figs. 13.1, 13.6, 13.7 and 13.9. Spectacular examples are the Cambrian explosion 540 million years ago, where a huge variety of life arose in a short time interval, and the Cretaceous–Tertiary boundary, where mammals replaced the dinosaurs as large animals.

In between the major "transitions" there were periods of quiescence, where species seemed to live in "the best of all worlds," with small risk of extinction. However, the pattern of life is more subtle than quietness versus worldwide on–off transitions. The record shown in Fig. 13.7 often has extinctions of a smaller size in the more quiet periods. Figure 13.8 shows that ecological events come on all scales, even cataclysmic events that involve most of the contemporary genera. Furthermore, one can see that larger events are gradually less frequent than smaller events. There is no "bump" or especially enhanced frequency for the largest-scale extinctions. In fact, the distribution of extinction size s is consistent with a scale-free distribution, as indicated by the fitted $1/s^{1.5}$ curve. This overall gradual decline of event size distributions indicates:

- That large and small events may be associated with similar types of underlying dynamics. If extinctions were always externally driven by events like, for example, asteroid impacts [748], one would expect a peak at the large events.
- The non-Gaussian probability distribution for extinction events shows that the species in the ecosystem do not suffer extinction independently of each other. This is consistent with co-evolution on the scale of the global ecosystem.

It may be instructive to look at macro-evolutionary data a little more finely grained, as in Fig. 13.9. Even more fine grained, Fig. 1.3 follows the family tree of ammonites from their origination 350 million years ago until their extinction 66 million years ago [16]. One observes many speciations, as well as extinctions. Occasionally, the whole

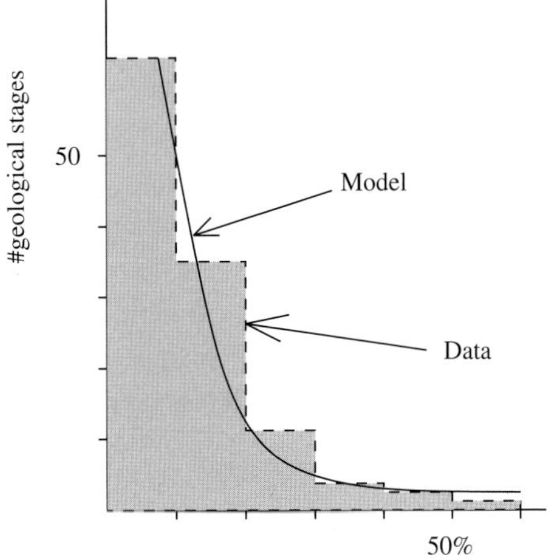

Figure 13.8 Histogram of family extinctions in the fossil record as recorded by Raup and Sepkoski [728]. The prediction, $p(s) \propto 1/s^{1.5}$ of the random-neighbor version of the BS model is marked as "model."

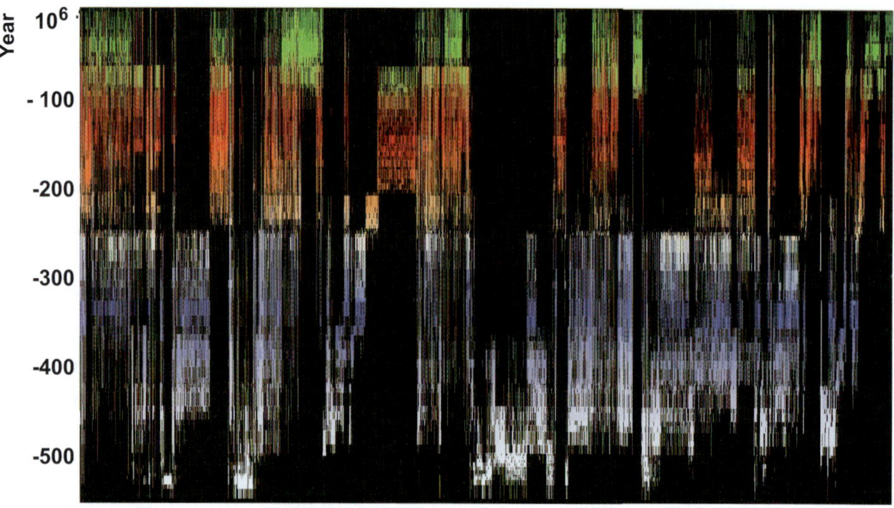

Figure 13.9 Extinction and origination of genera, colored according to their origination time. Thus genera that orginate before the Permian are marked blue, whereas more younger genera are marked red or green. The vertical axis is from 550 million years ago to the present. The horizontal axis marks about 30 000 genera positioned such that neighbors tend to be more closely related. Notice that very few of the old (blue) genera survive across the Permian extinction at 250 Mybp.

tree is even exposed to near total extinctions. However, as long as a single species survives, one can see that it subsequently diversifies and regains large diversity. Finally at the extinction event 66 million years ago, not a single ammonite survived, and as a consequence these group of animals are now only known as fossils.

What we cannot see on the family tree are the interactions between the species, nor can we see which other species these ammonites interacted with. Overall, one observes dynamics of coherent stasis alternating with rapid extinctions across all species, as also seen in Fig. 13.9. The observed genera level dynamics are far from what one would expect from random asynchronous extinction and origination.

To model the observed macro-evolutionary pattern we start with units on the size of the main players on this scale. We call these "agents" in the model for species. A species of course consists of many individual organisms, and the dynamics of a species represent the coarse-grained view of the dynamics of these entire populations. Thus, whereas population dynamics may be influenced by some sort of fitness, we here assume that species dynamics are governed by stabilities. In the language of fitness landscapes, we picture the species dynamics in terms of jumps over fitness barriers:

Population dynamics   *Barrier dynamics*

$$\xrightarrow{\text{large timescale}}$$

Survival of the fittest   *Evolve the least stable*

Stability of a species will depends on the particular environment and the type of fluctuations at hand. In terms of the fitness concept, changes in the environment may change the fitness landscape [18]. However, even if fitness landscapes are abandoned as meaningless,[3] the stability of a species retains meaning.

Given the basic evolutionary unit of a species, we characterize this with one number B_i. This number is the stability of the species on a timescale much longer than the time needed to fill a given ecological niche. An ecosystem of species then consists of N numbers B_i that each represent a species. Each of these species are connected to a few other species, with links that could be predation, collaboration, or niche maintenance [18].

13.3.2   Bak–Sneppen model

For simplicity, let us first assume that the numbers B_i, $i = 1, 2, ...N$ are placed on a line, mimicking a one-dimensional model ecosystem. At each time step, one changes the least stable of these species. As the stability is defined by its context, the fitness of a given species is a function of the species it interacts with, and accordingly the neighboring species will also change their stabilities. The co-evolutionary updating rule for the agent-based model then reads [749, 750]:

At each step, the smallest of $\{B_i\}_{i=1,N}$ are located. For this, as well as its nearest neighbors, one replaces their B value by new random numbers in $[0, 1]$.

This is traditionally referred to as the Bak–Sneppen (BS) model. It is the simplest model that exhibits a phenomenon called self-organized criticality [751]. As the system evolves, the smallest of B_i is eliminated. After a transient period, a statistically stationary distribution of B is obtained. For a system size $N \to \infty$ this distribution is a step function, where the selected minimum $B_{\min}$ is always below B_c. As a consequence the distribution of B is constant above B_c. For the dimension $d = 1$ by updating the two nearest neighbors,[4] one obtains a self-organized threshold $B_c = 0.6670$, see Fig. 13.10. The right-hand panel illustrates how the minimum B sites change in the "species space" as the system evolves. One observes highly correlated activity.

The right-hand panel of Fig. 13.10 shows a "space–time" map of the minimum B sites over the considered time interval. Whenever the lowest barrier is found among the three updates surrounding the previous minimum, the active site performs a random walk. The figure shows that this is what happens most frequently. When the site of the lowest barrier value moves by more than one lattice spacing, it most frequently backtracks in subsequent updates. Importantly, activity tends to stay localized and forms a sequence of changes in the same region of the model ecosystem. Thus evolution is reinforced locally, bridging punctuated equilibrium in single-species evolution [744, 745] to larger- scale quantum evolution and origination of new taxonomic groups [725, 726].

---

[3] As discussed earlier in this chapter, the populations of a species mostly fluctuate without any direction, implying that the time-averaged growth is zero. Thus, strictly speaking, the fitness is zero for all well-established species.

[4] Notice that the overall behavior of the model is robust to the choices of neighbors. For example, one may choose one of the neighbors at random.

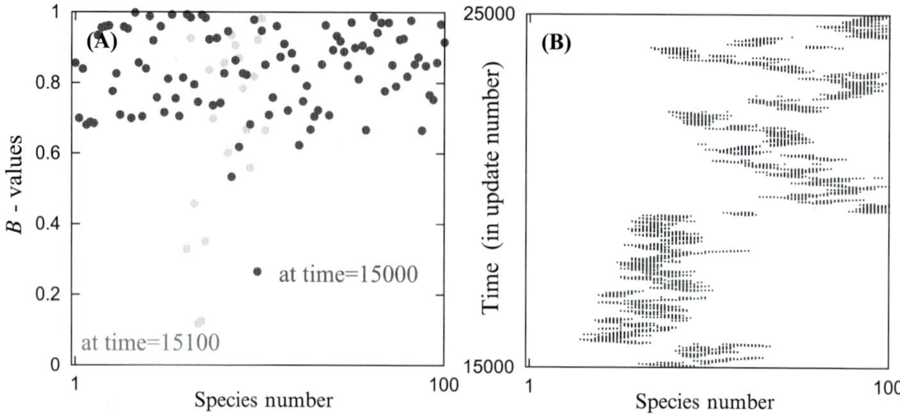

Figure 13.10 (A) An example of a snapshot distribution of barriers B_i in space. One can observe that sites with $B > B_c = 0.6670$ are distributed randomly in space. In contrast, sites with $B < B_c$ are highly correlated and tend to remain within a small region. (B) Space–time plot of the activity in a steady state where "time" is counted by the number of updates. At each step the site with minimum B is shown with a +.

The obtained correlation of evolutionary activity relies on a self-organization that demands time and allows the evolving system to develop towards a dynamic attractor state exhibiting scale-free correlations in both space and time. This attractor state is therefore critical, and the algorithm is one of a class of models that let a system self-organize toward such criticality.

13.3.3   The random neighbor model and its solution

To understand how the threshold in B emerges we consider a simpler random-neighbor version of the BS model [752, 753]. Here at each update one changes the site with minimum B, as well as one other randomly selected B (see Fig. 13.11). Because of the absence of spatial correlations between the Bs, this simpler model can be solved analytically [753].

Starting a simulation with an initial distribution of $B_i \in [0, 1]$ that is random and uniform, then the smallest of these B_i is first to get eliminated. This leads to a systematic depletion of small B values. After a transient period, one obtains a statistically stationary distribution of Bs. For $N \to \infty$ this distribution is a step function, where the selected minimum $B_{\min}$ is always at or below B_c. This in turn implies that species with $B > B_c$ cannot be selected as the minimum, and therefore are only changed because they were selected as the random neighbor, irrespective of their actual B value. Therefore the distribution of B is constant above B_c. At each step the dynamics then select one B below B_c and the other B above B_c. As the two newly assigned Bs are assigned uniform random values in $[0, 1]$, the condition for a statistically stationary distribution of the number of species in the interval $[0, B_c]$ is:

$$-1 + 2B_c = 0 \Rightarrow B_c = \frac{1}{2} \tag{13.3}$$

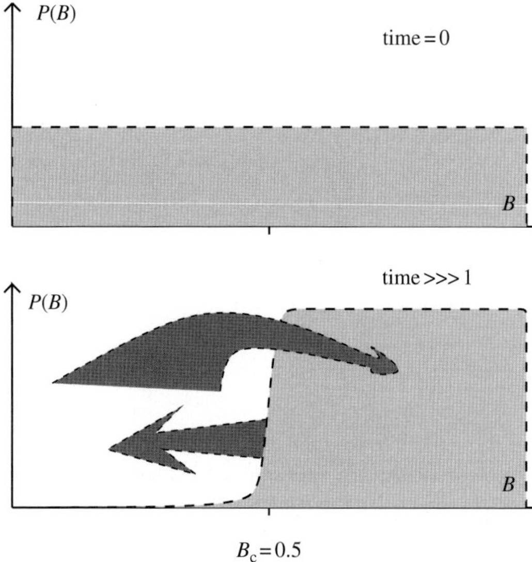

Figure 13.11 Distribution of barriers/fitness in the random neighbor version of the BS model, where one takes the minimum B and one random B, and replaces them with a new random $B \in [0 : 1]$. The B distribution reaches a steady state where the minimum selected below $B_c = 1/2$ and distributed everywhere in $[0, 1]$ provides the same net transport into the interval $[B_c, 1]$, and the B selected at random the opposite transport of species from $[B_c, 1]$ to $[0, B_c]$.

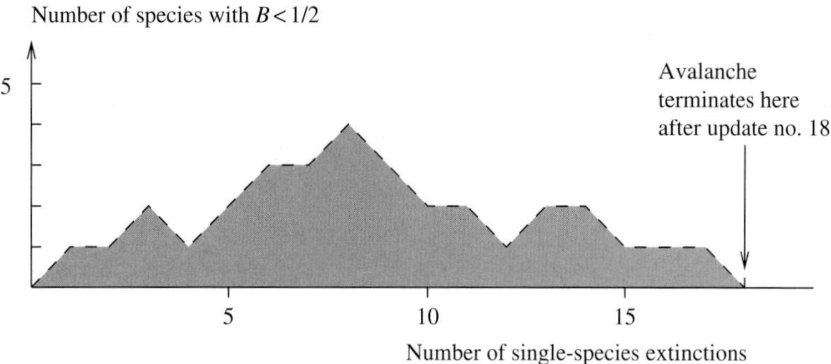

Figure 13.12 Evolution in the number of sites with $B < 1/2$ in the random-neighbor version of the BS model. At any time there is equal probability of increasing or decreasing the number of active species a, thus defining a random walk of this number. The first return of the random walk to zero defines an avalanche that terminates when all $B > 1/2$. When this occurs the system is so stable that the next change will occur rarely, but anywhere along the one-dimensional ecosystem. The probability of the first returns of random walks $\propto 1/updates^{3/2}$.

The time series of the minimum B exhibits correlations. An avalanche is defined as the number of steps s between two subsequent selections of minimum $B > B_t$. The number n of Bs below $B_t = B_c$ exhibits a random walk and the size of the avalanche is determined by the number of updates s before this random walk returns to zero:

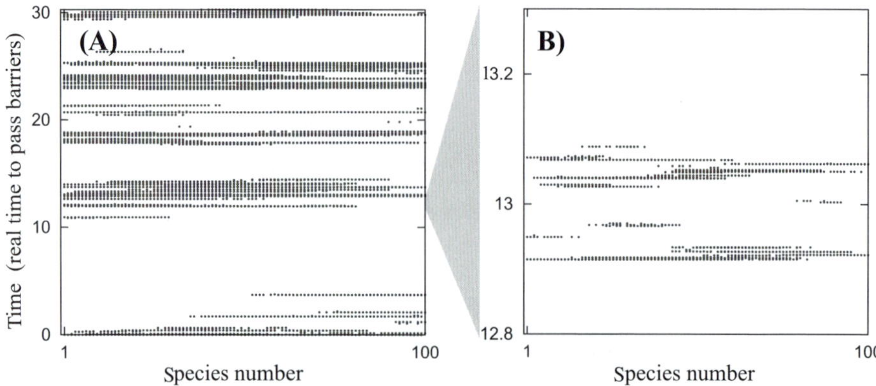

Figure 13.13 Space–time plot of the activity in the model. Each update is shown as a black mark. On the time resolution of the plot, the avalanches appear as almost horizontal lines. The increased magnification on the right shows that there are avalanches within avalanches. The calculation was done at mutation rate $\mu = 0.02$.

$P(s) \propto s^{-3/2}$. This is the famous distribution of waiting times in the gambler's ruin problem (details in the Appendix, scaling argument in footnote 5).

To compare the predicted $1/s^{3/2}$ distribution of the model with the macro-evolutionary data (Fig. 13.8) assume the number of active replacement of B's is proportional to the size of one coordinated extinction event.[6]

13.3.4   Timescale separation $\Rightarrow$ extreme dynamics and avalanches

Importantly, the BS model is defined in terms of updating of the site with the global minimum of B. This selection implicitly assumes a separation of timescales in the dynamics, which in fact also allows us to naturally separate avalanches when all B_i are above the the self-organized threshold.

---

[5] The first return distribution can also be deduced from a scaling argument for random walks in one dimension, using that this has to be scale-invariant because random walks repeat themselves on all scales. The distribution of first returns is accordingly of the form $P_{\text{first-return}}(t) \propto 1/t^{\sigma}$, where $\sigma = 3/2$ is determined by considering the time interval $[0, T]$, and the division of this into first returns [754]:

$$T = \langle t \rangle_T \cdot (\text{no. of returns in } T) = \langle t \rangle_T \cdot T^{1/2}$$

Here the number of returns is counted from the probability that the random walker is at $x = 0$ at any time $t < T$, given that it was at $x = 0$ at $t = 0$. This probability is $\propto 1/\sqrt{t}$, reflecting the density distribution for a random walk. Integrating this over $[0, T]$ then gives $\sqrt{T}$ in the above equation. When inserting the average first return within $[0, T]$, $\langle t \rangle_T = \int_0^T t/t^{\tau} \, dt$, one obtains:

$$T = T^{1/2} T^{2-\tau} \Rightarrow P_{\text{first-return}}(t) \propto 1/t^{3/2}$$

This is then also the avalanche size distribution in the random-neighbor BS model.

[6] Notice that in one dimension, the avalanche distribution would instead be $1/s^{1.1}$, whereas all dimensions $d > 3$, as well as any random network, would give the same distribution as the random-neighbor model, i.e. $1/s^{1.5}$.

The extremal dynamics can be seen as the $\mu \to 0$ limit of the following local model [755]:

At each step of size dt:
Select each $\{B_i\}_{i=1,N}$ with probability $\propto e^{-B_i/\mu}\, dt$. This selection defines a list of active sites. For members of this list, replace them, as well as their nearest neighbors, by new random numbers in $[0, 1]$.

Here μ represents an attempt rate for microscopic evolutionary changes, and is proportional to the mutation rate per generation.

For each barrier B_t below the self-organized critical threshold B_c, a "B_t avalanche" starts when the first selected $B_{\min}$ is below B_t and terminates when a selected $B_{\min}$ is above B_t. In the *local* formulation, all activity within a B_t avalanche occurs practically instantly, when seen on a timescale of the order of $\exp(B_t/\mu)$. This statement may be reiterated for the larger avalanches associated with $B_{t2} > B_t$, thereby defining a hierarchy of avalanches within avalanches. Thus one may view the avalanche-within an-avalanche picture as burst-like activity on different timescales (see Fig. 13.13). These timescales may be set by associating each step of the algorithm with a time interval:

$$\Delta t \propto \frac{1}{\sum_i e^{-B_i/\mu}} \quad \sim \quad e^{B_{\min}/\mu} \quad \text{for} \quad B_c - B_{\min} >> \mu \tag{13.4}$$

Here the last approximation uses that the distribution of barriers below B_c is scarce.

For $\mu \to 0$ the evolutionary avalanches may mimic the boundaries between geological periods, where all $B_i > B_c$ and the overall ecosystem is in a quasi-static period until an avalanche is initiated by a spontaneous mutation of one of the species at the threshold of stability. The duration of the stasis period is then set by the stability B_c, $t \sim e^{B_c/\mu}$, whereas the disturbance at the boundary will cascade as a critical branching process influencing a total number of species s with probability:

$$P_0(s) \propto s^{-\tau} \tag{13.5}$$

Here the exponent $\tau = 3/2$ for ecological networks with dimension $d \stackrel{>}{=} 4$ [754], which compares well with the histogram of extinction events shown in Fig. 13.8.

Apart from integrating small and large extinction events into one combined framework, the model predicts:

(1) Each evolutionary avalanche consists of subavalanches on smaller scales. Thus when we analyze the fossil data on more fine-grained time (and space) levels, we should expect to find each extinction event subdivided into smaller extinction events. Noticeably, such correlations between extinctions may be examined by more fine-grained data [756].

(2) Time separation between evolutionary events of a given lineage will be power-law distributed, with long periods of stasis that are sometimes broken by a sequence of multiple small jumps.

(3) Co-evolution allows for large evolutionary meanderings. Evolutionary barriers that seem impossible to pass at a stasis period get circumvented by changes in fitness landscapes due to sequences of co-evolution adaptations (Fig. 13.14).

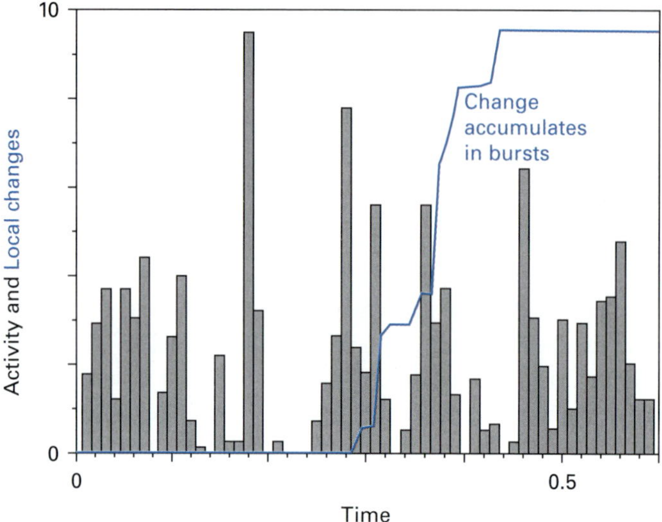

Figure 13.14 Global activity in the model (gray) plotted together with accumulated change at one position in the model ecosystem, using the finite-mutation-rate version [750, 755]. The figure illustrates that the individual species change most when the global ecosystem also exhibits big activity. Figure from [750]

Figure 13.15 Separated timescales in the ecological and paleontological record show up in long periods with geological depositions of similar types of sediments, interrupted by sharp demarcations before a new geological period starts. The left-hand panel shows an example of such a mountain landscape in the south-western USA. The right-hand panel shows a simulation of the activity in the co-evolution model with separation of timescales (between local adaptation and real "inventions"). The BS model is here simulated for $\mu = 0.005$ and $N = 140$. Left-hand picture courtesy of S. Semsey.

Accordingly, the model may well serve as a provocative perspective on macro-evolution by its emphasis on lack of needs for external events to create the sharp boundaries in the paleontological record. Figure 13.15 shows an example of how this record is visible in mountain landscapes, emphasizing that the boundaries between layers may be compared to actvity in an evolutionary model with separated timescales.

Although the above model is perhaps the simplest, with the ambition of addressing self-organization of the largest scale of ecosystems, it is not the only attempt [757, 758]. Other interesting approaches to describe facets of large-scale life systems include the Gaia hypothesis by Lovelock [7] and the coupled genetic network models of Kauffman [18], as well as attempts to understand early life through emerging and collapsing autocatalytic networks by Jain and Krishna [759] and Segre and Lancet [760].

13.3.5   Questions

13.3.1   Simulate the BS model for 100 species placed along a line in a variant of the model where only one of the neighbors is updated at each step. Plot the selected B_{min} as a function of time, as well as the maximum of all previous B_{min}s. How do the minima of B change as time progresses toward a steady state (look at the envelope defined as the maximum over all B_{min} at earlier times)?

13.3.2   Consider the externally driven version of an evolution based on stability of species [761]. Implement a system of $N = 1000$ species. Assign a random number B in [0, 1] to each species. At each timestep select an external noise x from a narrow distribution $p(x) \propto \exp(-x/\sigma)$, $\sigma = 0.1$. At each time step replace all $B < x$ with new random numbers $\in [0; 1]$ and, in addition, select one random species and set its B to a new random number $\in [0; 1]$. Simulate this model.

13.4   Summary

- Evolution is a historic process with individuals exposed to selection and random events on all scales.
- Co-evolution is expected to be important, but is only convincingly documented between phages and bacteria.
- Disasters happen, on all scales, perhaps again demonstrating that life depends on life that depends on life.

Appendix

Langevin versus Fokker–Planck equation

As mobility is associated with how fast a particle can be dragged through a viscous medium, there is a fundamental relationship between mobility μ and molecular noise, quantified by the diffusion constant D. If x fulfills the Langevin equation:

$$\frac{\mathrm{d}x}{\mathrm{d}t} = -\mu \cdot \frac{\mathrm{d}V}{\mathrm{d}x} + \sqrt{2D} \cdot \eta(t) \tag{1}$$

with $\langle \eta(t'')\eta(t') \rangle = \delta(t'' - t')$, one can construct a probability distribution for x at a given time t:

$$P(x,t)\mathrm{d}x = \text{probability} \ \ \text{for} \ \ x \in [x, x+\mathrm{d}x] \ \ \text{at} \ \ \text{time} \ \ t \tag{2}$$

By using physical insight the two terms in the Langevin equation can be recast in terms of loss and gain of particles within $[x, x+\mathrm{d}x]$, $P(x,t) \cdot \mathrm{d}x$. $P(x,t)$ will evolve according to the Fokker–Planck equation:

$$\frac{\mathrm{d}P(x,t)}{\mathrm{d}t} = -\frac{\mathrm{d}}{\mathrm{d}x}J \tag{3}$$

where the current J is given by:

$$J = -\mu \cdot P \cdot \frac{\mathrm{d}V}{\mathrm{d}x} - D \cdot \frac{\mathrm{d}P(x,t)}{\mathrm{d}x} \tag{4}$$

At equilibrium, $P = \text{constant}$, implying that $J = 0$ and thus:

$$P(x,t) = P(x) \propto e^{-\mu V(x)/D} \tag{5}$$

which can only be proportional to $e^{-V(x)/k_\mathrm{B}T}$ if $\mu = D/k_\mathrm{B}T$. This famous equation by Einstein states that the mobility (μ) is proportional to the diffusion constant D. The diffusion constant D has dimensions of a mean free path times a typical (thermal) velocity. The diffusive part of the Fokker–Planck equation describes how an initial localized particle spreads out in flat potentials to a Gaussian with spread $\sigma \propto \sqrt{2Dt}$ that simply follows from the central limit theorem. The convective part of the Fokker–Planck equation states that the particle moves downhill, with a speed proportional to both $\mathrm{d}V/\mathrm{d}x$ and the mobility μ.

Kramers' equation

Escape from a potential well is an old and important problem in physics and chemistry, as well as for a number of larger-scale biological problems related to the stability of genetic switches and punctuated macro-evolution in abstract fitness landscapes. We here present the derivation proposed by Kramers [762] for escape from a potential well.

Consider a one-dimensional potential well, as in Fig. 1, where point A is at the bottom of the well and point B is somewhere outside the well. Following Kramers we consider the stationary situation where the current leaking out from the well is insignificant. Thus, the current J in the corresponding Fokker–Planck equations is a constant number that is independent of position x and the corresponding P is constant in time. The position-independent current can be rewritten as:

$$J = -\frac{D}{k_B T} P \frac{\mathrm{d}}{\mathrm{d}x} V - D \frac{\mathrm{d}P}{\mathrm{d}x} = -D \cdot e^{-V/k_B T} \cdot \frac{\mathrm{d}}{\mathrm{d}x}\left(P \cdot e^{V/k_B T}\right) \tag{6}$$

When rewritten and integrated from point A to point B:

$$J \cdot e^{V/k_B T} = -D \cdot \frac{\mathrm{d}}{\mathrm{d}x}\left(P \cdot e^{V/k_B T}\right) \tag{7}$$

or

$$J = -D \frac{\left[Pe^{V/k_B T}\right]_A^B}{\int_A^B e^{V/k_B T}\mathrm{d}x} \tag{8}$$

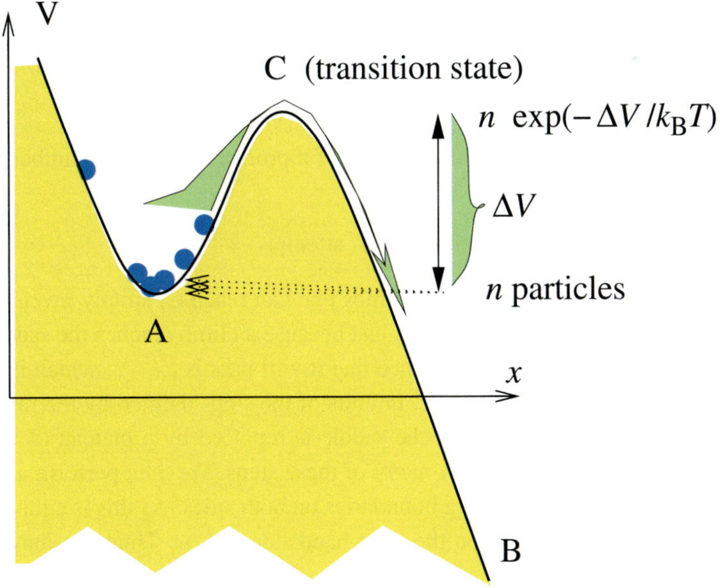

Figure 1 Escape over a one-dimensional potential, where a particle is confined at A, but allowed to escape over barrier C to the outside of the well. In the figure we show a number of particles, to illustrate the approach by Kramers.

where the quasi-stationary condition states that $P_B \sim 0$ and $P_A \approx$ local equilibrium value:

$$J = D \frac{P_A e^{V_A/k_B T}}{\int_A^B e^{V/k_B T} dx} \tag{9}$$

The value of P_A can be estimated from using a harmonic approximation around the minimum A, and setting P_A equal to the corresponding maximum of the corresponding Gaussian density profile. Thus, around point A:

$$V(x) \approx V_A + \frac{1}{2} \frac{d^2 V}{dx^2} (x - x_A)^2 = V_A + \frac{1}{2} k (\delta x)^2 \tag{10}$$

For a particle with mass m, this is a harmonic oscillator with frequency $\omega = \sqrt{k/m}$ where k is a "spring constant" provided by the *potential* V. The peak density P_A is given by normalization of $\exp(-k\delta x^2/(2k_B T))$ (everything has to be counted as if there is one particle in the potential well that can escape). The escape rate (J per particle):

$$r = \frac{\omega_A \tau}{m \cdot 2} \sqrt{\frac{m k_B T}{2\pi}} \frac{e^{V_A/k_B T}}{\int_A^B e^{V/k_B T} dx} \tag{11}$$

where $\tau = 2\mu m = 2 Dm/k_B T$ from Eqs. (5.2)–(5.5). The remaining integral is calculated by a saddle point around its maximum, i.e. around the barrier top at point C:

$$\int_A^B dx \, e^{V/k_B T} = e^{V_C/k_B T} \int_{-\infty}^{\infty} dx \, e^{-k_C^2 \delta x^2/(2k_B T)} = \frac{\sqrt{2\pi k_B T}}{\sqrt{k_C}} e^{V_C/k_B T} \tag{12}$$

which, with $k_C = m\omega_C^2$ gives the final escape rate for overdamped motion:

$$r = \frac{\omega_A}{2\pi} \cdot \omega_C \frac{\tau}{2} \cdot \exp\left(-\frac{V_C - V_A}{k_B T}\right) \tag{13}$$

This equation can be interpreted in terms of a product between a number of attempted climbs:

$$\text{number of attempts} = \frac{\omega_A}{2\pi} \tag{14}$$

multiplied by the fraction of these climbs that can reach C, simply given by the Boltzmann weight $e^{-(V_C - V_A)/k_B T}$. Finally, just because a climb reaches the saddle point, it is not given that it will pass. The chance that it will pass is $\omega_C \cdot \tau$, which is equal to one divided by the width of the saddle, in units of the steps defined by the random kicking frequency $2/\tau$; i.e. imagine that the saddle is replaced by a plateau of $w = 2/(\omega_C \tau)$ steps, and we enter the first (left-most) of these steps. We then perform a random walk over the plateau, with absorbing boundaries on both sides. As this is equivalent to a fair game, the chance to escape on the right-hand side is $1/w$. Thus one may interpret the overdamped escape as:

$$
\begin{aligned}
r = & \; (\text{attempts to climb}) \\
& \cdot (\text{chance to reach top given it attempts}) \\
& \cdot (\text{chance to pass top given it reached top})
\end{aligned}
\tag{15}
$$

and one can immediately see that a higher viscosity, meaning a lower τ, implies that the escape rate diminishes. This is not surprising, as a higher viscosity means that accordingly everything goes slower, including the escape.

First passage for random walks

Consider the first passage time for a random walker, defined as the time for the first visit to position x starting at position 0. Characterizing the walk with diffusion constant D, the distribution $P_x(t)$ for the first passage time t to position x is:

$$P_x(t) = \frac{x}{\sqrt{2\pi D} \cdot t^{3/2}} \cdot \exp\left(-\frac{x^2}{4Dt}\right) \tag{16}$$

We here prove this equation, based on a derivation that was presented in [763].
Considering the accumulated probability:

$$\mathcal{P}_x(t) = \int_0^t P_x(t')\mathrm{d}t' \tag{17}$$

equal the probability that the walk reached x before the time t. This probability is also equal to the probability that the walk passed x at some time prior to t. Thus it is equal to the probability that the *maximum* excursion of the walk in the time interval $[0; t]$ is larger than x.

The distribution of the *maximum* of the walk up to time t is related to the distribution of the endpoint at time t. For each endpoint $y \geq x$, there are in fact two walks that go beyond x at some time before t, the original path and a *mirror* path, defined as the path that follows the original path to the first passage of x, but thereafter is reflected in x, (see also Fig. 2). The first of these paths reaches position $\geq x$, the second mirror path does not. Therefore:

$$\mathcal{P}_x(t) = P(\text{max excursion in } [0; t] \geq x) = 2P(\text{end position } is \geq x)$$

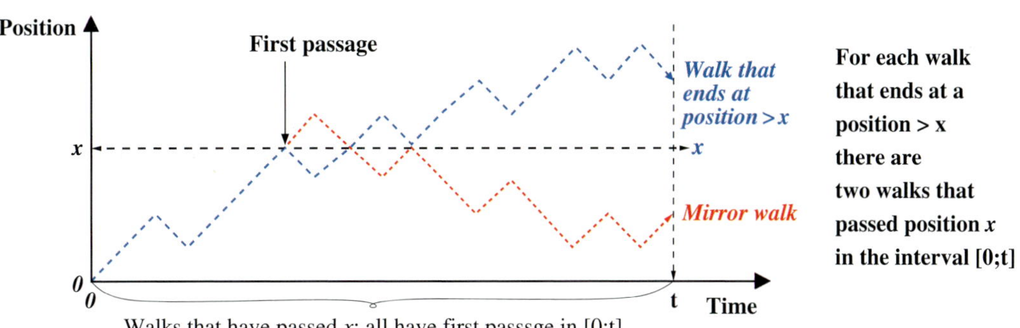

Figure 2 A random walk starting at position $x = 0$ at time $= 0$ and ending at some position $\geq x$. The figure illustrates that for each walk that ends at position $\geq x$ at time t, there are exactly two walks that have passed x at earlier times.

The distribution of positions at time t is given by the normal distribution of the random walk after time t:

$$P(\text{end position is } \geq x) = \frac{1}{\sqrt{4\pi Dt}} \int_x^\infty dy \, e^{-y^2/4Dt} \tag{18}$$

giving

$$\mathcal{P}_x(t) = 2\frac{1}{\sqrt{4\pi Dt}} \int_x^\infty dy \, e^{-y^2/4Dt} \tag{19}$$

The actual probability is given by the differential of this accumulated probability:

$$\begin{aligned}
P_x(t) &= \frac{d}{dt}\mathcal{P}_x(t) \\
&= 2\frac{1}{\sqrt{4\pi D}}\frac{d}{dt}\left(\frac{1}{\sqrt{t}}\int_x^\infty dy \, e^{-y^2/4Dt}\right)
\end{aligned} \tag{20}$$

which by substituting $v = 2Dt/y^2$, provides us with the distribution in Eq. (16), which shows the famous first-return scaling that is valid for large times compared to the time to reach x:

$$P_{x\sim0}(t) \propto \frac{1}{t^{3/2}} \tag{21}$$

References

[1] C. Darwin. *On the Origin of Species by Means of Natural Selection, or The Preservation of Favoured Races in the Struggle for Life*. John Murray, London, 1859.

[2] E. Schrödinger. *What is Life?* Cambridge University Press, Cambridge, 1944.

[3] D. Raup. *Extinction: Bad Genes or Bad Luck?* University of Princeton, Princeton, 1992.

[4] S. J. Gould. *Wonderful Life, The Burgess Shale and the Nature of History*. Penguin, New York, 1991.

[5] H. Meinhardt. *The Algorithmic Beauty of Sea Shells*. Springer-Verlag, Berlin, Heidelberg, New York, 1995.

[6] S. Kauffman. *The Origins of Order*. Oxford University Press, Oxford, 1993.

[7] J. Lovelock. *The Ages of Gaia*. Bantam Books / W.W. Norton & Company Inc., New York, 1990.

[8] M. Eigen. *Steps Towards Life*. Oxford University Press, Oxford, 1992.

[9] B. McClintock. Induction of instability at selected loci in maize. *Genetics*, 38:579–599, 1953.

[10] R. M. May. Biological populations with nonoverlapping generations: stable points, stable cycles, and chaos. *Science*, 186(4164):645–647, 1974.

[11] Mitchell J. Feigenbaum. Quantitative universality for a class of nonlinear transformations. *J. Stat. Phys.*, 19(1):25–52, 1978.

[12] D. L. Royer. CO_2 forced climate thresholds during the phanerozoic. *Geochim. Cosmochim. Acta*, 70:5665–5675, 2006.

[13] K. Sneppen and G. Zocchi. *Physics in Molecular Biology*. Cambridge University Press, Cambridge, 2005.

[14] C. N. Parkinson. Parkinsons law. *Economist*, p. 19, Nov. 1955.

[15] T. R. Malthus. *An Essay on the Principle of Population*. J. Johnson, in St. Paul's Church-yard, London, 1798.

[16] N. Eldredge. *Life Pulse, Episodes from the Story of the Fossil Record*. Facts on File Publications, New York, 1987.

[17] P. Bak and K. Sneppen. Punctuated equilibrium and criticality in a simple model of evolution. *Phys. Rev. Lett.*, 71:4083–4086, 1993.

[18] K. J. Kauffman, P. Prakash and J. S. Edwards. Advances in flux balance analysis. *Curr. Opin. in Biotech.*, 14:491–496, 2003.

[19] P. Bak and M. Paczuski. Complexity, contingenicity, and criticality. *Proc. Natl. Acad. Sci. USA*, 92:6689–6696, 1995.

[20] F. Jacob. Evolution and tinkering. *Science*, 196:1161–1166, 1977.

[21] O. L. Miller Jr., A. Barbara, A. Hamkalo and C. A. Thomas Jr. Visualization of bacterial genes in action. *Science*, 169:392–395, 1970.

[22] D. Godsell. *The Machinery of Life*. Springer Verlag, Berlin, Heidelberg, New York, 1992.

[23] C. B. Anfinsen and E. Haber. Studies on the reduction and re-formation of protein disulphide bonds. *J. Biol. Chem.*, 236:1361–1363, 1961.

[24] E. van Nimwegen. Scaling laws in the functional content of genomes. *Trends Genet.*, 19:479, 2003.

[25] A. B. Arnvig, S. Pedersen and K. Sneppen. Thermodynamics of heat shock response. *Phys. Rev. Lett.*, 84:3005–3008, 2000.

[26] R. FitzHugh. Mathematical models of threshold phenomena in the nerve membrane. *Bull. Math. Biophys.*, 17:257–278, 1955.

[27] A. Goldbeter and D. E. Koshland Jr. An amplified sensitivity arising from covalent modification in biological systems. *Proc. Natl. Acad. Sci. USA*, 78:6840–6844, 1981.

[28] X. Wei, S. K. Ghosh, M. E. Taylor *et al.* Viral dynamics in human immunodeficiency virus type 1 infection. *Nature*, 373:117–122, 1995.

[29] J. D. Barry and C. M. R. Turner. The dynamics of antigenic variation and growth of African trypanosomes. *Parasitol. Today*, 7:207–211, 1991.

[30] K. Sneppen, L. Lizana, M. H. Jensen, S. Pigolotti and D. Otzen. Modeling proteasome dynamics in Parkinson's disease. *Phys. Biol.*, 6:036005, 2009.

[31] J. B. Axelsen, S. Bernhardsson and K. Sneppen. One hub-one process: a tool based view on regulatory network topology. *BMC Syst. Biol.*, 2:25, 2008.

[32] R. Dawkins. *The Blind Watchmaker*. Norton & Company Inc., New York, 1986.

[33] O. G. Berg, R. B. Winter and P. H. von Hippel. How do genome-regulatory proteins locate their DNA target sites? *Trends Biochem. Sci.*, 7:52–55, 1982.

[34] M. Kirschner and T. Mitchison. Dynamic instability of microtubule growth. *Nature*, 312:237, 1984.

[35] M. B. Elowitz, A. J. Levine, E. D. Siggia and P. S. Swain. Stochastic gene expression in a single cell. *Science*, 297:1183–1186, 2002.

[36] E. M. Ozbudak, M. Thattai, I. Kurtser, A. D. Grossman and A. van Oudenarden. Regulation of noise in the expression of a single gene. *Nat. Genet.*, 31:69–73, 2002.

[37] H. Maamar, A. Raj and D. Dubnau. Noise in gene expression determines cell fate in *Bacillus subtilis*. *Science*, 317:526–529, 2007.

[38] S. Maslov and K. Sneppen. Well-temperate phage. arXiv:1308.1646[q-bio.PE].

[39] L. A. Meyers and J. J. Bull. Fighting change with change: adaptive variation in an uncertain world. *Trends Ecol. Evol.*, 17:551–557, 2002.

[40] J-W. Veening, W. K. Smits and O. Kuipers. Bistability, epigenetics and bet-hedging in bacteria. *Annu. Rev. Microbiol*, 62:193–210, 2008.

[41] E. Meron. Pattern formation in excitable media. *Phys. Rep.*, 218:1–66, 1992.

[42] D. A. Smith and M. A. Geeves. Strain-dependent cross-bridge cycle for muscle. *Biophys. J.*, 69:524–537, 1995.

[43] A. Hansen, M. H. Jensen, K. Sneppen and G. Zocchi. Modelling molecular motors as folding- unfolding cycles. *Europhys. Lett.*, 50:120, 2000.

[44] J. J. Hopfield. Kinetic proofreading: a new mechanism for reducing errors in biosynthetic processes requiring high specificity. *Proc. Natl. Acad. Sci. USA*, 71:4135–4139, 1974.

[45] B. Kramer, W. Kramer and H. J. Fritz. Different base/base mismatches are corrected with different efficiencies by the methyl-directed DNA mismatch-repair system of *E. coli*. *Cell*, 38:879–887, 1984.

[46] T. W. McKeithan. Kinetic proofreading in T-cell receptor signal transduction. *Proc. Natl. Acad. Sci. USA*, 92:5041–5046, 1995.

[47] B. Goldstein, J. R. Faeder and W. S. Hlavacek. Mathematical and computational models of immune-receptor signalling. *Nat. Rev. Immunol.*, 4:445–456, 2004.

[48] H. R. Drew, R. M. Wing, T. Takano *et al.* Structure of a b-dna dodacamer, conformation and dynamics. *Proc. Natl Acad. Sci.*, 78:2179, 1981.

[49] P. B. Moore and H. Shi. The crystal structure of yeast phenylalanine tRNA at 1.93AA resolution: a classic structure revisited. *RNA*, 6:1091, 2000.

[50] D. H. Ohlendorf, D. E. Tronrud, and B. W. Matthews. Refined structure of cro repressor protein from bacteriophage lambda suggests both flexibility and plasticity. *J. Mol. Biol.*, 280:129, 1998.

[51] F. H. C. Crick, F. R. S. Leslie Barnett, S. Brenner and R. J. Watts-Tobin. General nature of the genetic code for proteins. *J. Mol. Biol.*, 192:1227–1232, 1961.

[52] S. Pedersen. *Escherichia coli* ribosomes translate *in vivo* with variable rate. *EMBO J.*, 3:2895–2898, 1984.

[53] N. Mitarai, K. Sneppen and S. Pedersen. Ribosome collisions and translation efficiency: optimization by codon usage and mRNA destabilization. *J. Mol. Biol.*, 382:236–245, 2008.

[54] T. D. Schneider. Strong minor groove base conservation in sequence logos implies DNA distortion or base flipping during replication and transcription initiation. *Nucleic Acids Res.*, 29:4881–4891, 2001.

[55] T. D. Schneider. Claude Shannon: biologist. The founder of information theory used biology to formulate the channel capacity. *IEEE Eng. Med. Biol. Mag.*, 25:30–33, 2006.

[56] R. K. Shultzaberger, Roberts L. R., Lyakhov I. G. *et al.* Correlation between binding rate constants and individual information of *E. coli* FIS binding sites. *Nucleic Acids Res.*, 35:5275–5283, 2007.

[57] S. Pedersen, S. Reeh and J. D. Friesen. Functional mRNA half lives in *E. coli*. *Molec. Gen. Genet.*, 166:329–336, 1978.

[58] M. Zuker. On finding all suboptimal foldings of an RNA molecule. *Science*, 249:48–52, 1989.

[59] I. L. Hofacker, W. Fontana, P. F. Stadler *et al.* Fast folding and comparison of RNA secondary structures. *Chem. Monthly*, 125:167–188, 1994.

[60] A. A. Zamyatnin. Amino acid, peptide and protein volume in solution. *Annu. Rev. Biophys. Bioeng*, 13:145–165, 1984.

[61] Y. Nozaki and C. Tanford. The solubility of amino acids and two glycine peptides in aqueous ethanol and dioxane solutions. *J. Biol. Chem.*, 246:2211, 1971.

[62] K. W. Plaxco, K. T. Simons and D. Baker. Contact order, transition state placement and the refolding rates of single domain proteins. *J. Mol. Biol.*, 277:985–994, 1998.

[63] S. Sundararaj, A. Guo, B. Habibi-Nazhad *et al.* The cybercell database (CCDB): a comprehensive, self-updating, relational database to coordinate and facilitate *in silico* modeling of *Escherichia coli*. *Nucleic Acids Res.*, 32((Database issue)):D293–D295, 2004.

[64] S. B. Zimmereman and S. O. Trach. Estimation of macromolecular concentrations in *E. coli*. *J. Mol. Biol.*, 222:599–620, 1991.

[65] Y. Taniguchi, P. J. Choi, G.-W. Li *et al.* Quantifying *E. coli* proteome and transcriptome with single-molecule sensitivity in single cells. *Science*, 329:533, 2010.

[66] P. Mongiovi and R. Milo. *openwetware.org/index.php?title = BioNumber_of_The_Month&oldid = 351581.*

[67] N. S. Hill, P. J. Buske, Y. Shi and P. A. Levin. A moonlighting enzyme links escherichia coli cell size with central metabolism. *PLoS Genet.*, 9:e1003663, 2013.

[68] S. F. Conti and M. E. Gettner. Electron microscopy of cellular division in *Escherichia coli*. *J. Bacteriol.*, 83:544, 1962.

[69] X. Wang, P. M. Llopis and D. Z. Rudner. Organization and segregation of bacterial chromosomes. *Nat. Rev. Genet.*, 14:191–203, 2013.

[70] R. Kavenoff and B. C. Bowen. Electron microscopy of membrane-free folded chromosomes from *Escherichia coli*. *Chromosoma (Berlin)*, 59:89–101, 1976.

[71] D. J. Jin and J. E. Cabrera. Coupling the distribution of RNA polymerase to global gene regulation and the dynamic structure of the bacterial nucleoid in *Escherichia coli*. *J. Struct. Biol.*, 156:284–291, 2006.

[72] F. Jacob, A. Ullman and J. Monod. The promoter, a genetic element necessary for the expression of an operon. *C.R. Acad. Sci. Paris*, 258:3125–3128, 1964.

[73] S. Liang, M. Bipatnath, Y. Xu *et al.* Activities of constitutive promoters in *Escherichia coli. J. Mol. Biol.*, 292:19–37, 1999.

[74] U. Vogel and K. F. Jensen. Effects of guanosine $3', 5'$-bidiphosphate (PPGPP) on rate of transcription elongation in isoleucine-starved *Escherichia coli. J. Biol. Chem.*, 269:16236–16241, 1994.

[75] H. Bremer and P. B. Dennis. Modulation of chemical composition and other parameters of the cell by growth rate in *Escherichia coli* and *Salmonella typhimurium*. In *Escherichia coli and Salmonella*, ed. F.C. Niedhardt, ASM Press, Washington, DC, 1996.

[76] N. Shepherd, P. Dennis and H. Bremer. Cytoplasmic RNA polymerase in *Escherichia coli. J. Bact.*, 183:2527–2534, 2001.

[77] G. Walter, W. Zilig, P. Palm and E. Fuchs. Initiation of DNA-dependent RNA synthesis and the effect of heparin on RNA polymerase. *Eur. J. Biochem.*, 3:194–201, 1967.

[78] W. R. MacClure and D. K. Hawley. Mechanism of activation of transcription initiation from the λ PRM promoter. *J. Mol. Biol*, 157:493–525, 1982.

[79] W. R. McClure. Mechanism and control of transcription initiation in procaryotes. *Ann. Rev. Biochem.*, 54:171–204, 1985.

[80] P. H. von Hippel, D. G Bear, W. D. Morgan and J. A. McSwiggen. *Ann. Rev. Biochem.*, 53:389–446, 1984.

[81] W. R. McClure. Rate-limiting steps in RNA chain initiation. *Proc. Natl. Acad. Sci. USA*, 77:5634–5638, 1980.

[82] M. Ptashne and A. Gann. *Genes and Signals*. Cold Spring Harbor Laboratory, New York 2001.

[83] B. M. Burmann, K. Schweimer, X. Luo *et al.* A NusE:NusG complex links transcription and translation. *Science*, 328:501–504, 2010.

[84] S. Proshkin, A. R. Rahmouni, A. Mironov and E. Nudler. Cooperation between translating ribosomes and RNA polymerase in transcription elongation. *Science*, 328:504–508, 2010.

[85] Y.-C. Xu, S. T. Liang, P. Dennis and H. Brenner. mRNA composition and control of bacterial gene expression. *J. Bacteriol.*, 182:3037–3044, 2000.

[86] R. J Arnold and J. P. Reilly. Observation of *Escherichia coli* ribosomal proteins and their posttranslational modifications by mass spectrometry. *Anal. Biochem.*, 269:105–12, 1999.

[87] J. Vind, M. A. Sørensen, M. D. Rasmussen and S. Pedersen. Synthesis of proteins in *Escherichia coli* is limited by the concentration of free ribosomes: Expression from reporter genes does not always reflect functional mRNA levels. *J. Mol. Biol.*, 231:678–688, 1993.

[88] Y. Asato. Control of ribosome synthesis during the cell division cycles on *E. coli* and *synechococcus. Curr. Issues Mol. Biol*, 7:109–118, 2005.

[89] C. L. Squires and D. Zaporojets. Proteins shared by the transcription and translation machines. *Ann. Rev. Microbiol.*, 54:775–798, 2000.

[90] M. Torres, C. Condon, J. M. Balada, C. Squires and C. L. Squires. Ribosomal protein S4 is a transcription factor with properties remarkably similar to NusA, a protein involved in both non-ribosomal and ribosomal RNA antitermination. *EMBO J.*, 20:3811–3820, 2001.

[91] A. D. Tadmor and T. Tlusty. A coarse grained biophysical model of *E. coli.* and its application to perturbation of the RNA operon copy number. *PloS Comput. Biol*, 4:1000038, 2008.

[92] A. Kuroda, K. Nomura, R. Ohtomo *et al.* Role of inorganic polyphosphate in promoting ribosomal protein degradation by the Lon protease in *E. coli. Science*, 293:705–708, 2001.

[93] L. Ciandrini, I. Stansfield and M. C. Romano. Role of the particles stepping cycle in an asymmetric exclusion process: a model of mRNA translation. *Phys. Rev. E*, 81(5):051904, 2010.

[94] S. Pedersen, P. L. Bloch, S. Reeh and F. C. Neidhardt. Patterns of protein synthesis in *E. coli*: a catalog of the amount of 140 individual proteins at different growth rates. *Cell*, 14:179–190, 1978.

[95] S. Ringquist, S. Shinedling, D. Barrick *et al.* Translation initiation in *Escherichia coli*: sequences within the ribosome-binding site. *Mol. Microbiol.*, 6:1219–1229, 1992.

[96] S. R. Kushner. mRNA decay in *Escherichia coli* comes of age. *J. Bacteriol*, 184:4658–4665, 2002.

[97] H. R. Ueda, S. Hayashi, S. Matsuyama *et al.* Universality and flexibility in gene expression from bacteria to human. *Proc. Nat. Acad. Sci. USA*, 101(11):3765–3769, 2004.

[98] S. Poshkin, A. R. Rahmouni, A. Mironov and E. Nudler. Cooperation between translating ribosomes and RNA polymerases in transcription elongation. *Science*, 328:504–508, 2010.

[99] J. A. Bernstein, A. B. Khodursky, P.-H. Lin *et al.* Global analysis of mRNA decay and abundance in *Escherichia coli* at single-gene resolution using two-color fluorescent DNA microarrays. *Proc. Natl. Acad. Sci. USA*, 99:9697–9702, 2002.

[100] J. Monod. The growth of bacterial cultures. *Ann. Rev. Microbiol.*, 3:371–394, 1949.

[101] M. Schaechter, O. Maaloe and N. O. Kjeldsen. Dependency on medium and temperature of cell size and chemical composition during balanced growth of salmonella typhimurium. *J. Gen. Microbiol.*, 19:592–606, 1958.

[102] M. Scott, C. W. Gunderson, E. Mateescu, Z. Zhang and T. Hwa. Interdependence of cell growth and gene expression: origins and consequences. *Science*, 330:1099–1102, 2010.

[103] M. Scott and T. Hwa. Bacterial growth laws and their applications. *Curr. Opin. Biotechnol.*, 22:559–565, 2011.

[104] Y. Wakamoto, A. Y. Grosberg and E. Kussell. Optimal lineage principle for age-structured populations. *Evolution*, 66:116–134, 2012.

[105] M. Ackermann, S. C. Stearns and U. Jenal. Senescence in a bacterium with asymmetric division. *Science*, 300(5627):1920–1920, 2003.

[106] E. J. Stewart, R. Madden, G. Paul and F. Taddei. Aging and death in an organism that reproduces by morphologically symmetric division. *PLoS Biol.*, 3(2):e45, 2005.

[107] F. C. Neidhardt and B. Magasanik. Studies on the role of ribonucleic acid in the growth of bacteria. *Biochim. Biophys. Acta*, 42:99–116, 1960.

[108] A. C. Redfield. On the proportions of organic derivations in sea water and their relation to the composition of plankton. In *James Johnstone Memorial Volume*, ed. R. J. Daniel, University Press of Liverpool, Liverpool, 177–199, 1934.

[109] S. C. Schultz, G. C. Shields and T. A. Steitz. Crystal structure of a CAP-DNA complex is bent by 90 degrees. *Science*, 253:1001–1007, 1993.

[110] A. B. Pardee, F. Jacob and J. Monod. The genetic control and cytoplasmic expression of inducibility in the synthesis of β-galactosidase by *E.coli*. *J. Mol. Biol.*, 1:165–178, 1959.

[111] E. Engelsberg, D. Sheppard, C. Squires and F. Menronk Jr. An analysis of "revertants" of a deletion mutant in the c gene of the l-arabinose gene complex in *Escherichia coli* b/r: isolation of initiator constitutive mutants (i^c). *J. Mol. Biol.*, 43:281–298, 1969.

[112] M. Ptashne. *A Genetic Switch: Gene Control and Phage Lambda.* Blackwell Scientific Publications, Palo Alto, CA, 1986.

[113] H. Nakanishi, N. Mitarai and K. Sneppen. Dynamical analysis on gene activity in the presence of repressors and an interfering promoter. *Biophys. J.*, 95:4228–4240, 2008.

[114] S. Gottesman and M. R. Maurizi. Regulation by proteolysis: energy-dependent proteases and their target. *Microbial Rev.*, 56:592, 1992.

[115] J. W. Little. The SOS regulatory system: control of its state by the level of RecA protease. *J. Mol. Biol.*, 167:791–808, 1983.

[116] N. Rosenfeld, M. B. Elowitz and U. Alon. Negative autoregulation speeds the response times of transcription networks. *J. Mol. Biol.*, 323:785–793, 2002.

[117] J. S. Cox and P. Walter. A novel mechanism for regulating activity of a transcription factor that controls the unfolded protein response. *Cell*, 87:391–404, 1996.

[118] J. B. Axelsen and K. Sneppen. Quantifying the benefits of translation regulation in the unfolded protein response. *Phys. Biol.*, 1:159, 2004.

[119] M. Behar, N. Hao, H. G. Dhlman and T. C. Elston. Mathematical and computational analysis of adaptation via feedback inhibition in signal transduction pathways. *Biophys. J.*, 93:806–813, 2007.

[120] E. Masse and S. Gottesman. A small RNA regulates the expression of genes involved in iron metabolism in *Escherichia coli. Proc. Natl. Acad. Sci. USA*, 99:4620–4625, 2002.

[121] D. H. Lenz, K. C. Mok, B. N. Lilley *et al.* The small RNA chaperone HFQ and multiple small RNAs control quorum sensing in *vibrio harveyi* and *vibrio cholerae. Cell*, 118:69–82, 2004.

[122] E. Levine, Z. Zhang, T. Kuhlman and T. Hwa. Quantitative characteristics of gene regulation mediated by small RNA. *PLoS Biol.*, 5:e229, 2007.

[123] N. Mitarai, A. M. C. Andersson, A. Krishna, S. Semsey and K. Sneppen. Efficient degradation and expression prioritization with small RNAs. *Phys. Biol.*, 4:164–171, 2007.

[124] N. Mitarai, J.-A. M Benjamin, S. Krishna *et al.* Dynamic features of gene expression control by small regulatory RNAs. *Proc. Natl. Acad. Sci. USA*, 106:10655–10659, 2009.

[125] P. Mehta, S. Goyal and N. Wingreen. A quantitative comparison of sRNA-based and protein-based gene regulation. *Mol. Syst. Biol.*, 4(1):221, 2008.

[126] H. Nakanishi, M. Pedersen, A. K. Alsing and K. Sneppen. Modeling of the genetic switch of bacteriophage TP901-1: a heteromer of CI and MOR ensures robust bistability. *JMB*, 394:15, 2009.

[127] K. Gerdes and E. Maisonneuve. Bacterial persistence and toxin-antitoxin loci. *Annu. Rev. Microbiol.*, 66:103–123, 2012.

[128] I. Cataudella, A. Trusina, K. Sneppen, K. Gerdes and N. Mitarai. Conditional cooperativity in toxin-antitoxin regulation prevents random toxin activation and promotes fast translational recovery. *Nuc. Acids Res.*, 40:6424, 2012.

[129] W. B. Whitman, D. C. Coleman and W. J. Wiebe. Prokaryotes: the unseen majority. *Proc. Natl. Acad. Sci. USA*, 95:6578–6583, 1998.

[130] C. A. Carlson and H. W. Ducklow. Oceanic bacterial production. In *Advances in Microbial Ecology*, ed. K. C. Marshall, vol. 12, Springer-Verlag, Berlin, Heidelberg, New York, 113–181, 1992.

[131] F. d'Herelles. Sur un microbe invisible antagoniste des bacilles dysentiriques. *Comptes rendus Acad. Sci. Paris*, 165:373–375, 1917.

[132] E. L. Ellis and M. Delbruck. The growth of bacteriophage. *J. Gen. Physiol.*, 22:365–384, 1939.

[133] M. Breibart and F. Rohwer. Here a virus, there a virus, everywhere the same virus. *Trends Microbiol.*, 13:278–284, 2005.

[134] C. A. Suttle. Marine viruses: major players in the global ecosystem. *Nat. Rev. Microbiol.*, 5:801–812, 2007.

[135] E. Lederberg. Lysogenicity in *Escherichia coli* strain K-12. *Microb. Gen. Bull.*, 1:5–8, 1950.

[136] H. Eisen, P. Brachet, L. P. da Silva and F. Jacob. Regulation of repressor expression in λ. *Proc. Natl. Acad. Sci. USA*, 66:855–862, 1970.

[137] M. Ptashne, A. Jeffrey, A. D. Johnson *et al.* How the lambda repressor and cro work. *Cell*, 1:1–11, 1980.

[138] A. D. Johnson, A. R. Poteete, G. Lauer *et al.* λ repressor and cro: components of an efficient molecular switch. *Nature*, 294:217–223, 1981.

[139] D. I. Friedman, E. R. Olson, C. Georgopoulos *et al.* Interactions of bacteriophage and host macromolecules in the growth of bacteriophage lambda. *Microbiol. Rev.*, 48:299–325, 1984.

[140] A. Arkin, J. Ross and H. H. McAdams. Stochastic kinetic analysis of developmental pathway bifurcation in phage λ-infected *Escherichia coli* cells. *Genetics*, 149:1633–1648, 1998.

[141] M. Ptashne. On the use of the word "epigenetic." *Curr. Biol.*, 17: R233–R236, 2007.

[142] Z. Toman, C. Dambly-Chaudihre, L. Tenenbaum and M. Radman. A system for detection of genetic and epigenetic alterations in *Escherichia coli* induced by DNA-damaging agents. *J. Mol. Biol.*, 186:97–105, 1985.

[143] A. Novick and M. Weiner. Enzyme induction as an all-or-none phenomenon. *Proc. Natl. Acad. Sci. USA*, 43:553–566, 1957.

[144] L. Wolpert and J. H. Lewis. Towards a theory of development. *Fed. Proc.*, 34:14–20, 1975.

[145] R. Thomas, A. Gathoye and L. Lambert. A complex control circuit: regulation of immunity in temperate bacteriophages. *Eur. J. Biochem.*, 71:211–227, 1976.

[146] R. Thomas. Laws for the dynamics of regulatory networks. *Int. J. Dev. Biol.*, 42:479–485, 1998.

[147] R. W. Hendrix and R. L. Duda. Bacteriophage lambda papa: not the mother of all lambda phages. *Science*, 258:1145–1148, 1992.

[148] C. Werts, V. Michel, M. Hofnung and A. Charbit. Adsorption of bacteriophage lambda on the LamB protein of *Escherichia coli* K-12: point mutations in gene J of lambda responsible for extended host range. *J. Bact.*, 176:941–947, 1994.

[149] S. Szmelcman and M. Hofnung. Maltose transport in *Escherichia coli* K-12: involvement of the bacteriophage lambda receptor. *J. Bacteriol.*, 124:112–118, 1975.

[150] P. Kourilsky. Lysogenization by bacteriophage lambda. I. Multiple infection and the lysogenic response. *Mol. Gen. Genet.*, 122:183–195, 1973.

[151] A. A. Pakula, V. B. Young and R. T. Sauer. Bacteriophage lambda cro mutations: effects on activity and intracellular degradation. *Proc. Natl. Acad. Sci. USA*, 83:8829–8833, 1986.

[152] A. Rattray, A. Altuvia, G. Mahajna, A. B. Oppenheim and M. Gottesman. Control of bacteriophage lambda CII activity by bacteriophage and host functions. *J. Bact.*, 159:238–242, 1984.

[153] P. Kourilsky. Lysogenization by bacteriophage lambda. II. Identification of genes involved in the multiplicity dependent process. *Biochimie*, 56:1511–1516, 1974.

[154] A. D. Kaiser and F. Jacob. Recombination between related temperate bacteriophages and the genetic control of immunity and prophage induction. *Virology*, 4:509, 1957.

[155] M. Ptashne. Isolation of the λ phage repressor. *Proc. Natl. Acad. Sci.*, 57:306–313, 1967.

[156] M. Dutreix, A. Bailone and R. Devoret. Efficiency of induction of prophage lambda mutants as a function of RecA alleles. *J. Bacteriol.*, 161:1080–1085, 1985.

[157] A. Trusina, K. Sneppen, I. B. Dodd, K. E. Shearwin and J. B. Egan. Functional alignment of regulatory networks: a study of temerate phages. *PloS Comp. Biol.*, 1:7, 2005.

[158] S. Gottesman, M. Gottesman, J. E. Shaw and M. L. Pearson. Protein degradation in *E. coli*: the ion mutation and bacteriophage λN and CII protein stability. *Cell*, 24:225–233, 1981.

[159] M. Glinkowska, J. Majka, W. Messer and G. Wegrzyn. The mechanism of bacteriophage λ pR promoter activity be *Escherichia coli* DnaA protein. *J. Biol. Chem.*, 278: 22250–22256, 2003.

[160] M. Chalfie, Y. Tu, G. Euskirchen, W. W. Ward and D. C. Prasher. Gene fluorescent protein as a marker for gene expression. *Science*, 263:802–805, 1994.

[161] A. Amir, O. Kobiler, A. Rokney, A. B. Oppenheim and J. Stavans. Noise in timing and precision of gene activity in a genetic cascade. *Mol. Sys. Biol.*, 3:1–10, 2007.

[162] K. Sergueev, D. Court, L. Reaves and S. Austin. *E. coli* cell cycle regulation by bacteriophage lambda. *J. Mol. Biol.*, 324:297–307, 2002.

[163] J. W. Roberts and C. W. Roberts. Proteolytic cleavage of bacteriophage lambda repressor in induction. *Proc. Natl. Acad. Sci., USA*, 72:147–151, 1975.

[164] A. Hotschild and M. Ptashne. Cooperative binding of λ repressor to sites separated by integral turns of the DNA helix. *Cell*, 44:681–687, 1986.

[165] P. J. Darling, J. M. Holt and G. K. Ackers. Coupled energetics of lambda cro repressor self-assembly and site specific DNA operator binding. *J. Mol. Biol.*, 302:625–638, 2000.

[166] J. W. Little, D. P. Shepley and D. W. Wert. Robustness of a gene regulatory circuit. *EMBO J.*, 18:4299–4307, 1999.

[167] E. L. Haseltine, V. Lam, J. Yin and J. B. Rawlings. Image-guided modeling of virus growth and spread. *Bull. Math. Biol.*, 70:1730–1748, 2008.

[168] S. Kannoly, R. Gallet and I.-N. Wang. Effects of bacteriophage traits on plaque formation. *BMC Microbiol.*, 11:181, 2011.

[169] W. Gilbert and D. Dressler. DNA replication: the rolling circle model. *Cold Spring Harb. Symp. Quant. Biol.*, 33:473–484, 1968.

[170] P. M. Lizardi, X. Huang, Z. Zhu *et al.* Mutation detection and single-molecule counting using isothermal rolling-circle amplification. *Nat. Gen.*, 19:225–232, 1998.

[171] I.-N. Wang, D. L. Smith and R. Young. The protein clocks of bacteriophage infections. *Ann. Rev. Microbiol.*, 54:799–825, 2000.

[172] S. Heilmann, K. Sneppen and S. Krishna. Sustainability of virulence in a phage-bacterial ecosystem. *J. Virology*, 84:3016–3022, 2010.

[173] G. Ackers, A. Johnson and M. Shea. Quantitative model for gene regulation by λ phage repressor. *Proc. Natl. Acad. Sci. USA*, 79:1129–1133, 1982.

[174] M. A. Shea and G. K. Ackers. The OR control system of bacteriophage lambda: a physical-chemical model for gene regulation. *J. Mol. Biol.*, 181:211–230, 1985.

[175] E. Aurell, S. Brown, J. Johansen and K. Sneppen. Stability puzzles in lambda phage. *Phys. Rev. E*, 65:51914, 2002.

[176] L. Bintu, N. E. Buchler, H. G. Garcia *et al.* Transcriptional regulation by the numbers: models. *Curr. Opin. Gen. Devel.*, 15(2):116–124, 2005.

[177] R. Philoips, J. Kondev and J. Theriot. *Physical Biology of The Cell*. Garland Science, Taylor & Francis Group, New York, 2009.

[178] P. Shkilnyj and G. B. Koudelka. Effects of salt shock on stability of λ^{imm434} lysogens. *J. Bact.*, 189:3115–3123, 2007.

[179] J. Reinitz and J. R. Vaisnys. Theoretical and experimental analysis of the phage lambda genetic switch implies missing levels of co-operativity. *J. Theoret. Biol.*, 145:295–318, 1990.

[180] I. B. Dodd, K. E. Shearwin, A. J. Perkins *et al.* Cooperativity in long-range gene regulation by the lambda ci repressor. *Genes Dev.*, 18:344–354, 2004.

[181] L. Reichardt and A. D. Kaiser. Control of λ repressor synthesis. *Proc. Natl. Acad. Sci. USA*, 68:2185–2189, 1971.

[182] T. Record, E. S. Courtenay, S. Cayley and H. J. Guttman. Biophysical compensation mechanisms buffering *E.coli* protein nucleic-acid interactions against changing environments. *Trends Biochem. Sci.*, 23:190–194, 1998.

[183] N. J. Trun and J. F. Marko. Architecture of a bacterial chromosome. *ASM News*, 64:276–283, 1998.

[184] T. A. Azam, A. Iwata, A. Nishimura, S. Ueda and A. Ishihama. Growth phase-dependent variation in protein composition of the *Escherichia coli* nucleoid. *J. Bacteriol.*, 181:6361–6370, 1999.

[185] J. Rouviere-Yaniv. Localization of the HU protein in the *E. coli* nucleoid. *Cold Spring Harbor Symp. Quant. Biol.*, 42:439–447, 1978.

[186] A. Travers. DNA-protein interactions: IHF – the master bender. *Curr. Biol.*, 7:R252–4, 1997.

[187] B. Revet, B. von Wilcken-Bergmann, H. Bessert, A. Barker and B. Muller-Hill. Four dimers of lambda repressor bound to two suitably spaced pairs of lambda operators form octamers and DNA loops over large distances. *Curr. Biol.*, 9:151–154, 1999.

[188] I. B. Dodd, A. J. Perkins, D. Tsemitsidis and J. B. Egan. Octamerization of lambda CI repressor is needed for effective repression of PRM and efficient switching from lysogeny. *Genes Dev.*, 15:3013–3022, 2001.

[189] C. Zurla, C. Manzo, D. Dunlap *et al.* Direct demonstration and quantification of long-range dna looping by the ? bacteriophage repressor. *Nucl. Acids Res.*, 37:2789–2795, 2009.

[190] J. Shimada and H. Yamakawa. Ring-closure probabilities for twisted wormlike chains. application to DNA. *Macromolecules*, 17(4):689–698, 1984.

[191] L. Han, H. G. Garcia, S. Blumberg *et al.* Concentration and length dependence of DNA looping in transcriptional regulation. *PLoS One*, 4(5):e5621, 2009.

[192] A. Yu. Grosberg, S. K. Nechaev and E. I. Shakhnovich. The role of topological constraints in the kinetics of collapse of macromolecules. *J. Phys. France*, 49:2095–2100, 1988.

[193] E. Leiberman-Aiden, N. L. van Berleum, L. Williams *et al.* Comprehensive mapping of long-range interactions reveals folding principles of the human genome. *Science*, 326:289–293, 2009.

[194] L. A. Mirny. The fractal globule as a model of chromatin architecture in the cell. *Chromosome Res.*, 19:37–51, 2011.

[195] J. K. Fisher, A. Bourniquel, G. Witz *et al.* Four-dimensional imaging of e. coli nucleoid organization and dynamics in living cells. *Cell*, 153:882–895, 2013.

[196] D. F. Senear, T. M. Lave, J. B. Ross *et al.* The primary self-assembly reaction of bacteriophage lambda ci repressor dimers is to octamer. *Biochem.*, 32:6179–6189, 1993.

[197] E. Rusinova, J. B. Ross, T. M. Laue, L. C. Sowers and D. F. Senear. Linkage between operator binding and dimer to octamer self-assembly of bacteriophage lambda CI repressor. *Biochem.*, 36:12994–13003, 1997.

[198] D. G. Priest, L. Cui, S. Kumar *et al.* Quantitation of the DNA tethering effect in long-range DNA looping *in vivo* and *in vitro* using the Lac and λ repressors. *PNAS*, 111:349–354, 2014.

[199] K. Norregaard, M. Andersson, K. Sneppen *et al.* DNA supercoiling enhances cooperativity and efficiency of an epigenetic switch. *Proc. Natl. Acad. Sci. USA*, 111: 17386–17391, 2013.

[200] G. D. Stormo and D. S. Fields. Specificity, free energy and information content in protein-DNA interactions. *TIBS*, 23:101–113, 1998.

[201] U. Gerland, J. D. Moroz and T. Hwa. Physical constraints and functional characteristics of transcription factor-DNA interaction. *Proc. Natl. Acad. Sci. USA*, 99:12015–12020, 2002.

[202] I. B. Dodd, K. E. Shearwin and K. Sneppen. Modelling transcriptional interference and DNA looping in gene regulation. *J. Mol. Biol.*, 369:1200–1213, 2007.

[203] M. B. Elowitz, M. G. Surette, P. E. Wolf, J. B. Stock and S. Leibler. Protein mobility in the cytoplasm of *Escherichia coli*. *J. Bacteriol.*, 181:197–203, 1999.

[204] K. M. Bendtsen, M. B. Jensen, A. May *et al.* Distinct dynamics of DNA repair proteins in the nucleoplasm, nucleoli and at sites of DNA damage. Submitted to *PLoS ONE*, 2014.

[205] C. Molenaar, K. Wiesmeijer, N. P. Verwoerd *et al.* Visualizing telomere dynamics in living mammalian cells using pna probes. *EMBO J.*, 22:6631–6641, 2003.

[206] P. Nelson. *Biological Physics: Energy, Information, Life*. W.H. Freeman and Company, New York, 2004.

[207] A. Einstein. Uber die von der molekularkinetischen theorie der warme geforderte bewegung von in ruhenden flussigkeiten suspendierten teilchen. *Annalen der Physik*, 322:549–560, 2005.

[208] R. Metzler and J. Klafter. The random walk guide to anomalous diffusion: fractional dynamics approach. *Phys. Rep.*, 339:1–77, 2000.

[209] M. von Smoluchowski. Zur kinetischen theorie der brownschen molekularbewegung und der suspensionen. *Annalen der Physik*, 326:756–780, 1906.

[210] P. von Hippel and H. Berg. Facilitated target location in biological systems. *J. Biol. Chem.*, 264:675, 1989.

[211] M. A. Surby and N. O. Reich. Facilitated diffusion of the EcoRI DNA methyltransferase is described by a novel mechanism. *Biochem.*, 35:2209–2217, 1996.

[212] J. Elf, G-W. Li and S. Xie. Probing transcription factor dynamics at the single-molecule level in a living cell. *Science*, 316:1191–1194, 2007.

[213] S. Adhya and M. Gottesman. Promoter occlusion: transcription through a promoter may inhibit its activity. *Cell*, 29:939, 1982.

[214] B. P. Callen, K. E. Shearwin and J. B. Egan. Transcriptional interference between convergent promoters caused by elongation over the promoter. *Mol. Cell*, 14:647–656, 2004.

[215] K. Sneppen, I. Dodd, K. E. Shearwin *et al.* A mathematical model for transcriptional interference by RNA polymerase traffic in *Escherichia coli. J. Mol. Biol.*, 346:399, 2005.

[216] H. Y. Wu, S. Shyy, J. C. Wang and L. F. Liu. Transcription generates positively and negatively supercoiled domains in the template. *Cell*, 53:433–440, 1988.

[217] H. Bremer and P. B. Dennis. Modulation of chemical composition and other parameters of the cell by growth rate in *Escherichia coli* and *salmonella typhimurium*. In *Escherichia coli and Salmonella*, ed. F. C. Niedhardt, ASM Press, Washington, DC, 2009.

[218] E. M. Ozbudak, M. Thattai, I. Kurtser, A. D. Grossman and A. van Oudenaarden. Regulation of noise in the expression of a single gene. *Nat. Gen.*, 31(1):69–73, 2002.

[219] M. Thattai and A. van Oudenaarden. Intrinsic noise in gene regulatory networks. *Proc. Natl. Acad. Sci. USA*, 98(15):8614–8619, 2001.

[220] I. Golding, J. Paulsson, S. M. Zawilski1 and E. C. Cox. Real-time kinetics of gene activity in individual bacteria. *Cell*, 123:1025–1036, 2005.

[221] N. Mitarai, I. B. Dodd, M. T. Crooks and K. Sneppen. The generation of promoter-mediated transcriptional noise in bacteriaa. *PLoS Comp. Biol.*, 4:e1000109, 2008.

[222] J. T. Kittleson, S. Cheung and J. C. Anderson. Rapid optimization of gene dosage in *E. coli* using dial strains. *J. Biol. Eng.*, 5:10, 2011.

[223] N. G. Van Kampen. *Stochastic Processes in Physics and Chemistry*, third edition. Elsevier, Amsterdam, 2007.

[224] H. M. Lim, D. E. Lewis, H. J. Lee, M. Liu and S. Adhya. Effect of varying the supercoiling of dna on transcription and its regulation. *Biochem.*, 42:10718–10725, 2003.

[225] M. L. Opel and G. W. Hatfield. DNA supercoiling-dependent transcriptional coupling between the divergent transcribed promoters of the ilvYC operon of *Escherichia coli* is proportional to promoter strength and transcription lengths. *Mol. Micro.*, 39:191–198, 2001.

[226] F. Jacob, D. Perrin, C. Sanchez and J. Monod. The operon: a group of genes with expression coordinated by an operator. *C. R. Acad. Sci.* (Paris), 250:1727–1729, 1960.

[227] K. Sneppen, S. Pedersen, S. Krishna, I. Dodd and S. Semsey. Economy of operon formation: cotranscription minimizes shortfall in protein complexes. *MBio.*, 1:e00177, 2010.

[228] J. Paulsson. Models of stochastic gene expression. *Phys. Life Rev.*, 2:157–175, 2005.

[229] K. T. Bæk, S. Svenningsen, H. Eisen, K. Sneppen and S. Brown. Single-cell analysis of lambda immunity regulation. *J. Mol. Biol.*, 334:363, 2003.

[230] J. Paulsson. Summing up noise in gene networks. *Nature*, 427:415–418, 2004.

[231] Y. Dublanche, K. Michalodimitrakis, N. Kummerer, N. Foglierini and L. Serrano. Noise in transcription negative feedback loops: simulation and experimental analysis. *Mol. Syst. Biol.*, 2:41, 2006.

[232] D. A. McQuarrie. Stochastic approach to chemical kinetics. *J. Appl. Prob.*, 4:413, 1967.

[233] D. T. Gillespie. A general method for numerically simulating the stochastic time evolution of coupled chemical reactions. *J. Comp. Phys.*, 22:403–434, 1976.

[234] D. T. Gillespie. Exact stochastic simulation of coupled chemical reactions. *J. Phys. Chem.*, 81:2340–2361, 1977.

[235] R. J. Allen, P. B. Warren and P. Rein ten Wolde. Sampling rare switching events in biochemical networks. *Phys. Rev. Lett.*, 94:018104, 2005.

[236] J. S. van Zon and P. Rein ten Wolde. Green's-function reaction dynamics: a particle-based approach for simulating biochemical networks in time and space. *J. Chem. Phys.*, 123:234910, 2005.

[237] R. J. Allen, D. Fraenkel and P. Rein ten Wolde. Forward flux sampling-type schemes for simulating rare events: efficiency analysis. *J. Chem. Phys.*, 124:194111, 2006.

[238] T. S. Gardner, C. R. Cantor and J. J. Collins. Construction of a genetic toggle switch in *Escherichia coli*. *Nature*, 403:339–342, 2000.

[239] I. B. Dodd, K. E. Shearwin and J. B. Egan. Revisited gene regulation in bacteriophage lambda. *Curr. Opin. Genet. Dev.*, 15:145–152, 2005.

[240] J. W. Little and C. B. Michalowski. Stability and instability in the lysogenic state of phage lambda. *J. Bacteriol.*, 192:6064–6076, 2010.

[241] I. Golding. Decision making in living cells: lessons from a simple system. *Annu. Rev. Biophys.*, 40:63–80, 2011.

[242] P. Langevin. On the theory of brownian motion. *C.R. Acad. Sci. (Paris)*, 146:530533, 1908.

[243] E. Aurell and K. Sneppen. Epigenetics as a first exit problem. *Phys. Rev. Lett.*, 88:51914, 2002.

[244] O. Kobiler, A. Rokney, N. Friedman *et al.* Quantitative kinetic analysis of the bacteriophage λ genetic network. *Proc. Natl. Acad. Sci. USA*, 102:4470–4475, 2005.

[245] A. C. Palmer, A. Ahlgren-Berg, J. B. Egan, I. B. Dodd and K. E. Shearwin. Potent transcriptional interference by pausing of RNA polymerases over a downstream promoter. *Mol. Cell*, 34:545–555, 2009.

[246] J. S. Weitz, Y. Mileyko, R. I. Joh and E. Q. Voit. Collective decision making in bacterial viruses. *Biophys. J.*, 95:2673–2680, 2008.

[247] M. Avlund, I. B. Dodd, K. Sneppen and S. Krishna. Minimal gene regulatory circuits that can count like bacteriophage lambda. *J. Bacteriol.*, 191:4487, 2009.

[248] M. Avlund, S. Krishna, S. Semsey, I. B. Dodd and K. Sneppen. Minimal gene regulatory circuits for lysis-lysogeny choice in the presence of noise. *PLoS One*, 5:e15037, 2010.

[249] G. Von Dassow and G. M. Odell. Design and constraints of the *drosophila* segment polarity module: robust spatial patterning emerges from intertwined cell state switches. *J. Exp. Zool.*, 294:179–215, 2002.

[250] P. Kourilsky and A. Knapp. Lysogenization by bacteriophage lambda. III. Multiplicity dependent phenomena occurring upon infection by lambda. *Biochimie*, 56:1517–1523, 1974.

[251] O. Brandman, J. E. Ferrell Jr., R. Li and T. Meyer. Interlinked fast and slow positive feedback loops drive reliable cell decisions. *Science*, 310:496–498, 2005.

[252] F. St-Pierre and D. Endy Determination of cell fate selection during phage lambda infection. *Proc. Natl. Acad. Sci. USA*, 105:20705–20710, 2008.

[253] L. Zeng, S. O. Skinner, C. Zong *et al.* Decision making at a subcellular level determines the outcome of bacteriophage infection. *Cell*, 141(4):682–691, 2010.

[254] P. Deighan and A. Hochschild. The bacteriophage λ Q anti-terminator protein regulates late gene expression as a stable component of the transcription elongation complex. *Mol. Microbiol.*, 63:911–920, 2007.

[255] M. Lehmann and K. Sneppen. Gene regulatory networks that count to 3. *JTB*, 329:15–19, 2013.

[256] N. Balaskas, A. Ribeiro, J. Panovska *et al.* Gene regulatory logic for reading the sonic hedgehog signaling gradient in the vertebrate neural tube. *Cell*, 148:273–284, 2012.

[257] H. Echolz, D. Court and L. Green. On the nature of *cis*-acting regulatory proteins and genetic organization in bacteriophage: The example of gene Q in bacteriophage λ. *Genetics*, 83:5–10, 1976.

[258] E. McFall. *Cis*-acting proteins. *J. Bacteriol.*, 167:429, 1986.

[259] D. W. Burt and W. J. Brammar. The *cis*-specificity of the Q-gene product of bacteriophage lambda. *Mol. Gen. Genet.*, 185:468–472, 1982.

[260] G. Lindahl. Bacteriophage p2: replication of the chromosome requires a protein which acts only on the genome that code for it. *Virology*, 17:522–533, 1966.

[261] A. Bird. DNA methylation patterns and epigenetic memory. *Genes Dev.*, 16:6–21, 2002.

[262] J. L. MacDonald, C. S. Y. Gin and A. J. Roskams. Stage-specific induction of dna methyltransferases in olfactory receptor neuron development. *Dev. Biol.*, 288:461–473, 2005.

[263] M. Braunstein, R. E. Sobel, C. D. Allis, B. M. Turner and J. R. Broach. Efficient transcriptional silencing in saccharomyces cerevisiae requires a heterochromatin histone acetylation pattern. *Mol. Cell Biol.*, 16:4349–4356, 1996.

[264] G. Thon and T. Friis. Epigenetic inheritance of transcriptional silencing and switching competence in fission yeast. *Genetics*, 145:685, 1997.

[265] S. I. Grewal and D. Moazed. Heterochromatin and epigenetic control of gene expression. *Science*, 301:798–802, 2003.

[266] E. Y. Xu, K. A. Zawadzki and J. R. Broach. Single-cell observations reveal intermediate transcriptional silencing states. *Mol. Cell*, 23:219–229, 2006.

[267] A. Chess, I. Simon, C. Howard and A. Richard. Allelic inactivation regulates olfactory receptor gene expression. *Cell*, 78:823–834, 1994.

[268] M. Kambere and R. Lane. Co-regulation of a large and rapidly evolving repertoire of odorant receptor genes. *BMC Neurosci.*, 8(Suppl.3):S2, 2007.

[269] S. H. Fuss, M. Omura and P. Mombaerts. Local and *cis* effects of the H element on expression of odorant receptor genes in mouse. *Cell*, 130:373–384, 2007.

[270] S. DeMaria and J. Ngai. The cell biology of smell. *J. Cell Biol.*, 191:443–452, 2010.

[271] T. Sugiyama, H. Can, A. Vedel, D. Moazed and S. I. Greval. RNA polymerase is an essential component of a self-enforcing loop coupling heterochromatin assembly to siRNA production. *Proc. Natl. Acad. Sci. USA*, 102:152–157, 2006.

[272] A. Magklara, A. Yen, B. M. Colquitt *et al.* An epigenetic signature for monoallelic olfactory receptor expression. *Cell*, 145:555–570, 2011.

[273] L. A. Boyer, K. Plath, J. Zeitlinger *et al.* Polycomb complexes repress developmental regulators in murine embryonic stemm cells. *Nature*, 441:349–353, 2006.

[274] T. I. Lee, R. G. Jenner, L. A. Boyer *et al.* Control of developmental regulators by polycomb in human embryonic stem cells. *Cell*, 125:301–313, 2006.

[275] M. Sauvageau and G. Sauvageau. Polycomb group genes: keeping stem cell activity in balance. *PLoS Biol.*, 6(4):e113, 2008.

[276] Y. B. Schwartz and V. Pirrotta. Polycomb complexes and epigenetic states. *Curr. Opin. Cell Biol.*, 20:266–273, 2008.

[277] R. Margueron, N. Justin, K. Ohno *et al.* Role of the polycomb protein EED in the propagation of repressive histone marks. *Nature*, 461:762–767, 2009.

[278] B. Schuettengruber, A. M. Martines, N. Livino and G. Cavalli. Thritorax group proteins: switching genes on and keeping them active. *Nat. Rev. Mol. Cell. Biol.*, 12:799–814, 2011.

[279] C. A. Davey, D. F. Sargent, A. W. Maeder K. Luger and T. J. Richmond. Solvent mediated interactions in the structure of the nucleosome core particle at 1.9 angstrøm resolution. *J. Mol. Biol.*, 319:1097–1113, 2002.

[280] T. E. Shrader and D. M. Crothers. Artificial nucleosome positioning sequences. *Proc. Natl. Acad. Sci. USA*, 86:7418–7422, 1989.

[281] P. T. Lowary and J. Widom. New DNA sequence rules for high affinity binding to histone octamer and sequence-directed nucleosome positioning. *J. Mol. Biol.*, 276:19–42, 1998.

[282] B. R. Cairns, Y. Lorch, Y. Li *et al.* Rsc; an essential abundant chromatin-remodeling complex. *Cell*, 87:1249–1260, 1996.

[283] M. Kruithof, F. T. Chien, A. Routh *et al.* Single-molecule force spectroscopy reveals a highly compliant helical folding for the 30-nm chromatin fiber. *Nat. Struct. Mol. Biol.*, 16:534, 2009.

[284] B. D. Reddy, Y. Wang, E. C. Higuchi *et al.* Elimination of a specific histone H3K14 acetyltransferase complex bypasses the RNAI pathway to regulate pericentric heterochromatin functions. *Genes Dev.*, 25:214–219, 2011.

[285] A. Johnson, G. Li, T. W. Sikorski *et al.* Reconstitution of heterochromatin-dependent transcriptional gene silencing. *Mol. Cell*, 35:769–781, 2009.

[286] J. Nakayama, J. C. Rice, B. D. Strahl, C. D. Allis and S. I. Grewal. Role of histone H3 lysine 9 methylation in epigenetic control of heterochromatin assembly. *Science*, 292:110–113, 2001.

[287] A. J. Bannister, P. Zegerman, J. F. Partridge *et al.* Selective recognition of methylated lysine 9 on histone H3 by the HP1 chromo domain. *Nature*, 410:120–124, 2001.

[288] T. A. Volpe, C. Kidner, I. M. Hall *et al.* Regulation of heterochromatic silencing and histone H3 lysine-9 methylation by RNAI. *Science*, 297:1833–1837, 2002.

[289] A. Verdel, S. Jia, S. Gerber *et al.* RNAI-mediated targeting of heterochromatin by the RITS complex. *Science*, 303:672–676, 2004.

[290] V. J. Petrie, J. D. Wuitschick, C. D. Givens, A. M. Kosinski and J. F. Partridge. RNA interference (RNAI)- dependent and RNAI-independent association of the CHP1 chromodomain protein with distinct heterochromatic loci in fission yeast. *Mol. Cell Biol.*, 25:2331–2346, 2005.

[291] T. Schalch, G. Job, V. J. Noffsinger *et al.* High-affinity binding of chp1 chromodomain to K9 methylated histone H3 is required to establish centromeric heterochromatin. *Mol. Cell*, 34:36–46, 2009.

[292] K. Zhang, K. Mosch, W. Fischle and S. I. Grewal. Roles of the CLR4 methyltransferase complex in nucleation, spreading and maintenance of heterochromatin. *Nat. Struct. Mol. Biol.*, 15:381–388, 2008.

[293] T. Yamada, W. Fischle, T. Sugiyama, C. D. Allis and S. I. Grewal. The nucleation and maintenance of heterochromatin by a histone deacetylase in fission yeast. *Mol. Cell*, 20:173–185, 2005.

[294] M. R. Motamedi, E. J. Hong, X. Li *et al.* HP1 proteins form distinct complexes and mediate heterochromatic gene silencing by nonoverlapping mechanisms. *Mol. Cell*, 32:778–790, 2008.

[295] T. Fisher, B. Cui, J. Dhakshnamoorthy *et al.* Diverse roles of hp1 proteins in heterochromatin assembly and functions in fision yeast. *Proc. Natl. Acad. Sci. USA.*, 106:8998–9003, 2009.

[296] S. I. Grewal, M. J. Bonaduce and A. J. Klar. Histone deacetylase homologs regulate epigenetic inheritance of transcriptional silencing and chromosome segregation in fission yeast. *Genetics*, 150:563–576, 1998.

[297] P. Bjerling, R. A. Silverstein, G. Thon *et al.* Functional divergence between histone deacetylases in fission yeast by distinct cellular localization and *in vivo* specificity. *Mol. Cell Biol.*, 22:2170–2181, 2002.

[298] G. D. Shankaranarayana, M. R. Motamedi, D. Moazed and S. I. Grewal. Sir2 regulates histone H3 lysine 9 methylation and heterochromatin assembly in fission yeast. *Curr. Biol*, 13:1240–1246, 2003.

[299] E. L. Gerace, D. Moazed and M Halic. The methyl transferase activity of Clr4Suv39h triggers RNAi independently of histone H3K9. *Mol. Cell*, 39:360–372, 2010.

[300] D. Moazed. Mechanisms for the inheritance of chromatin states. *Cell*, 146:510–518, 2011.

[301] E. H. Bayne, S. A. White, A. Kagansky *et al.* Stc1: a critical link between RNAI and chromatin modification required for heterochromatin integrity. *Cell*, 140:666–667, 2010.

[302] A. Hayashi, M. Ishida, R. Kawaguchi *et al.* Heterochromatin protein 1 homologue Swi6 acts in concert with Ers1 to regulate RNAI-directed heterochromatin assembly. *Proc. Natl. Acad. Sci. USA*, 109:6159–6164, 2012.

[303] M. Rougemaille, S. Braun, S. Coyle *et al.* Ers1 links HP1 to RNAI. *Proc. Natl. Acad. Sci. USA*, 109:11258–11263, 2012.

[304] F. Lan, M. Zaratiequi, J. Villén *et al.* S. pombe LSD1 homologs regulate heterochromatin propagation and euchromatic gene transcription. *Mol. Cell*, 26:89–101, 2007.

[305] M. Opel, D. Lando, C. Bonilla *et al.* Genome wide studies of histone demethylation catalysed by the fission yeast homologs of mammalian LSD1. *Plos One*, 2:e386, 2007.

[306] I. B. Dodd, M. A. Micheelsen, K. Sneppen and G. Thon. Theoretical analysis of epigenetic cell memory by nucleosome modification. *Cell*, 129:813–822, 2007.

[307] G. J. Hoppe, J. C. Tanny, A. D. Rudner *et al.* Steps in assembly of silent chromatin in yeast: Sir3-independent binding of a Sir2/Sir4 complex to silencers and role for Sir2-dependent deacetylation. *Mol. Cell. Biol.*, 22:4167–4180, 2002.

[308] J. S. Smith, C. B. Brachmann, L. Pillus and J. D. Boeke. Distribution of a limited Sir2 protein pool regulates the strength of yeast RDNA silencing and is modulated by Sir4p. *Genetics*, 149:1205–1219, 1998.

[309] M. Sedighi and A. M. Sengupta. Epigenetic chromatin silencing: bistability and front propagation. *Phys. Biol.*, 4:246–255, 2007.

[310] P. D. Kaufman and O. J. Rando. Chromatin as a potential carrier of heritable information. *Curr. Opin. Cell Biol.*, 22:284–290, 2010.

[311] A. Angel, J. Song, C. Dean and M. Howard. A polycomb-based switch underlying quantitative epigenetic memory. *Nature*, 476:105–108, 2011.

[312] K. H. Hansen, A. P. Bracken, D. Pasini *et al.* A model for transmission of the H3K27me3 epigenetic mark. *Nat. Cell Biol.*, 10:1291–300, 2008.

[313] J. R. Danzer and L. L. Wallrath. Mechanisms of HP1-mediated gene silencing in *drosophila*. *Development*, 131:3571–80, 2004.

[314] J. Wysocka, T. Swigot, T. A. Milne *et al.* WDR5 associates with histone H3 methylated at K4 and is essential for H3 K4 methylation and vertebrate development. *Cell*, 121:859–872, 2005.

[315] S. I. Grewal and S. C. Elgin. Heterochromatin: new possibilities for the inheritance of structure. *Curr. Opin. Genet. Dev.*, 12:178–187, 2002.

[316] H. Renauld, O. M. Aparicio, P. D. Zierath *et al.* Silent domains are assembled continuously from the telomere and are defined by promoter distance and strength, and by Sir3 dosage. *Genes Dev.*, 7:1133–1145, 1993.

[317] E. Ising. Beitrag zur Theorie des Ferromagnetismus. *Z. Phys.*, 31:253–258, 1925.

[318] D. Poland and H. A. Scheraga. Phase transitions in one dimension and the helix-coil transition in polyamino acids. *J. Chem. Phys.*, 45:1456, 1966.

[319] M. A. Micheelsen, N. Mitarai, K. Sneppen and I. B. Dodd. Theory for the stability and regulation of epigenetic landscapes. *Phys. Biol.*, 7:026010, 2010.

[320] C. H. Waddington. *The Strategy of the Genes: A Discussion of Some Aspects of Theoretical Biology*. Allen & Unwin, Sydney, 1957.

[321] K. Sneppen, M. A. Micheelsen and I. B. Dodd. Ultrasensitive gene regulation by positive feedback loops in nucleosome modification. *Mol. Syst. Biol.*, 4:182, 2008.

[322] M. Vignali, D. J. Steeger, K. E. Neely and J. L. Workman. Distribution of acetylated histones resulting from GAL4-VP16 recruitment of SAGA and NuA4 complexes. *EMBO J.*, 19:2629–2640, 2000.

[323] M. Carey, Y. S. Lin, M. R. Green and M. Ptashne. A mechanism for synergistic activation of a mammalian gene by gal4 derivatives. *Nature*, 345:361–364, 1990.

[324] C. M. Ajo-Franklin, D. A. Drubin, J. A. Eslein *et al.* Rational design of memory in eukaryotic cells. *Genes Dev.*, 21:2271–2276, 2007.

[325] G. Berrozpe, G. O. Bryant, K. Warpinski and M. Ptashne. Regulation of a mammalian gene bearing a cpg island promoter and a distal enhancer. *Cell Rep.*, 4:445–453, 2013.

[326] I. B. Dodd and K. Sneppen. Barriers and silencers: a theoretical toolkit for control and containment of nucleosome-based epigenetic states. *J. Mol. Biol.*, 414:624–637, 2011.

[327] L. Ringrose, S. Chabanis, P. O. Angrand, C. Woodroofe and A. F. Stewart. Quantitative comparison of dna looping *in vitro* and *in vivo*: chromatin increases effective DNA flexibility at short distances. *EMBO J.*, 18:6630–6641, 1999.

[328] K. Weiss and R. T. Simpson. High-resolution structural analysis of chromatin at specific loci: *Saccharomyces cerevisiae* silent mating type locus HML alpha. *Mol. Cell. Biol.*, 18:5392–5403, 1998.

[329] A. Ravindra, K. Weiss and R. T. Simpson. High resolution structural analysis of chromatin at specific loci: *Saccharomyces cerevisiae* silent mating-type locus HMRA. *Mol. Cell. Biol.*, 19:7944–7950, 1999.

[330] S. Venkatasubrahmanyam, W. W. Hwang, M. D. Meneghini, A. H. Tong and H. D. Madhani. Genomewide, as opposed to local, antisilencing is mediated redundantly by the euchromatic factors Set1 and H2A. *Proc. Natl. Acad. Sci. USA*, 104:16609–16614, 2007.

[331] A. Turing. The chemical basis of morphogenesis. *Phil. Trans. B*, 237:37–72, 1952.

[332] A. Gierer and H. Mainhardt. A theory of biological pattern formation. *Kybernetik*, 12:30–39, 1972.

[333] A. K. Alsing and K. Sneppen. Differentiation of developing olfactory neurons analysed in terms of coupled epigenetic landscapes. *Nucl. Acids Res.*, 41:4755–4764, 2013.

[334] P. Qasba and R. R. Reed. Tissue and zonal-specific expression of an olfactory receptor transgene. *J. Neurosci.*, 18:227–236, 1998.

[335] B. M. Shykind. Gene switching and the stability of odorant receptor gene choice. *Cell*, 117:801–815, 2004.

[336] S. Serizawa. Negative feedback regulation ensures the one receptor-one olfactory neuron rule in mouse. *Science*, 302:2088–2094, 2003.

[337] J. W. Lewcock and R. R. Reed. A feedback mechanism regulates monoallelic odorant receptor expression. *Proc. Natl. Acad. Sci. USA*, 101:1069–1074, 2004.

[338] K. Sneppen and I. B. Dodd. A simple histone code opens many paths to epigenetics. *PLoS Comp. Biol.*, 8:e1002643, 2012.

[339] K. Sneppen and N. Mitarai. Multistability with a mixed metastable state. *Phys. Rev. Lett.*, 109:100602, 2012.

[340] A. P. Bird. CpG islands as gene markers in the vertebrate nucleus. *Trends Genet.*, 3:343–347, 1987.

[341] M. Monk, M. Boubelik and S. Lehnert. Temporal and regional changes in DNA methylation in the embryonic, extraembryonic and germ cell lineages during mouse embryo development. *Development*, 99:371–382, 1987.

[342] R. Goyal, R. Reinhardt and A. Jeltsch. Accuracy of DNA methylation pattern preservation by the Dnmt1 methyltransferase. *Nucl. Acids Res.*, 34:1182–1188, 2006.

[343] A. Jeltsch. On the enzymatic properties of Dnmt1. *Epigenetics*, 1:63–66, 2006.

[344] H. Cedar and Y. Bergman. Linking DNA methylation and histone modification: patterns and paradigms. *Nat. Rev., Genet.*, 10:295–304, 2009.

[345] M. Okano, D. W. Bell, D. A. Haber and E. Li. DNA methyltransferases Dnmt3a and Dnmt3b are essential for de novo methylation and mammalian development. *Cell*, 99:247–257, 1999.

[346] Y. F. He, B.-Z. Li, Z. Li *et al.* Tet-mediated formation of 5-carboxylcytosine and its excision by TDG in mammalian DNA. *Science*, 333:1303–1307, 2011.

[347] L. B. Sontag, W. C. Lorincz and G. Luebeck. Dynamics, stability and inheritance of somatic DNA methylation imprints. *J. Theoret. Biol.*, 242:890–899, 2006.

[348] K. Williams, J. Christensen and K. Helin. DNA methylation: Tet proteins guardians of CpG islands? *EMBO Rep.*, 13:28–35, 2011.

[349] K. Williams, J. Christensen, M. T. Pedersen *et al.* Tet1 and hydroxymethylcytosine in transcription and DNA methylation fidelity. *Nature*, 473:343–348, 2011.

[350] S. K. Ooi, C. Qui, E. Bernetein *et al.* DNMT3L connects unmethylated lysine 4 of histone H3 to de novo methylation of DNA. *Nature*, 448:714–717, 2007.

[351] F. Mohn, M. Weber, M. Rebhan *et al.* Lineage-specific polycomb targets and de novo DNA methylation define restriction and potential of neuronal progenitors. *Mol. Cell*, 30:755–766, 2008.

[352] H. Cedar and Y. Bergman. Linking DNA methylation and histone modification: patterns and paradigms. *Genetics*, 10:295–304, 2009.

[353] S. Semsey, S. Krishna, K. Sneppen and S. Adhya. Signal integration in the galactose network of *Escherichia coli. Mol. Microbiol.*, 65:465, 2007.

[354] S. Krishna, L. Orosz, K. Sneppen, S. Adhya and S. Semsey. Relation of intracellular signal levels and promoter activities in the GAL regulon of *Escherichia coli. J. Mol. Biol.*, 391:671, 2009.

[355] S. Semsey, A. M. C. Andersson, S. Krishna *et al.* Genetic regulation of fluxes: Iron homeostasis of *Escherichia coli. Nucl. Acids Res.*, 34:4960–4967, 2006.

[356] A. Amir, S. Meshner, T. Beatus and J. Stavans. Damped oscillations in the adaptive response of the iron homeostasis network of *E. coli. Mol. Microbiol.*, 76:428–436, 2010.

[357] S. V. Aksenov. Dynamics of the inducing signal for the SOS regulatory system in *Escherichia coli* after ultraviolet irradiation. *Math. Biosci*, 157:269–286, 1999.

[358] N. Friedman, S. Vardi, M. Ronen, U. Alon and J. Stavans. Precise temporal modulation in the response of the SOS DNA repair network in individual bacteria. *PloS Biol.*, 3:1261–1267, 2005.

[359] S. Krishna, S. Maslov and K. Sneppen. UV-induced mutagenesis in the *Escherichia coli* SOS response: a quantitative model. *PLoS Comp. Biol.*, 3:e41, 2007.

[360] J. C. Gerhart and A. C. Pardee. The enzymology of control by feedback inhibition. *J. Biol. Chem.*, 237:891–896, 1961.

[361] B. Goodwin. Oscillatory behavior in enzymatic control processes. *Adv. Enzyme Reg.*, 3:425–428, 1965.

[362] R. D. Bliss, P. R. Painter and A. G. Marr. Role of feedback inhibition in stabilizing the classical operon. *J. Theor. Biol.*, 97:177–93, 1982.

[363] A. Goldbeter and D. E. Koshland Jr. Ultrasensitivity in biochemical systems controlled by covalent modification. Interplay between zero-order and multistep effects. *J. Biol. Chem.*, 259:14441–14447, 1984.

[364] J. E. Ferrell. Jr. Tripping the switch fantastic: how a protein kinase cascade can convert graded inputs into switch-like outputs. *Trends Biochem. Sci.*, 21:460–466, 1996.

[365] P. Ruoff and L. Rensing. The temperature-compensated Goodwin model simulates many circadian clock properties. *J. Theoret. Virol.*, 179:275–285, 1996.

[366] N. Barkai and S. Leibler. Robustness in simple biochemical networks. *Nature*, 387:913–917, 1997.

[367] A. Becskei and L. Serrano. Engineering stability in gene networks by autoregulation. *Nature*, 405:590–593, 2000.

[368] M. B. Elowitz and S. Leibler. A synthetic oscillatory network of transcriptional regulators. *Nature*, 403:335–338, 2000.

[369] J. Hasty, J. Pradines, M. Dolnik and J. J. Collins. Noise-based switches and amplifiers for gene expression. *Proc. Natl. Acad. Sci. USA*, 97:2075–2080, 2000.

[370] S. W. Omholt, E. Plahte, L. Øyehaug and K. Xiang. Gene regulatory networks generating the phenomena of additivity, dominance and epistasis. *Genetics*, 155:969–980, 2000.

[371] R. L. Bar-Or, R. Maya, L. A. Segel *et al.* Generation of oscillations by the p53-MDM2 feedback loop: a theoretical and experimental study. *Proc. Natl. Acad. Sci. USA*, 97:11250–11255, 2000.

[372] H. Hirata, S. Yoehivra, T. Ohtsulea *et al.* Oscillatory expression of the bhlh factor hes1 regulated by a negative feedback loop. *Science*, 298:840–843, 2002.

[373] D. Gonze, J. Halloy and A. Goldbeter. Robustness of circadian rhythms with respect to molecular noise. *Proc. Natl. Acad. Sci. USA*, 99:673–678, 2002.

[374] J. R. Ferrell, Jr. Self-perpetuating states in signal transduction: positive feedback, double-negative feedback and bistability. *Curr. Opin. Cell Biol.*, 14:140–148, 2002.

[375] A. Hoffmann, A. Levchenko, M. L. Scott and D. Baltimore. The IkappaB-NF-kappaB signaling module: temporal control and selective gene activation. *Science*, 298:1241–1245, 2002.

[376] G. Tiana, M. H. Jensen and K. Sneppen. Time delay as a key to apoptosis induction in the p53 network. *Eur. Phys. J. B*, 29:135–140, 2002.

[377] M. H. Jensen, G. Tiana and K. Sneppen. Sustained oscillations and time delays in gene expression of protein HES1. *FEBS Lett.*, 541:176–177, 2003.

[378] J. Lewis. Autoinhibition with transcriptional delay: a simple mechanism for the zebrafish somitogenesis oscillator. *Curr. Biol.*, 13:1398–408, 2003.

[379] A. Trusina, F. R. Papa and C. Tang. Rationalizing translation attenuation in the network architecture of the unfolded protein response. *Proc. Natl. Acad. Sci. USA.*, 105:20280–20285, 2008.

[380] K. Sneppen, S. Krishna and S. Semsey. Simplified models of biological networks. *Ann. Rev. Biophys.*, 39:43–59, 2010.

[381] J. J. Tyson and B. Novak. Functional motifs in biochemical reaction networks. *Ann. Rev. Phys. Chem.*, 61:219–240, 2010.

[382] S. S. Shen-Orr, R. Milo, S. Mangan and U. Alon. Network motifs in the transcriptional regulation of *Escherichia coli*. *Nat. Genet.*, 22(2002), 2002.

[383] U. Alon. Network motifs: theory and experimental approaches. *Nat. Rev. Genet.*, 8(6):450–461, 2007.

[384] S. Krishna, M. H. Jensen and K. Sneppen. Minimal model of spiky oscillations in NF-κB signalling. *Proc. Natl. Acad. Sci. USA*, 103:10840–10845, 2006.

[385] D. E. Nelson, A. E. C. Ihekwaba, M. Elliot *et al.* Oscillations in NF-κB signalling control the dynamics of gene expression. *Science*, 306:704–708, 2004.

[386] L. Ashall C. A. Horton, D. E. Nelson *et al.* Pulsatile stimulation determines timing and specifity of NF-κB-dependent transcription. *Science*, 324:242–246, 2009.

[387] M. Mandal and R. R. Breaker. Gene regulation by riboswitches. *Nat. Rev. Mol. Cell Biol.*, 5:451–463, 2004.

[388] M. E. Wall, W. S. Hlavacek and M. A. Savagcau. Design of gene circuits: lessons from bacteria. *Nat. Rev. Genet.*, 5:34–42, 2004.

[389] S. Semsey, S. Krishna, J. Erdossy, P. Horvath and L. Orosz. Dominant negative autoregulation limits steady-state repression levels in gene networks. *J. Bacteriol.*, 191:4487–4491, 2009.

[390] S. L. Svenningsen, C. M. Waters and B. L. Bassler. A negative feedback loop involving small RNAS accelerates *Vibrio cholerae*'s transition out of quorum-sensing mode. *Genes Dev.*, 22:226–238, 2008.

[391] I. Ventre, A. L. Goodman, I. Vallet-Gely *et al.* Multiple sensors control reciprocal expression of pseudomonas aeruginosa regulatory rna and virulence genes. *Proc. Natl Acad. Sci.*, 103:171–176, 2006.

[392] S. Krishna, A. M. C. Andersson, S. Semsey and K. Sneppen. Structure and function of negative feedback loops at the interface of genetic and metabolic networks. *Nucl. Acids Res.*, 34:2455, 2006.

[393] T. Lamark, T. P. Røknes, J. McDougall *et al.* The complex bet promoters of escherichia coli: regulation by oxygen (ArcA), choline (BetI), and osmotic stress. *J. Bacteriol.*, 178:1655–1662, 1996.

[394] L. M. Meng M. Kilstrup and P. Nygaard Autoregulation of purR repressor synthesis and involvement of purR in the regulation of purB, purC, purL, purMN and guaBA expression in *Escherichia coli*. *Eur. J. Biochem.*, 187:373–379, 1990.

[395] A. Z. Ansari, J. E. Bradner and T. V. O'Halloran. DNA-bend modulation in a repressor-to-activator switching mechanism. *Nature*, 374:371–375, 1995.

[396] S. Oehler E. R. Eismann, H. Krämer and B. Müller-Hill. The three operators of the lac operon cooperate in repression. *EMBO J.*, 9:973–979, 1990.

[397] D. B. Straus, W. A. Walter and C. A. Gross. The heat shock response of *E. coli* is regulated by changes in the concentration of σ^{32}. *Nature*, 329:348–351, 1987.

[398] E. Guisbert, C. Herman, C. Z. Lu and C. A. Gross. A chaperone network controls the heat shock response in *E. coli*. *Genes Dev.*, 18:2812–2821, 2004.

[399] M. Inoue, N. Mitarai and A. Trusina. Circuit architecture explains functional similarity of bacterial heat shock responses. *Phys. Biol*, 9:066003, 2012.

[400] G. Lahav, N. Rosenfeld, A. Sigal *et al.* Dynamics of the p53-Mdm2 feedback loop in individual cells. *Nature Genetics*, 36:147, 2004.

[401] E. Batchelor, A. Loewer, C. Mock and G. Lahav. Stimulus-dependent dynamics of p53 in single cells. *Mol. Syst. Biol.*, 7:488, 2011.

[402] S. Tay, J. J. Hughey, T. K. Lee *et al.* Single-cell NF-κB dynamics reveal digital activation and analogue information processing. *Nature*, 466:267–271, 2010.

[403] J. Stricker S. Cookson, M. R. Bennett *et al.* A fast, robust and tunable synthetic gene oscillator. *Nature*, 456:516–520, 2008.

[404] G. Tiana, S. Krishna, S. Pigolotti, M. H. Jensen and K. Sneppen. Oscillations and temporal signalling in cells. *Phys. Biol.*, 4:R1–R17, 2007.

[405] T. Kobayashi, L. Chen and K. Aihara. Modeling genetic switches with positive feedback loops. *J. Theoret. Biol.*, 221(3):379–399, 2003.

[406] P. Francois and V. Hakim. Design of genetic networks with specified functions by evolution *in silico*. *Proc. Natl. Acad. Sci. USA*, 101:580–585, 2004.

[407] M. Pedersen and K. Hammer. The role of MOR and the CI operator sites on the genetic switch of the temperate bacteriophage TP901-1. *JMB*, 384:577–589, 2008.

[408] Y. T. Maeda and M. Sano. Regulatory dynamics of synthetic gene networks with positive feedback. *J. Mol. Biol.*, 359(4):1107–1124, 2006.

[409] J. E. Ferrell Jr. Self-perpetuating states in signal transduction: positive feedback, double-negative feedback and bistability. *Curr. Opin. Cell Biol.*, 14(2):140–148, 2002.

[410] D. J. Rodda, J. L. Chew, L. H. Lim *et al.* Transcriptional regulation of Nanog by Oct4 and Sox2. *J. Biol. Chem.*, 280:24731–24737, 2005.

[411] V. Chickarmane, V. Oliriu and C. Peterson. Probing the role of stochasticity in a model of the embryonic stem cell: heterogeneous gene expression and reprogramming efficiency. *BMC Syst. Biol.*, 6:98, 2012.

[412] J. Shu, C. Wu, Y. Wu *et al.* Induction of pluripotency in mouse somatic cells with lineage specifiers. *Cell*, 153:963–975, 2013.

[413] J. Wang, D. N. Levasseur and S. H. Orkin. Requirement of Nanog dimerization for stem cell self-renewal and pluripotency. *Proc. Natl. Acad. Sci. USA*, 105:6326–6331, 2008.

[414] P. Navarro N. Testuccia, D. Colby *et al.* OCT4/SOX2-independent Nanog autorepression modulates heterogeneous Nanog gene expression in mouse ES cells. *EMBO J.*, 31:4547–4562, 2012.

[415] A. Yates, R. Callard and J. Stark. Combining cytokine signalling with T-bet and GATA-3 regulation in Th1 and Th2 differentiation: a model for cellular decision making. *J. Theoret. Biol.*, 231:181–196, 2004.

[416] O. Clinquin and J. Demongeot. High dimensional switches and the modeling of cellular differentiation. *J. Theoret. Biol.*, 233:391–411, 2005.

[417] J. C. Furusawa and K. Kaneko. A dynamical-systems view of stem cell biology. *Science*, 338:215–217, 2012.

[418] B. A. Edgar, G. M. Odell and G. Schubiger. A genetic switch, based on negative regulation, sharpens stripes in *drosophila* embryos. *Dev. Genet.*, 10:124–142, 1989.

[419] H. L. Ashe and J. Briscoe. The interpretation of morphogen gradients. *Development*, 133:385–394, 2006.

[420] N. Balaskas, A. Ribeiro, J. Panovska *et al.* Gene regulatory logic for reading the sonic hedgehog signalling gradient in the vertebrate neural tube. *Cell*, 148:273–284, 2012.

[421] M. Babu and S. A. Teichmann. Evolution of transcription factors and the gene regulatory network in *Escherichia coli*. *Nucl. Acids Res.*, 31:1234–1244, 2003.

[422] V. Anantharaman, E. V. Koonin and L. Aravind. Regulatory potential, phyletic distribution and evolution of ancient, intracellular small-molecule-binding domains. *J. Mol. Biol.*, 307:1271–92, 2001.

[423] J. M. Berg, J. L. Tymoczko and L. Stryer. *Biochemistry*, 5th edition W.H. Freeman, New York, 2002.

[424] P. Wong, S. Gladney and J. D. Keasling. Mathematical model of the lac operon: inducer exclusion, catabolite repression, and diauxic growth on glucose and actose. *Biotech. Prog.*, 13:132–143, 1997.

[425] S. Krishna, S. Semsey and K. Sneppen. Combinatorics of feedback in cellular uptake and metabolism of small molecules. *Proc. Natl. Acad. Sci. USA*, 104:20815–20819, 2007.

[426] E. M. Ozbudak, M. Thattai, H. N. Lim, B. I. Shraiman and A. van Oudenadden. Multistability in the lactose utilization network of *Escherichia coli*. *Nature*, 427:737–740, 2004.

[427] B.-K. Cho, C. L. Barrett, E. M. Knight, Y. S. Park and B. Ø Palsson. Genome-scale reconstruction of the lrp regulatory network in escherichia coli. *Proc. Natl. Acad. Sci. USA*, 105:19462–19467, 2008.

[428] B.-K. Cho, S. Federowicz, Y.-S. Park, K. Zegler and B. Ø Palsson. Deciphering the transcriptional regulatory logic of amino acid metabolism. *Nat. Chem. Biol.*, 8:65–71, 2011.

[429] S. Kondo and R. Asai. A reaction-diffusion wave on the skin of the marine angelfish *Pomacanthus*. *Nature*, 376(6543):765–768, 1995.

[430] S. Ishihara, K. Fujimoto and T. Shibata. Cross talking of network motifs in gene regulation that generates temporal pulses and spatial stripes. *Genes to Cells*, 10(11):1025–1038, 2005.

[431] S. Sawai, P. A. Thomason and E. C. Cox. An autoregulatory circuit for long-range self-organization in dictyostelium cell populations. *Nature*, 433(7023):323–326, 2005.

[432] J. B. Chang and J. E. Ferrell, Jr. Mitotic trigger waves and the spatial coordination of the xenopus cell cycle. *Nature*, 500(7464):603–607, 2013.

[433] Alexander Aulehla and Randy L Johnson. Dynamic expression of lunatic fringe suggests a link between notch signaling and an autonomous cellular oscillator driving somite segmentation. *Dev. Biol.*, 207(1):49–61, 1999.

[434] O. Pourquié. The segmentation clock: converting embryonic time into spatial pattern. *Science*, 301(5631):328–330, 2003.

[435] Y. Masamizu, T. Ohtsuka, Y. Takashima *et al.* Real-time imaging of the somite segmentation clock: revelation of unstable oscillators in the individual presomitic mesoderm cells. *Proc. Nat. Acad. Sci. USA*, 103(5):1313–1318, 2006.

[436] E. Ben-Jacob, O. Schochet, A. Tenenbaum *et al.* Generic modelling of cooperative growth patterns in bacterial colonies. *Nature*, 368:46–49, 1994.

[437] T. Sams, K. Sneppen, M. H. Jensen *et al.* Morphological instabilities in a growing yeast colony: experiment and theory. *Phys. Rev. Lett.*, 79:313–316, 1997.

[438] P. Yde, B. Mengel, M. H. Jensen, S. Krishna and A. Trusina. Modeling the NFκB mediated inflammatory response predicts cytokine waves in tissue. *BMC Syst. Biol.*, 5:115, 2011.

[439] B. M. Sager. Propagation of traveling waves in excitable media. *Genes Dev.*, 10:2237–2250, 1996.

[440] M. Zhabotinsky. Periodic liquid phase reactions. *Biofizika*, 9:329, 1964.

[441] T. Matsusaka, K. Fujikawa, Y. Nishio *et al.* Transcription factors NF-IL6 and NF-κB synergistically activate transcription of the inflammatory cytokines, interleukin 6 and interleukin 8. *Proc. Natl. Acad. Sci. USA*, 90:10193–10197, 1993.

[442] M. Nian, P. Lee and P. Liu Khaper. Inflammatory cytokines and postmyocardial infarction remodelling. *Circ. Res.*, 94:1543–1553, 2004.

[443] S. Sick, S. Reinker, J. Timmer and T. Schlake. Wnt and DKK determine hair follicle spacing through a reaction-diffusion mechanism. *Science*, 314:1447–1450, 2006.

[444] P. K. Maini, R. E. Baker and M. Chuong C. The Turing model comes of molecular age. *Science*, 314(5804):1397–1398, 2006.

[445] S. Kondo and T. Miura. Reaction-diffusion model as a framework for understanding biological pattern formation. *Science*, 329:1616–1620, 2010.

[446] D. Sprinzak, A. Lakhampal, L. LeBon *et al.* *Cis*-interactions between Notch and Delta generate mutually exclusive signalling states. *Nature*, 465:86–90, 2010.

[447] Y. Kuramoto and T. Tsuzuki. Persistent propagation of concentration waves in dissipative media far from thermal equilibrium. *Prog. Theoret. Phys.*, 55:356–369, 1976.

[448] G. Sivashinsky. Nonlinear analysis of hydrodynamic instability in laminar flames i. derivation of basic equations. *Acta Astron.*, 4:1177–1206, 1977.

[449] G. Sivashinsky and D. Michelson. On irregular wavy flow of a liquid film down a vertical plane. *Prog. Theoret. Phys.*, 63:2112–2114, 1980.

[450] J. M. Hyman and B. Nicolaenko. The Kuramoto-Sivashinsky equation: a bridge between PDE's and dynamical systems. *Physica D*, 18:113–126, 1986.

[451] P. Collet, J.-P. Eckmann, H. Epstein and J. Stubbe. Analyticity for the Kuramoto-Sivashinsky equation. *Physica D*, 67:321–326, 1993.

[452] K. Sneppen, J. Krug, M. Jensen, C. Jayaprakash and T. Bohr. Dynamic scaling and crossover analysis for the Kuramoto-Sivashinsky equation. *Phys. Rev. A*, 46:R7351–R7354, 1992.

[453] J. Berg and M. Lassig. Local graph alignment and motif search in biological networks. *Proc. Natl. Acad. Sci. USA*, 101:14689–14694, 2004.

[454] R. Sharan and T. Ideker. Modeling cellular machinery through biological network comparison. *Nat. Biotech.*, 24:427–433, 2006.

[455] S. V. Rajagopala, S. Casjens and P. Uetz. The protein interaction map of bacteriophage lambda. *BMC Microbiol.*, 11:213, 2011.

[456] A. Trusina, M. Rosvall and K. Sneppen. Communication boundaries in networks. *Phys. Rev. Lett.*, 94:238701, 2004.

[457] P. Erdös and A. Rényi. On the evolution of random graphs. *Publ. Math. Inst. Hung. Acad. Sci.*, 5:1760, 1960.

[458] M. E. J. Newman, S. H. Strogatz and D. J. Watts. Random graphs with arbitrary degree distributions and their applications. *Phys. Rev. E*, 64:026118, 2001.

[459] R. D. Luce and A. D Perry. A method of matrix analysis of group structure. *Psychometrika*, 14:95–116, 1949.

[460] P. W. Holland and S. Leinhardt. Transitivity in structural models of small groups. *Comparative Group Studies*, 2:107–124, 1971.

[461] D. J. Watts and S. H. Strogatz. Collective dynamics of small-world networks. *Nature*, 393:409–410, 1998.

[462] L. C. Freeman. A set of measures of centrality based on betweenness. *Sociometry*, 40:35–41, 1977.

[463] L. C. Freeman. Centrality in social networks conceptual clarification. *Social Networks*, 1:215–239, 1978/1979.

[464] M. W. Hahn and A. D. Kahn. Comparative genomics of centrality and essentiality in three eukaryotic protein-interaction networks. *Mol. Biol. Evol.*, 22:603–806, 2005.

[465] M. P. Joy, A. Brock, D. E. Ingber and S. Huang. High-betweenness proteins in the yeast protein interaction network. *J. Biomed. Biotechnol.*, 2005:96–103, 2005.

[466] H. Yu, P. M. Kim, E. Sprecher, V. Trifanov and M. Gerstein. The importance of bottlenecks in protein networks: correlation with gene essentiality and expression dynamics. *PloS. Comp. Biol.*, 3:713–720, 2007.

[467] M. Girvan and M. E. J. Newman. Community structure in social and biological networks. *Proc. Natl. Acad. Sci. USA*, 99:7821–7826, 2002.

[468] K. A. Eriksen, I. Simonsen, S. Maslov and K. Sneppen. Modularity and extreme edges of the internet. *Phys. Rev. Lett.*, 90:148701, 2003.

[469] K. A. Eriksen, I. Simonsen, S. Maslov and K. Sneppen. Modularity and extreme edges of the internet. *Physica A*, 336:163, 2004.

[470] M. Rosvall and C. T. Bergstrøm. Maps of random walks on complex networks reveal community structure. *Proc. Natl. Acad. Sci. USA*, 105:1118–1123, 2008.

[471] H. Rieger and J. D. Noh. Random walks on complex networks. *Phys. Rev. Lett.*, 92:118701, 2004.

[472] M. E. J. Newman. A measure of betweenness centrality based on random walks. *Social Networks*, 27:39–54, 2005.

[473] A.-L. Barabasi and R. Albert. Emergence of scaling in random networks. *Science*, 509:286, 1999.

[474] H. Jeong, B. Tombor, R. Albert, Z. N. Oltvai and A.-L. Barabasi. The large scale organization of metabolic networks. *Nature*, 407:651–654, 2000.

[475] H. Ma and A-P. Zeng. Reconstruction of metabolic networks from genome data and analysis of their global structure for various organisms. *Bioinformatics*, 19:270–277, 2003.

[476] P. Minnhagen and S. Bernhardsson. The blind watchmaker network: scale-freeness and evolution. *PLoS One*, 3:e1690, 2008.

[477] R. Cohen, K. Erez, D. ben Avraham and S Havlin. Resilience of the internet to random breakdowns. *Phys. Rev. Lett.*, 85:4626–4628, 2000.

[478] K. Christenseni, R. Donangelo, B. Koiler and K. Sneppen. Evolution of random networks. *Phys. Rev. Lett.*, 81:2380, 1998.

[479] M. Newman. Spread of epidemic disease on networks. *Phys. Rev. E*, 66:016128, 2002.

[480] M. E. J Newman. The structure and function of complex networks. *SIAM Rev.*, 45:167–256, 2002.

[481] L. D. Valdez, C. Buono, P. A. Macri and L. A. Braunstein. Social distancing strategies against disease spreading. *arXiv*, 1308, 2009.

[482] A. Trusina, S. Maslov, P. Minnhagen and K. Sneppen. Hierarchy and anti-hierarchy in real and scale free networks. *Phys. Rev. Lett.*, 92:178702, 2004.

[483] J. B. Axelsen, S. Bernhardsson, M. Rosvall, K. Sneppen and A. Trusina. Degree landscapes in scale-free networks. *Phys. Rev. E*, 74:036119, 2006.

[484] S. Maslov and K. Sneppen. Specificity and stability in topology of protein networks. *Science*, 296:910, 2002.

[485] T. Ito, T. Chiba, R. Ozawa *et al.* A comprehensive two-hybrid analysis to explore the yeast protein interactome. *Proc. Natl. Acad. Sci. USA*, 98:4569, 2001.

[486] P. Uetz, L. Giot, G. Cagney *et al.* A comprehensive analysis of protein-protein interactions in *Saccharomyces cerevisiae*. *Nature*, 403:623–627, 2000.

[487] R. Albert. Scale-free networks in cell biology. *J. Cell Sci.*, 118:4947–4957, 2005.

[488] C. K. Stover, X. Q. Pham, A. L. Erwin *et al.* Complete genome sequence of *Pseudomonas aeruginosa* PAO1, an opportunistic pathogen. *Nature*, 406:959–964, 2000.

[489] E. van Nimwegen. Scaling laws in the functional content of genomes. *Trends Genet.*, 19:479, 2003.

[490] N. Molina and E. van Nimwegen. Scaling laws in functional genome content across prokaryotic clades and lifestyles. *Trends Genet.*, 25:243–247, 2009.

[491] S. Maslov, K. Sneppen and A. Zaliznyak. Detection of topological patterns in complex networks: correlation profile of the internet. *Physica A*, 333:529-540, 2004.

[492] R. Milo S. Itzkovitz, N. Kashtan *et al.* Superfamilies of evolved and designed networks. *Science*, 303:1538–1542, 2004.

[493] R. Milo, S. S. Shen-Orr, S. Itzkovitz, N. Kashtan and U. Alon. Network motifs: simple building blocks of complex networks. *Science*, 298:824–827, 2002.

[494] S. Mangan and U. Alon. Structure and function of the feed-forward loop network motif. *Proc. Natl. Acad. Sci. USA*, 100:11980–11985, 2003.

[495] M. Costanzo, A. Baryshnikova, J. Bellay *et al.* The genetic landscape of a cell. *Science*, 327:425–431, 2010.

[496] S. Maslov and K. Sneppen. Computational architecture of the yeast regulatory network. *Phys. Biol.*, 2:94, 2005.

[497] S. Maslov, S. Krishna, T. Y. Pang and K. Sneppen. Toolbox model of evolution of prokaryotic metabolic networks and their regulation. *Proc. Natl. Acad. Sci. USA*, 106:9743–9748, 2009.

[498] G. Calderelli, A. Capocci, P. De Los Rios and M. A. Munoz. Scale-free networks from varying vertex intrinsic fitness. *Phys. Rev. Lett.*, 89:258702, 2002.

[499] E. J. Deeds, O. Ashenberg and E. I. Shakhnovich. A simple physical model for scaling in protein-protein interaction networks. *Proc. Natl. Acad. Sci. USA*, 103:311–316, 2006.

[500] S. Maslov, K. Sneppen, K. A. Eriksen and K.-K. Yan. Upstream plasticity and downstream robustness in evolution of molecular networks. *BMC Evol. Biol.*, 4:9, 2004.

[501] N. Kleckner. Transposable elements in prokaryotes. *Ann. Rev. of Genet.*, 15:341–404, 1981.

[502] M. G. Kidwell and D. Lisch. Transposable elements as sources of variation in animals and? plants. *Proc. Natl. Acad. Sci. USA*, 94:7704–7711, 1997.

[503] A. C. Burke, C. E. Nelson, B. A. Morgan and C. Tabin. Hox genes and the evolution of vertebrate axial morphology. *Development*, 121:333–346, 1995.

[504] F. Galis. Why do almost all mammals have seven cervical vertebrae? Developmental constraints, hox genes, and cancer. *J. Exp. Zool.*, 285:19–26, 1999.

[505] J. Zákány, C. Fromental-Ramain, X. Warot and D. Duboule. Regulation of number and size of digits by posterior Hox genes: a dose-dependent mechanism with potential evolutionary implications. *Proc. Natl. Acad. Sci. USA*, 94:13695–13700, 1997.

[506] T. I. Lee, N. J. Rinaldi, F. Robert *et al.* Transcription regulatory networks in *Saccharomyces cerevisiae*. *Science*, 298:799–804, 2002.

[507] A. Wagner. The yeast protein interaction network evolves rapidly and contains few redundant duplicate genes. *Mol. Biol. Evol.*, 18:1283–1292, 2001.

[508] Z. Gu, D. Nicolae, H. H-S. Lu and W. Li. Rapid divergence in expression between duplicate genes inferred from microarray data. *Trends Genet.*, 18:609–613, 2002.

[509] A. Wagner. Evolution of gene networks by gene duplications: a mathematical model and its implications on genome organization. *Proc. Natl. Acad. Sci. USA*, 91:4387, 1994.

[510] A. Bhan, D. J. Galas and T. G. Dewey. A duplication growth model of gene expression networks. *Bioinformatics*, 18:1486–1493, 2002.

[511] A. Vazquez, A. Flammini, A. Maritan and A. Vespignani. Modeling of protein interaction networks. *Complexus*, 21:38–44, 2002.

[512] R. V. Sole, R. Pastor-Satorras, E. Smith and T. B. Kepler. Model of large-scale proteome evolution. *Adv. Compl. Syst.*, 5:43–54, 2002.

[513] S. A. Teichmann and M. Babu. Gene regulatory network growth by duplication. *Nat. Genet.*, 36:492–496, 2004.

[514] J. Berg, M. Lassig and A. Wagner. Structure and evolution of protein interaction networks: a statistical model for link dynamics and gene duplications. *BMC Evol. Biol.*, 4:51, 2004.

[515] J. Enemark and K. Sneppen. Analyzing a stochastic model for evolving regulatory networks by unbiased gene duplication. *JSTAT*, P11007, 2007.

[516] T. Yamada, M. Kanehisa and S. Goto. Extraction of phylogenetic network modules from metabolic network. *BMC Bioinform.*, 7:130, 2006.

[517] T. Y. Pang and S. Maslov. Universal distribution of component frequencies in biological and technological systems. *Proc. Natl. Acad. Sci.* USA, 110:6235–6239, 2013.

[518] M. Babu and S. A. Teichmann. Evolution of transcription factors and the gene regulatory network in *Escherichia coli*. *Nucl. Acids Res.*, 31:1234–1244, 2003.

[519] Y. I. Wolf, I. B. Rogozin, N. V. Grishin and E. V. Koonin. Genome trees and the tree of life. *Trends Genet.*, 472–479:18, 2002.

[520] C. Pal, B. Pabb and M. Lercher. Adaptive evolution of metabolic networks by horizontal gene transfer. *Nat. Genet.*, 37:1372–1375, 2005.

[521] D. Medini, C. Donati, H. Tettelin, V. Masignani and R. Rappuoli. The microbial pan-genome. *Curr. Opin. Genet. Dev.*, 15:589–594, 2005.

[522] H. Tettelin, V. Masignani, M. J. Cieslewicz *et al.* Genome analysis of multiple pathogenic isolates of *Streptococcus agalactiae*: implications for the microbial "pan-genome". *Proc. Natl. Acad. Sci. USA*, 102:13950–13955, 2005.

[523] M. Kanehisa and S. Goto. Kegg: Kyoto encyclopedia of genes and genomes. *Nucl. Acids Res.*, 28:27–30, 2000.

[524] T. Handorf, O. Ebenhoh and R. Heinrich. Expanding metabolic networks: scopes of compounds, robustness, and evolution. *J. Mol. Evol.*, 61:498–512, 2005.

[525] T. Handorf and O. Ebenhoh. Metapath online: a web server implementation of the network expansion algorithm. *Nucl. Acids Res.*, 35:613–618, 2007.

[526] T. Y. Pang and S. Maslov. A toolbox model of evolution of metabolic pathways on networks of arbitrary topology. *PLoS Comp. Biol.*, 7(5):e1001137, 2011.

[527] J. B. Axelsen, S. Krishna and K. Sneppen. The cost and capacity of signaling in the *Escherichia coli* protein reaction network. *J. Stat. Mech.*, P01018, 2008.

[528] M. J. Weickerta and S. Adhya. The galactose regulon of *Escherichia coli*. *Mol. Microbiol.*, 167:245–251, 1993.

[529] S. A. Kauffman and S. Johnsen. Coevolution to the edge of chaos: coupled fitness landscapes, poised states, and coevolutionary avalanches. *J. Theor. Biol.*, 149:467–505, 1991.

[530] E. H. Davidson, J. P. Rast, P. Oliveri *et al.* A provisional regulatory gene network for specification of endomesoderm in the sea urchin embryo. *Dev. Biol.*, 246:162–190, 2002.

[531] K. J. Kauffman, P. Prakash and J. S. Edwards. Advances in flux balance analysis. *Curr. Opin. Biotech.*, 14:491–496, 2003.

[532] N. Buchler, U. Gerland and T. Hwa. On schemes of combinatorial transcription control. *Proc. Natl. Acad. Sci. USA*, 100:5136–5141, 2003.

[533] R. Hermsen, S. Tans and P. Rein ten Wolde. Transcriptional regulation by competing transcription factor modules. *PLoS Comp. Biol.*, 2:e164, 2006.

[534] K. J. Peterson and E. H. Davidson. Regulatory evolution and the origin of the bilaterians. *Proc. Natl. Acad. Sci. USA*, 97:4430–4433, 2000.

[535] E. H. Davidson, J. P. Rast, P. Oliveri *et al.* A genomic regulatory network for development. *Science*, 295(5560):1669–2002, 2002.

[536] K. Young. Yeast two-hybrid: so many interactions, (in) so little time. *Biol. Reprod.*, 58:302–311, 1998.

[537] J. Joung, E. Ramm and C. Pabo. A bacterial two-hybrid selection system for studying protein-DNA and protein-protein interactions. *Proc. Natl. Acad. Sci. USA*, 97:7382–7387, 2000.

[538] S. Maslov, K. Sneppen and I. Ispolatov. Spreading out of perturbations in reversible reaction networks. *New J. Phys.*, 9:273–283, 2007.

[539] C. Stark, B.-J. Breitkreutz, A. Breitkreutz *et al.* BioGRID: a general repository for interaction datasets. *Nucl. Acids Res.*, 34:D535, 2006.

[540] S. Ghaemmaghami, W. K. Huh, K. Bower *et al.* Global analysis of protein expression in yeast. *Nature*, 425:737–741, 2003.

[541] G. Burnett and E. P. Kennedy. The enzymatic phosphorylation of proteins. *J. Biol. Chem.*, 211:969–980, 1954.

[542] C. Y. Huang and J. E. Ferrell Jr. Ultrasensitivity in the mitogen-activated protein kinase cascade. *Proc. Natl. Acad. Sci. USA*, 93:10078–10083, 1996.

[543] T. Shibata and K. Fijimoto. On the relation between fluctuation and response in biological systems. *Proc. Natl. Acad. Sci. USA.*, 100:14086–14090, 2003.

[544] T. Shibata and K. Fijimoto. Stochastic signal processing and transduction in chemotactic response of eukaryotic cells. *Biophys. J.*, 93:11–20, 2007.

[545] W. Ma, A. Trusina, W. Lim, H. El-Samad and C. Tang. Defining network topologies that can achieve biochemical adaptation. *Cell*, 138:760, 2009.

[546] U. Alon. An *Introduction to Systems Biology: Design Principles of Biological Circuits.* Chapman & Hall/CRC, 2006.

[547] T.-M. Yi, Y. Huang, M. I. Simon and J. Doyle. Robust perfect adaptation in bacterial chemotaxis through integral feedback control. *Proc. Natl. Acad. Sci. USA*, 97:4649–4653, 2000.

[548] M. G. Surette, U. Alon, N. Barkai and S. Leibler. Robustness in bacterial chemotaxis. *Nature*, 397:168–171, 1998.

[549] J. Adler and W.-W. Tso. Decision-making in bacteria: chemotactic response of *Escherichia coli* to conflicting stimuli. *Science*, 184:1292–1294, 1974.

[550] Howard C. Berg. *E. coli in Motion.* Springer-Verlag, New York, 2003.

[551] T. S. Shimizu, N. Le Novere, M. D. Levin *et al.* Molecular model of a lattice of signalling proteins involved in bacterial chemotaxis. *Nat. Cell Biol.*, 2:792–796, 2000.

[552] C. M. Waters and B. L. Bassler. Quorum sensing: Cell-to-cell communication in bacteria. *Annu. Rev. Cell Dev. Biol.*, 21:319–346, 2005.

[553] T. S. Shimizu, S. V. Aksenov and D. Bray. A spatially extended stochastic model of the bacterial chemotaxis signalling pathway. *J. Mol. Biol.*, 329:291–309, 2003.

[554] B. A. Mello and Y. Tu. Effects of adaptation in maintaining high sensitivity over a wide range of backgrounds for *Escherichia coli* chemotaxis. *Biophys. J.*, 92:2329–2337, 2007.

[555] V. Sourjik and W. W. Tso. Receptor clustering and signal processing in *E. coli* chemotaxis. *Trends Microbiol.*, 12:569–576, 2004.

[556] D. A. Fell and A. Wagner. The small world of metabolism. *Nat. Biotech.*, 18:1121–1122, 2000.

[557] S. Schuster, T. Pfeiffer, F. Moldenhower, I. Koch and T. Dandekat. Exploring the pathway structure of metabolism: decomposition into subnetworks and application to mycoplasma pneumoniae. *Bioinformatics*, 18:351–361, 2002.

[558] H. W. Ma and A. P. Zeng. The connectivity structure, giant strong component and centrality of metabolic networks. *Bioinformatics*, 19:1423–1430, 2003.

[559] P. Holme and M. Huss. Currency and commodity metabolites: their identification and relation to the modularity of metabolic networks. *IET Syst. Biol.*, 1:280–285, 2007.

[560] P. Gerlee, L. Lizana and K. Sneppen. Pathway identification by network pruning in the metabolic network of *Escherichia coli. Bioinformatics*, 25(3282), 2009.

[561] J. M. Savinell and B. O. Palsson. Network analysis of intermediary metabolism using linear optimization. *J. Theor. Biol.*, 154:455–473, 421–454, 1992.

[562] P. O. Palsson. *Systems Biology: Properties of Reconstructed Networks*. Cambridge University Press, Cambridge, 2006.

[563] M. W. Covert, C. H. Schilling and B. O. Palsson. Regulation of gene expression in flux balance models of metabolism. *J. Theor. Biol.*, 213:73–78, 2002.

[564] J. D. Orth, I. Thiele and B. Palssson. What is flux balance analysis. *Nat. Biotech.*, 28:245–248, 2010.

[565] A. Varma and B. Palsson. Stoichiometry flux balance models quantitatively predict growth and metabolic by-product secretion in wild-type *Escherichia coli* W3110. *Appl. Environ. Microbiol.*, 60:3724–3731, 1994.

[566] D. G. Luenberger. *Linear and Nonlinear Programming*. Addison-Wesley, Cambridge, MA 1984.

[567] J. von Neumann. The general and logical theory of automata. In *Cerebral Mechanics in Behaviour: the Hixon Symposium*, ed. L. A. Jeffres. John Wiley & Sons, New York, 1–31, 1951.

[568] T. C. Schelling. Models of segregation. *Am. Econ. Rev.*, 59:488–493, 1969.

[569] S. Wolfram. Statistical mechanics of cellular automata. *Rev. Mod. Phys.*, 55:601–644, 1983.

[570] D. Helbing, I. Farkas and T. Vicsek. Simulating dynamical features of escape panic. *Nature*, 407:487–490, 2000.

[571] E. Bonabeau. Agent-based modeling: methods and techniques for simulating human systems. *Proc. Natl. Acad. Sci. USA*, 99:7280–7287, 2002.

[572] E. Bonabeau, M. Dorigo and G. Theraulaz. Inspiration for optimization from social insect behaviour. *Nature*, 406:39–42, 2000.

[573] N. Wiener and A. Rosenblueth. The mathematical formulation of the problem of conduction of connected excitable elements, specifically the cardiac muscle. *Arch. Inst. Cardiol. Mex.*, 16:205–265, 1946.

[574] M. Greenberg and S. P. Hastings. Spatial patterns for discrete models of diffusion in excitable media. *SIAM J. Appl. Math.*, 54:515–523, 1978.

[575] S. W. Gangstad, C. W. Feldager, J. Juul and A. Trusina. Noisy transcription factor NF-κB oscillations stabilize and sensitize cytokine signaling in space. *Phys. Rev. E*, 87:022702, 2013.

[576] K. Yanagita. *Kagyuko*. Toko Shoin, Tokyo, 1930.

[577] L. Lizana, N. Mitarai, K. Sneppen and H. Nakanishi. Modeling the spatial dynamics of culture spreading in the presence of cultural strongholds. *Phys. Rev. E*, 83:066116, 2011.

[578] K. Sneppen, A. Trusina, M. H. Jensen and S. Bornholdt. A minimal model for multiple epidemics and immunity spreading. *PloS One*, 5:e13326, 2010.

[579] F. Uekermann and K. Sneppen. Spreading of multiple epidemics with cross immunization. *Phys. Rev. E*, 86:036108, 2012.

[580] Y. Iwasa, M. Naleamaru and S. A. Levin. Allelopathy of bacteria in a lattice population: competition between colicin-sensitive and colicin-producing strains. *Evol. Ecol.*, 12(7):785–802, 1998.

[581] M. S. Steinberg. Does differential adhesion govern self-assembly processes in histogenesis? Equilibrum configurations and the emergence of a hierarchy among populations of embryonic stem cells. *J. Exp. Zool.*, 173:395–433, 1970.

[582] P. B. Armstrong. Cell sorting out: the self-assembly of tissues *in vitro*. *Info. Healthcare*, 24:119–149, 1989.

[583] I. D. Chase. Dynamics of hierarchy formation: the sequential development of dominance relation. *Behaviour*, 80:218–240, 1982.

[584] E. Bonabeau, G. Theraulaz and J. L. Deneubourg. Phase diagram of a model of self-organizing hierarchies. *Physica A*, 217:373, 1995.

[585] M. Rosvall and K. Sneppen. Modeling self-organization of communication and topology in social networks. *Phys. Rev. E*, 74:16108, 2006.

[586] M. Rosvall and K. Sneppen. Reinforced communication and social navigation generate groups in model networks. *Phys. Rev. E*, 79:026111, 2009.

[587] H. Spencer. *The Principles of Psychology*. Longmans, London, 1855.

[588] H. A. Simon. On a class of skew distribution functions. *Biometrika*, 42:425–440, 1955.

[589] D. de Solla Price. A general theory of bibliometric and other cumulative advantage processes. *J. Am. Soc. Inform. Sci.*, 27:292, 1976.

[590] S. Bornholdt and K. H. Ebel. World wide web scaling exponent from Simon's 1955 model. *Phys. Rev. E*, 64:035104(R), 2001.

[591] V. Pareto. La legge della domanda. *Giornale degli Economisti*, 10(59–68):691–700, 1895.

[592] G. U. Yule. A mathematical theory of evolution, based on the conclusions of Dr. J. C. Willis. *Phil. Trans. R. Soc. L. B*, 213:2187, 1924.

[593] G. K. Zipf. *Human Behavior and the Principle of Least Effort*. Addison-Wesley, Cambridge, MA, 1949.

[594] J. Juul, K. Sneppen and J. Mathiesen. Labyrinthine clustering in a spatial rock-paper-scissors ecosystem. *Phys. Rev. E*, 87:042702, 2013.

[595] G. Hardin. The tragedy of the commons. *Science*, 162:1243–1248, 1968.

[596] C. R. Johnson and I. Seinen. Selection for restraint in competitive ability in spatial competition systems. *Proc. Biol. Sci.*, 269:655–663, 2002.

[597] M. Frean and E. R. Abraham. Rock-scissors-paper and the survival of the weakest. *Proc. R. Soc. L. B*, 268:1323–1327, 2001.

[598] B. Kerr, M. A. Riley, M. W. Feldman and B. J. M. Bohannan. Local dispersal promotes biodiversity in a real-life game of rock-paper-scissors. *Nature*, 418:171–174, 2002.

[599] B. Kerr, C. Neuhauser and B. J. M. Bohannan. Local migration promotes competitive restraint in a host-pathogen "tragedy of the commons." *Nature*, 442:75–78, 2006.

[600] T. Reichenbach, M. Mobilia and E. Frey. Coexistence versus extinction in the stochastic cyclic Lotka-Volterra model. *Phys. Rev*, 74:E051907, 2006.

[601] T. Reichenbach, M. Mobilia and E. Frey. Noise and correlations in a spatial population model with competition. *Phys. Rev. Lett.*, 99:238105, 2007.

[602] E. Frey. Evolutionary game theory: theoretical concepts and applications to microbial communities. *Physica A*, 389:4265–4298, 2010.

[603] T. L. Czaran, R. F. Hoekstra and L. Pagie. Chemical warfare between microbes promotes biodiversity. *Proc. Natl. Acad. Sci. USA*, 99:786–790, 2002.

[604] B. C. Kirkup and M. A. Riley. Antibiotic-mediated antagonism leads to a bacterial game of rock-paper-scissors *in vivo*. *Nature*, 428:412–414, 2004.

[605] J. Juul, K. Sneppen and J. Mathiesen. Clonal selection prevents tragedy of the commons when neighbors compete in a rock-paper-scissors game. *Phys. Rev. E*, 85:061924, 2012.

[606] J. R. Nahum, B. N. Harding and B. Kerr. Evolution of restraint in a structured rock-paper-scissors community. *Proc. Nat. Acad. Sci. USA*, 108:10831–10838, 2011.

[607] C. M. Eshelman R. Voule, J. L. Stewart *et al.* Unrestricted migration favours virulent pathogens in experimental metapopulations: evolutionary genetics of a rapacious life history. *Phil. Trans. R. Soc. B*, 365:2503–2513, 2010.

[608] J. Jackson and L. Buss. Alleopathy and spatial competition among coral reef invertebrates. *Proc. Nat. Acad. Sci. USA*, 72:5160, 1975.

[609] M. C. Boerlijst and P. Hogeweg. Spiral wave structure in pre-biotic evolution: hypercycles stable against parasites. *Physica D*, 48:17–28, 1991.

[610] M. Eigen and P. Schuster. The hypercycle. a principle of natural self-organisation. Part B: The abstract hypercycle. *Naturwissenschaften*, 65:7–41, 1978.

[611] R. Axelrod. *The Evolution of Cooperation*. Basic Books, New York, 1984.

[612] S. Bornholdt, M. H. Jensen and K. Sneppen. Emergence and decline of scientific paradigms. *Phys. Rev. Lett.*, 106:058701, 2011.

[613] F. Rohwer and R Edwards. The phage proteomic tree: a genome-based taxonomy system for phage. *J. Bacteriol.*, 184:4529–4535, 2002.

[614] R. W. Hendrix, M. C. M. Smith, R. N. Burns *et al.* Evolutionary relationships among diverse bacteriophages and prophages: all the world's a phage. *Proc. Natl. Acad. Sci. USA*, 96:2192–2197, 1999.

[615] G. Lima-Mendez, J. Van Helden, A. Toussaint, and R. Leplae. Reticulate representation of evolutionary and functional relationships between phage genomes. *Mol. Biol. Evol.*, 25:762–777, 2008.

[616] M. L. Coleman, M. B. Sullivan, A. C. Martiny *et al.* Genomic islands and the ecology and evolution of *Prochlorococcus*. *Science*, 311:1768–1770, 2006.

[617] D. Lindell, J. D. Jaffe, M. L. Coleman *et al.* Genome-wide expression dynamics of a marine virus and host reveal features of co-evolution. *Nature*, 449:83–86, 2007.

[618] P. Youderian. Genetic, physical, and restriction map of bacteriophage ϕ80. In *DNA: Insertion Elements, Pasmids and Episomes*, ed. A. I. Shapiro, J. A. Shapiro and S. L. Adhya. Cold Spring Harbor Laboratory, New York, 1977.

[619] Campbell A. M. Conditions for the existence of bacteriophage. *Evolution*, 15:153–165, 1960.

[620] M. L. Pedulla, M. E. Ford, J. M. Houtz *et al.* Origins of highly mosaic mycobacteriophage genomes. *Cell*, 18:171–182, 2003.

[621] F. Rohwer. Global phage diversity. *Cell*, 113:141, 2003.

[622] C. M. Jessup and S. E. Forde. Ecology and evolution in microbial systems: the generation and maintenance of diversity in phage-host interactions. *Res. Microbiol*, 159:382–389, 2008.

[623] J. S. Weitz, H. Hartman, and S. A. Levin. Coevolutionary arms races between bacteria and bacteriophage. *Proc. Natl. Acad. Sci. USA*, 102:9535–9540, 2005.

[624] D. A. Marvin. Filamentous phage structure, infection and assembly. *Curr. Opin. Struct. Biol.*, 8:150–158, 1998.

[625] K. S. Makarova, Y. I. Wolf, and E. V. Koonin. Comparative genomics of defense systems in archaea and bacteria. *Nucl. Acids Res.*, 41:4360–4377, 2013.

[626] R. Korona, B. Korona and R. Levin. Sensitivity of naturally occuring coliphages to type I and type II restriction and modification. *J. Gen. Microbiol.*, 139:1283–1290, 1993.

[627] C. Kessler and V. Manta. Specificity of restriction endonucleases and DNA modification methyltransferases: a review. *Gene*, 92:1–248.

[628] W. Arber. Host specificity of DNA produced by *Escherichia coli*. V. The role of methionine in the production of host specificity. *J. Mol. Biol*, 11:247–256, 1965.

[629] A. Rambach and P. Tiollais. Bacteriophage lambda having EcoRI endonuclease sites only in the nonessential region of the genome. *Proc. Natl. Acad. Sci. USA*, 71:3927–3930, 1974.

[630] R. Korona and B. R. Levin. Phage-mediated selection and the evolution and maintenance of restriction-modification. *Evolution*, 47:565–575, 1993.

[631] Y. Ishino, H. Shinagawa, K. Makino, M. Amemura and A. Nakata. Nucleotide sequence of the IAP gene, responsible for alkaline phosphatase isozyme conversion in *Escherichia coli*, and identification of the gene product. *J. Bacteriol.*, 169:5429–5433, 1987.

[632] R. Barrangou, C. Fremaux, H. Deveau, *et al.* CRISPR provides acquired resistance against viruses in prokaryotes. *Science*, 315:1709–1712, 2007.

[633] P. Horvath and R. Barrangou. CRISPR/Cas, the immune system of bacteria and archaea. *Science*, 327:167–170, 2010.

[634] S. J. Labrie, J. E. Samson and S. Moineau. Bacteriophage resistance mechanisms. *Nat. Rev. Microbiol.*, 8:317–327, 2010.

[635] R. Sorek, E. V. Koonin and P. Hugenholtz. CRISPR: A widespread system that provides acquired resistance against phages in bacteria and archaea. *Nat. Rev. Microbiol.*, 6:181–186, 2008.

[636] E. V. Koonin and Y. I. Wolf. Is evolution darwinian or/and lamarckian? *Biol. Direct*, 4:42, 2009.

[637] L. M. Childs, N. L. Held, M. J. Young, R. J. Whitaker and J. S. Weitz. Multi-scale model of crispr-induced coevolutionary dynamics: diversification at the interface of Lamarck and Darwin. *Evolution*, 66:2015–2029, 2012.

[638] A. Stern and R. Sorek. The phage-host arms race: shaping the evolution of microbes. *Bioessays Rev.*, 33:43–51, 2011.

[639] N. L. Held and R. J. Whitaker. Viral biogeography revealed by signatures in sulfolobus islandicus genomes. *Environ. Microbiol.*, 11:457–466, 2009.

[640] J. O. Haerter and K. Sneppen. Spatial structure and lamarckian adaptation explain extreme genetic diversity at crispr locus. *mBio*, 3:4, 2012.

[641] B. J. M. Bohannan and R. E. Lenski. Linking genetic change to community evolution: insights from studies of bacteria and bacteriophages. *Ecol. Lett.*, 3:362–377, 2000.

[642] B. Koskella, D. M. Lin, A. Bucklin and J. Thompson. The cost of evolving resistance in heterogenous environments. *Proc. R. Soc. B*, rspb20112259, 2011.

[643] R. Hazan, B. Sat and H. Engelberg-Kulka. *Escherichia coli* mazEF-mediated cell death is triggered by various stressful conditions. *J. Bacteriol.*, 186:3663–3669, 2004.

[644] M. Koga, Y. Otsuka, S. Lemire and T. Yonesaki. *Escherichia coli* rnlA and rnlB compose a novel toxin-antitoxin system. *Genetics*, 187:123–130, 2011.

[645] H. Sberro, A. Leavitt, R. Kiro *et al.* Discovery of functional toxin/antitoxin systems in bacteria by shotgun cloning. *Mol. Cell*, 50:1–13, 2013.

[646] M. C. Chopin, A. Chopin and E. Bidenko. Phage abortive infection in lactococci: variations on a theme. *Curr. Opin. Microbiol.*, 8:473–479, 2005.

[647] D. H. Parma, M. Snyder, S. Sobolevski, M. Nawroz, E. Brody and L. Gold. The Rex system of bacteriophage lambda: tolerance and altruistic cell death. *Genes Dev.*, 6:497–510, 1992.

[648] G. E. Christie and R. Calendar. Interactions between satellite bacteriophage p4 and its helpers. *Annu. Rev. Genet.*, 24:465–490, 1990.

[649] B. H. Lindqvist, G. Deho and R. Calender. Mechanisms of genome propagation and helper exploitation by satellite phage p4. *Microbiol. Mol. Biol. Rev.*, 57:683–702, 1993.

[650] N. Friedman, S. Vardi, M. Ronen, U. Alon and J. Stavans. Precise temporal modulation in the response of the SOS DNA repair network in individual bacteria. *PloS Biol.*, 3:1261–1268, 2005.

[651] J. Heinrich, M. Velleman and H. Schuster. The tripartite immunity system of phages p1 and p7. *FEMS Microbiol. Rev.*, 17:121–126, 1995.

[652] J. Nesper, J. Blass, M. Fountoulakis and J. Reidl. Characterization of the major control region of *Vibrio cholerae* bacreriophage k139: immunity, exclusion, and integration. *J. Bact.*, 181:2902–2913, 1999.

[653] E. V. Koonin and Y. I. Wolf. Evolution of microbes: a paradigm shift in evolutionary ecology? *Front. Cell. Infect. Microbiol.*, 2:119, 2012.

[654] P. Forterre and D. Prangishvili. The billion-year war between ribosome- and capsid-encoding organisms (cells and viruses) as the major source of evolutionary novelties. *Ann. NY Acad. Sci.*, 1178:65–77, 2009.

[655] D. H. Kruger and T. A. Bickle. Bacteriophage survival: multiple mechanisms for avoiding deoxyribonucleic acid restriction systems of their host. *Microbiol. Rev.*, 47:345–360, 1983.

[656] T. A. Bickle and D. H. Kruger. Biology of DNA restriction. *Microbiol. Rev.*, 57:434–450, 1993.

[657] P. A. Waite-Rees, C. J. Keating, L. S. Moran *et al.* Characterization and expression of the escherichia coli mrr restriction system. *J. Bact.*, 173:5207–5219, 1991.

[658] K. D. Seed, D. W. Lazinski, S. B. Calderwood and A. Camilli. A bacteriophage encodes its own CRISPR/CAS adaptive response to evade host innate immunity. *Nature*, 494:489–491, 2013.

[659] R. E. Lenski. Experimental studies of pleiotropy and epistasis in *Escherichia coli* variation in competitive fitness among mutants resistant to virus t4. *Evolution*, 42:425–432, 433–440, 1988.

[660] T. F. Thingstad and R. Lignell. Theoretical models for the control of bacterial growth rate, abundance, diversity and carbon demand. *Aquat. Microb. Ecol.*, 13:19–27, 1997.

[661] H. T. P. Williams. Phage-induced diversification improves host evolvability. *BMC Evol. Biol.*, 13:17, 2013.

[662] J. Lederberg, L. L. Cavalli and E. M. Lederberg. Sex compatibility in *Escherichia coli*. *Genetics*, 37(6):720–730, 1952.

[663] D. C. Arthur, A. F. Ghetu, M. J. Gubbins, *et al.* FinO is an RNA chaperone that facilitates sense-antisense RNA interactions. *EMBO J.*, 22(23):6346–55, 2003.

[664] M. Achtman, N. Willetts and A. J. Clark. Beginning a genetic analysis of conjugational transfer determined by the F factor in *Escherichia coli* by isolation and characterization of transfer-deficient mutants. *J. Bacteriol.*, 106:529–538, 1971.

[665] S. N. Cohen, A. C. Y. Chang and L. Hsu. Nonchromosomal antibiotic resistance in bacteria: genetic transformation of *Escherichia coli* by R-factor DNA. *Proc. Natl. Acad. Sci. USA*, 69:2110–2114, 1972.

[666] K. Moebus and H. Nattkemper. Bacteriophage sensitivity pattern among bacteria isolated from marine waters. *Helgol. Meeresunters*, 34:375–385, 1981.

[667] C. O. Flores, S. Valverde and J. S. Weitz. Multi scale structure and geographic drivers of cross-infection within marine bacteria and phages. *ISME J.*, 1–13, 2012.

[668] K. Moebus and H. Nattkemper. Taxonomic investigations of bacteriophage sensitive bacteria isolated from marine waters. *Helgol. Meeresunters*, 36:357–373, 1983.

[669] M. Vos, P. J. Birkett, E. Birch, R. I. Griffits and A. Bucklin. Local adaptation of bacteriophages to their bacterial hosts in soil. *Science*, 325:833–833, 2009.

[670] B. Koskella, J. N. Thompson, G. M. Preston, and A. Bucklin. Local biotic environment shapes the spatial scale of bacteriophage adaptation to bacteria. *Am. Nat*, 177:440–451, 2011.

[671] B. R. Levin, J. Antonovics, and H. Sharma. Frequency-dependent selection in bacterial populations. *Proc. R Soc. Lond. B*, 319:459–472, 1988.

[672] A. J. Lotka. Contribution to the theory of periodic reactions. *J. Phys. Chem.*, 14:271–274, 1910.

[673] M. De Paepe and F. Taddei. Viruses life history: Towards a mechanistic basis of a trade-off between survival and reproduction among phages. *PLoS Biol*, 4:e193, 2006.

[674] B. R. Levin, F. M. Stewart, and L. Chao. Resource-limited growth, competition, and predation: A model and experimental studies with bacteria and bacteriophage. *The American Naturalist*, 111:3–24, 1977.

[675] P. F. Verhulst. Notice sur la loi que la population poursuit dans son accroissement. *Correspondance Methematique et Physique*, 10:113–121, 1838.

[676] J. Hu, K. Miyanaga, and Y. Tanji. Diffusion of bacteriophages through artificial biofilm models. *Prog Biotechnol.*, pages 1–8, 2011.

[677] V. Volterra. Variazioni e fluttuazioni del numero d'individui in specie animali conviventi. *Mem. Acad. Lincei Roma*, 2:31–113, 1926.

[678] M. Middelboe. Bacterial growth rate and marine virus-host dynamics. *Microbia Ecol.*, 40:114–124, 2000.

[679] C. Brussaard. Viral control of phytoplankton populations - a review. *Eukaryotic Microbiology*, 51:125–138, 2004.

[680] S. Sillankorva, R. Oliveira, M. Vieira, I. Sutherland, and J. Azeredo. Pseudonomas fluorescens infection by bacteriophage $\phi s1$ the influence of temperature, host growth and media. *FEMS Microbiol. Lett.*, 241:13–20, 2004.

[681] J. Weitz and J. Dushoff. Alternative stable states in host-phage dynamics. *Theoretical Ecology*, 1:13–19, 2008.

[682] R. T. Noble and J. A. Fuhrman. Virus decay and its causes in coastal waters. *Appl. Environ. Microbiol.*, 63:77–83, 1997.

[683] Z. Wang and N. Goldenfeld. Fixed points and limit cycles in the population dynamics of lysogenic viruses and their hosts. *Phys. Rev. E*, 82:011918, 2010.

[684] S. Heilman, K. Sneppen, and S. Krishna. Coexistence of phage and bacteria on the boundary of self-organized refuges. *Proc. Nat. Acad. Sci. USA*, 109:12828–12833, 2012.

[685] S. A. Levin. Community equilibria and stability, and an extension of the competitive exclusion principle. *Am. Naturalist*, 413–423, 1970.

[686] L. F. Jover, M. H. Cortez and J. S. Weitz. Mechanisms of multi-strain coexistence in host–phage systems with nested infection networks. *J. Theoret. Biol.*, 332:65–77, 2013.

[687] J. O. Haerter, N. Mitarai and K. J. Sneppen. Phage and bacteria support mutual diversity in a narrowing staircase of coexistence. ISME doi10.1038 2014.

[688] U. Bastilla, M. Lassig, S. Manrubia, and A. Valleriani. Decision making in living cells: Lessons from a simple system. *J. Thor. Biol.*, 235:521–530, 2005.

[689] S. E. Luria and M. Delbruck. Mutations of bacteria from virus sensitivity to virus resistance. *Genetics*, 28:491–511, 1943.

[690] J. O. Haerter, A. Trusina, and K. Sneppen. Bacterial crispr defence system buffers phage diversity. *Journal of Virology*, 85:10554, 2011.

[691] A. Wallace. On the law which has regulated the introduction of new species. *Annals and Magazine of Natural History*, 16:184–196, 1855.

[692] A. Bolotin, B. Quinquis, A. Sorokin, and S. D.Ehrlich. Clustered regularly interspaced short palindrome repeats (crisprs) have spacers of extrachromosomal origin. *Microbiology*, 151:2551–2561, 2005.

[693] P. Horvath, D. A. Romero, A. C. Côuté-Monvoisin *et al.* Diversity, activity, and evolution of crispr loci in streptococcus thermophilus. *J. Bacteriol*, 190:1401–1412, 2008.

[694] L. Van Valen. A new evolutionary law. *Evolutionary Theory*, 1:1, 1973.

[695] A. Lipton and A. Weissbach. The bacteriophages. *American Journal of Medicine*, 46:264–274, 1969.

[696] B. A. Wiggins and M. Alexander. Minimum bacterial density for bacteriophage replication: Implications for significance of bacteriophages in natural ecosystems. *Applied and Environmental Microbiology*, 49:19–23, 1985.

[697] A. Buckling and P. B. Rainey. Antagonistic coevolution between a bacterium and a bacteriophage. *Proc. R. Soc. Lond. B*, 269:931–936, 2002.

[698] J. S. Weitz, H. Hartman, and S. A. Levin Coevolutionary arms races between bacteria and bacteriophage. *Proc. Natl. Acad. Sci. USA*, 102:9535–9540, 2005.

[699] L. C. Coberly, W. Wei, K. Y. Sampson *et al.* Space, time, and host evolution facilitate coexistence of competing bacteriophages: Theory and experiment. *Am. Nat.*, 173:E121–E138, 2009.

[700] M. Lieb. The establishment of lysogenicity in e. coll. *J. Bact.*, 65:642–651, 1953.

[701] K. Khemayan, T. Pasharawipas, O. Puiprom *et al.* Unstable lysogeny and pseudolysogeny in vibrio harveyi siphovirus-like phage. *Appl. Environ. Microbiol.*, 72:1355–1363, 2006.

[702] R. A. Weisberg and J. A. Gallant. Dual function of the λ prophage repressor. *Journal of Mol. Biology*, 25:537–544, 1967.

[703] M. Avlund, I. B. Dodd, S. Semsey, K. Sneppen, and S. Krishna. Why do phage play dice? *J. of Virology*, 83:11416–11420, 2009.

[704] P. Haccou and Y. Iwasa. Optimal mixed strategies in stochastic environments. *Theoretical Population Biology*, 47(2):212–243, 1995.

[705] C. T. Bergstrøm and M. Lachman. Shannon information and biological fitness. *Information Theory Workshop, IEEE*, 0-7803-8720-1:50–54, 2004.

[706] J. L. Kelly. A new interpretation of information rate. *Bell System Technical Journal*, 35:917–926, 1956.

[707] E. Kussell, R. Kishony, N-Q. Balaban, and S. Leibler. Bacterial persistence – a model of survival in changing environments. *Genetics*, 169:1807–1814, 2005.

[708] E. Kussell and S. Leibler. Phenotypic diversity, population growth, and information in fluctuating environments. *Science*, 309:2075–2078, 2005.

[709] S. Maslov and K. Sneppen. Well temperate phage. *preprint, arXiv:1308.1646*, 2013.

[710] Maslov S and Zhang Y-C. Optimal investment strategy for risky assets. *International Journal of Theoretical and Applied Finance*, 1:377–387, 1998.

[711] M. Marsili, S. Maslov, and Y. Zhang. Dynamical optimization theory of a diversified portfolio. *Physica A*, 253:403–418, 1998.

[712] R. May. Will a large complex system be stable? *Nature*, 238:413–414, 1972.

[713] M. Hall, K. Christensen, S. A. Collobiano, and H. J. Jensen. Time-dependent extinction rate and species abundance in a tangled-nature model of biological evolution. *Phys. Rev. E*, 66:011904, 2002.

[714] E. Rezende, J. E. Lavabre, P. R. Guimarães *et al.* Non-random coextinctions in phylogenetically structured mutualistic networks. *Nature*, 448(7156):925–928, 2007.

[715] C. J. Melián, J. Bascompte, P. Jordano, and V. Krivan. Diversity in a complex ecological network with two interaction types. *Oikos*, 118(1):122–130, 2009.

[716] J. Bascompte and R. V. Solé. Rethinking complexity: modelling spatiotemporal dynamics in ecology. *Trends in Ecology & Evolution*, 10(9):361–366, 1995.

[717] J. Mathiesen, N. Mitarai, K. Sneppen, and A. Trusina. Ecosystems with mutually exclusive interactions self organize into a state of high diversity. *Phys. Rev. Lett.*, 107:188101, 2011.

[718] R. T. Paine. Food web complexity and species diversity. *The American Naturalist*, 100:65–75, 1966.

[719] R. T. Paine. A note on trophic complexity and community stability. *The American Naturalist*, 103:91–93, 1969.

[720] J. P. Wright, C. G. Jones, and A. S. Flecker. An ecosystem engineer, the beaver, increases species richness at the landscape scale. *Oecologia*, 132(1):96–101, 2002.

[721] J. A. Estes and J. F. Palmisano. Sea otters: Their role in structuring nearshore communities. *Science*, 185:1058–1060, 1974.

[722] N. Mitarai, J. Mathiesen, and Kim Sneppen. Emergence of diversity in a model ecosystem. *Phys. Rev. E*, 86:011929, 2012.

[723] R. A. Fisher. *The Genetical Theory of Natural Selection*. Clarendon, Oxford, 1930.

[724] S. Wright. Character change, speciation, and the higher taxa. *Evolution*, 36:427–443, 1982.

[725] G. G. Simpson. *Tempo and Mode in Evolution*. Columbia University Press, New York, 1944.

[726] G. G. Simpson. *The Major Features of Evolution*. Columbia University Press, New York, 1953.

[727] G. P. Wagner and L. Altenberg. Perspective: Comple adaptations and the evolution of evolvability. *Evolution*, 50:967–976, 1996.

[728] J. J. Sepkoski. Ten years in the library: new data confirm paleontological patterns. *Paleobiology*, 19:43–51, 1993.

[729] S. Bornholdt, K. Sneppen, and H. Westphal. Longevity of orders is related to the longevity of their constituent genera rather than genus richness. *Theory in Biosciences*, 128(2):75–83, 2009.

[730] B. Burlando. The fractal geometry of evolution. *J. Theor. Biol.*, 163:161, 1993.

[731] S. Conway. *The Crucible of Creation: The Burgess Shale and the Rise of Animals*. Oxford University Press, Oxford, England, 1998.

[732] J. W. Valentine, D. Jablonski, and D. H. Erwin. Fossils, molecules and embryos: new perspectives on the cambrian explosion. *Development*, 126:851–859, 1999.

[733] D. M. Raup and J. J. Sepkoski Jr. Mass extinctions and the marine fossil record. *Science*, 215:1501, 1992.

[734] M. Kimura. A simple method for estimating evolutionary rates of base substitutions through comparative studies of nucleotide sequences. *J. Mol. Evol.*, 16:111–120, 1980.

[735] M. Eigen, J. McCaskill, and P. Schuster. The molecular quasi-species. *Advances in Chemical Physics*, LXXV:149–161, 1989.

[736] M. Martell, J. I. Esteban, J. Quer *et al.* Hepatitis c virus (hcv) circulates as a population of different but closely related genomes: quasispecies nature of hcv genome distribution. *J. Virol.*, 66:3225–3229, 1992.

[737] L. Peliti. Quasispecies evolution in general mean-field landscapes. *Europhys. Lett*, 57:745, 2002.

[738] L. Peliti. Introduction to the statistical theory of Darwinian evolution. *arXiv:cond-mat/*, 9712027, 1997.

[739] B. Derrida and L. Peliti. Evolution in a flat fitness landscape. *Bulletin of Mathematical Biology*, 53:355–382, 1991.

[740] E. van Nimwegen, J. P. Crutchfield, and M. Huynen. Neutral evolution of mutational robustness. *Proc. Natl Acad. Sci. USA*, 96:9716–9720, 1999.

[741] P. Schuster, W. Fontana, P. F. Stadler, and I. L. Hofacker. From sequences to shapes and back: a case study in rna secondary structures. *Proc. Roy. Soc. London B*, 255:279–284, 1994.

[742] S. Bornholdt, and K. Sneppen. Robustness as an evolutionary principle. *Proc. Roy. Soc. London B*, 267:2281–2286, 2000.

[743] R. Lande. Expected time for random genetic drift of a population between stable phenotypic states. *Proc. Natl Acad. Sci. USA*, 82:7641–7645, 1985.

[744] N. Eldredge and S. J. Gould. Punctuated equilibriua: An alternative to phyletic gradualism. In T. J. M Schopf, J. M. Thomas, and S. Francisco, editors, *Models in Paleobiology*. Freeman and Cooper, 1972.

[745] S. J. Gould and N. Eldredge. Punctuated equilibrium comes of age. *Nature*, 366:223–227, 1993.

[746] S. F. Elena, V. S. Cooper, and R. E. Lenski. Punctuated evolution caused by selection of rare beneficial mutations. *Science*, 272:1802, 1996.

[747] J. Cela-Conde and F. J. Ayala. Genera of the human lineage. *Proc. Natl Acad. Sci. USA*, 100:7684–7689, 2003.

[748] L. W. Alvarez. Mass extinctions caused by large solid impacts. *Physics Today*, 40:24–33, 1987.

[749] P. Bak and K. Sneppen. Punctuated equilibrium and criticality in a simple model of evolution. *Phys. Rev. Lett.*, 71:4083, 1993.

[750] K. Sneppen, P. Bak, H. Flyvbjerg, and M. H. Jensen. Evolution as a self-organized critical phenomenon. *Proc. Natl. Acad. Sci. USA*, 92:5209–5213, 1995.

[751] P. Bak, C. Tang, and K. Wiesenfeld. Self-organized criticality: an exlanation of 1/f noise. *Phys. Rev Lett.*, 59:381–384, 1987.

[752] H. Flyvbjerg, K. Sneppen, and P. Bak. Mean field model for a simple model of evolution. *Phys. Rev. Lett.*, 71:4087, 1993.

[753] J. de Boer, B. Derrida, H. Flyvbjerg, A. D. Jackson, and T. Wettig. Simple model of self-organized biological evolution. *Phys. Rev. Lett.*, 73:906–909, 1994.

[754] M. Paczuski, S. Maslov, and P. Bak. Avalanche dynamics in evolution, growth, and depinning models. *Phys. Rev. E*, 53:414–443, 1996.

[755] K. Sneppen. Extremal dynamics and punctuated co-evolution. *Physica A*, 221:168, 1995.

[756] Y. G. Jin, Y. Wang, W. Wang *et al.* Pattern of marine mass extinction near the permina-triassic boundary in south china. *Science*, 289:432–436, 2000.

[757] R. V. Sole and S. C. Manrubia. Extinction and self-organized criticality in a model of large-scale evolution. *Phys. Rev. E*, 54:R42–R45, 1996.

[758] R. V. Sole, S. C. Manrubia, M. Benton, S. Kauffman, and P. Bak. Criticality and scaling in evolutionary ecology. *Trends in Ecology and Evolution,* 14:156–160, 1999.

[759] S. Jain and S. Krishna. Autocatalytic sets and the growth of complexity in an evolutionary model. *Phys. Rev. Lett.*, 81:5684, 1998.

[760] D. Segre and D. Lancet. Composing life. *EMBO Reports*, 1:217–222, 2000.

[761] M. E. J. Newman and K. Sneppen. Avalanches, scaling and coherent noise. *Phys. Rev. E*, 54:6226–6231, 1996.

[762] H. A. Kramers. Brownian motion in a field of force and the diffusion model of chemical reactions. *Physica*, 7:284–304, 1940.

[763] K. Jacobs. *Stochastic Processes fro Physicists*. Cambridge University Press, Cambridge, 2010.

[764] J. O. Haerter, *et al. Nucleic Acid Research* glet 1235, 2013.

Index

$k_B T$ physics, 12, 17

abortive infection, 247
activation
    regulatory dynamics, 42
activator, 209
adaptation, 216
    Barkei–Leibler model, 218
    buffer, 217
    E. coli chemotaxis, 218
    RNA as a buffer, 43
ageing, 34
    of information, 228
    species does not get old, 281
agent-based model, 226
    co-evolution, 289
    communication and social climbing, 233
    diversity with competition, 273
    excitable media, 226
    information hierarchy, 232
    nucleosome-mediated epigenetics, 127, 128
    rock–paper–scissors, 238
    Schelling model, 230
    segregation, 231
    self-organized criticality, 289
    social climbing, 233
    social hierarchies, 233
    social segregation, 233
    tragedy of the commons, 238
    word spreading, 229
α-helix, 36
amino acid, 20, 21
    hydrophobicity, 20
    mass, 17
    size, 20
anti-immune state
    Cro domination, 111
anti-terminator
    N, 53
    Q, 54
ATP, 10
avalanches
    co-extinctions, 292

inside avalanche, 292
of avalanches, 292

bacteria, see prokaryotes, 259
bacterial immunity, 246
    to phage is local, 261, 263
bet-hedging, 7, 9
    fitness loss for virulent mutant, 271
    temperate phages, 268, 269, 278
binding energies, 59, 64
binding energy
    entropy battle, 14, 70, 74, 76
bistability, 129
    ecosystem model, 276
    cis, 124, 144
    phage λ switch, 111
Boltzmann weight, 14, 63, 298

CAP, see also CRP, 37
cell growth, 2
    morphological instability, 173
    Verhulst equation, 253
cell individuality, 9, 95, 127
central dogma, 4
    with protein feedback, 37
    with RNA feedback, 44
chaos from determinstic equations, 174
chemical binding
    $1\,M \sim 1\,nm^3$, 59
    counting exercise, 59
chemotaxis
    random walk, 221
CI, 121
    binding energies, 64
    Cro, 52
    degradation and RecA activation, 54
    immunity, superinfection, 51
    positive feedback, 65, 66
    repressor of Cro, 51
    tetramerization, 53
CII, 53
    allows for counting, 121
    degradation rate, 51

infection trajectory, 117
    predicted infection trajectory, 120
cis
    A in phage ϕX174, 123
    A in phage P2, 123
    epigenetics in, 124, 143
    Q in phage λ, 124
    regulation in, 124
clear plaques, 57
clonal selection, 239
    colony as a unit in ABM, 262
    mediocre killer phages, 240
closed complex, 25
clr4, a read–write enzyme, 126
co-evolution
    avalanches, 292
    Bak–Sneppen model, 289
    help barrier passing, 294
    phage λ uses HflB in *E. coli*, 51
    phage λ uses RecA in *E. coli*, 52
    phages and bacteria, 245, 248
    punctuated equilibrium, 292
    separated timescales, 292
    transitions in geology, 294
codons, 4, 17
    mass, 17
coefficient of variation (see CV), 95
colisins, 231, 236
combined umbrella motifs, 166
communication, 7
    signaling across tissue, 226
    social climbing, 233
    social coherence, 230, 234
    social hierarchies, 233
    social segregation, 233
complementary base pairs, 5, 18
computation
    E. coli, 6
    λ phage count to two, 118
    somite segmentation count to 31, 122
    spinal cord count to three, 122
consensus sequence
    OR and OL in λ, 53
    promoters, 25
    Shine–Delgarno, 29
contingencies
    macro-evolution, 4, 288
    phage λ as a model organism, 49
co-operativity
    dimerization, 38, 39
    Hill coefficient, 40
    needed for epigenetics, 130, 159
    recruitment as a two-step process, 131
copying, 1, 2, 3
counting to three, 12
counting to two, 118

coupled feedbacks, 142
covalent bond
    binding energy, 13
    keeping DNA together, 18
    protein backbone, 20
    RNA backbone, 20
    separation of energy scales, 6
CpG islands, 149
CRISPR
    bacterial immunity, 246
    ecosystem model, 261
    gain and loss of immunity, 262
Cro
    anti-immune state, 111, 112
    binding energies, 64
    in early decision, 121
    properties, 51
    stochastic production in lysogen, 108
CRP, 37
    hub regulator, see also CAP, 208
    in galactose network, 151
CV, coefficient of variation, 95
    CI in λ lysogen, 113
    of species lifetimes, 281
    plasmid copy numbers, 95
cycles, 11, 235
    $\rightarrow$ patchiness, 276
    ecosystems, 273
    positive feedback, 277
    species diversity, 277

dalton, 17
Darwin, 2, 279
developmental decisions
    counting to three, 122
    somite segmentation, 122
    stem cells, 161
    temperate phages, 51
diffusion, 82
    across an *E. coli* cell, 83
    constant and mobility, 80
    constant and random walks, 81
    constant in cytoplasm, 80
    equation, 82
    inside eukaryotes, 80
    limited on rate, 85
    time to locate a binding site, 84
    time to locate a binding site in cell, 83
diffusion constant, 79
diffusion equation
    relation to random walk, 81
    solution, 82
    solve with absorbing boundary, 84
dimerization, 39
dissipation timescale, 81
dissociation constant, 37

distribution
  gene expression, 30
distributions
  binomial, 179
  exponential, 13, 179
  Gaussian, 179
  Poisson, 179
  power law, 179
  scale-free, 185
  scale-free, see power law, 179
diversity
  biological, sustainability, 272
  sustainability, 285
DNA, 5
  structure, 17
  backbone, 18
  base pairing, 18
  base pairs, 18
    binding energy, 18
  double helix, 19
  major grove, 19
    information, 19
  minor groove, 19
  persistence length, 18
DNA looping
  gene regulation, 71
  in nucleosome-mediated epigenetics,
    128, 139
DNA methylation, 246
DNA trafficking, 90
DNAP, 90

E. coli, 6
  content, 22, 23, 32
  limits on growth, 33
  network
    modular, 207
  network degree distribution, 207
  reaction network, 207
  ribosome fraction, 33
  variations in growth, 34
E. coli chemotaxis, 218
ecosystems
  cyclic relationships, 273
  diversity or exponential growth, 272
  keystone species, 272
elongation initiation, 26
enzyme kinetics, 212
epigenetic landscapes
  coupled, 145
  regulated, 137
epigenetic states
  dynamics of, 133
  recruitments per generation, 133
epigenetics
  bistability, 129
  co-operativity, 131, 136

DNA methylation, 149
  in S. cerevisiea, 124, 147
  Kramers escape formalism, 115
  λ phage, 50
  need non-local recruitment, 132
  nucleosome-mediated, 124, 128
    three-state model, 128
    two-state model, 135, 136
  Pombe, 127
  positive feedback, 52, 128
  production and decay, 114
  stable and unstable fixed points, 114
  toolkit, 139
epigenetics states
  phage λ, 111
escape over a barrier, 297
eukaryote
  macro-evolution, 280, 281
eukaryotes
  cell size, 123
  inflammation response in mammals, 170
  multicellular differentiation, 122, 161
  nucleosomes and gene ragulation, 123
  S. cerevisiae, 151
  stem cells, 161
event-based simulation
  see Gillespie algorithm, 106
evolution
  adaptive walks, 283
  co-evolution, 293
  extinction avalanches, 293
  extinction by asteroid impacts, 287
  fitness, 282, 283
  macro-scale, 280
  neutral, 282
  Red Queen hypothesis, 281
  separated timescales, 288, 293
  timescales, 280
evolvability
  co-evolutionary avalanches, 294
  evolving in itself, 279, 283
  transposons, 3
excitable media, 168
  agent-based model, 226
  forest fire, 171
  inflammation, 171
exponential distribution
  waiting time distribution, 96
extinctions on geological scales, 287

F plasmid, 249
facilitated target location, 86
Fano factor, 96
feedback
  examples, 151–153
  excitable media, 168
  homeostasis, 154

local inhibition, 173
mixed, 154, 162
negative, 7, 153
    oscillations, 155
    time delay, 157
negative, stress response, 155
pattern formation, 172
positive, 126, 142, 159
    enzymatic, 160
    mixed, 160
    switch, 159, 160
    transcriptional, 160
positive global, negative local, 168
positive local, negative global, 142, 172
time delay, 157
transcription and small molecules, 154, 162
wave propagation, 170
first return of random walks, 292, 299
fitness
    hill climbing, 282
    landscape with barriers, 283
    landscapes, 283
    landscapes, Fisher theory, 284
    landscapes, S. Wright, 282
    limitations, 277, 285
    loss of virulent mutant, 271
    neutral landscapes, 283
    variable environment, 268
fitness landscapes
    barriers, 288
    interacting landscapes, 289
    neutral, 282
fixed points, 130
fluctuation–dissipation theorem, 80, 296
flux
    balance, 221, 223
    metabolites, 162
Fokker–Planck equation, 296, 297
forest fire
    excitable medium, 171
    with fire that needs to be extinguished, 236
fractal globule, 140
    absolute scale, 75
    distance dependent contacts, 74, 140
    E. coli chromosome, 75

GAIA, 1, 280, 281
gambler's ruin, 292
gene expression, 5, 8, 10, 152, 179
    distribution, 30
    stochastic, 9, 121
gene regulation
    DNA loop, 71
    indirect, 138
    logic gates, 209
    nucleosome mediated, 138
genetic code, 17

genetic switch, 52
    λ phage, 49
    Pombe, 127
geology and biological history, 280, 294
GFP
    cell variations of *S. pombe*, 127
    expression from plasmids, 95
    noise in λ phage lysogen, 113
Gillespie algorithm
    coarse-grained, 108
    event-based simulation, 106
    proteins from one mRNA, 108
globular protein, 21
GO-annotation of proteins, 197
GTP, 10

Hill coefficient, 40
history
    contingencies in macro-evolution, 4
    geological record, 294
    separation of timescales, 2
hydrogen bond
    binding energy, 17
hydrogen bonds
    base-pair matching, 18
    between DNA strands, 5
    binding energy, 13
    protein structure, 17
    weak binding perpendicular to backbone, 6

inflammation
    agent-based models, 226
    excitable media, 171, 226
information, 7
    value and age, 228
    waves, 227
isomerization, 26

J-factor, 73
    entropy cost, 76
    free energy, 73

$kcal \, mol^{-1}$, 17
keystone species, 272
Kramers' equation, 297
Kuhn length, 74
Kuramoto–Sivashisnsky equation, 174

λ
    RexA and RexB, 247
λ phage, 49
    anti-immune state, 109
    CI, 54, 64, 109, 121
    CII, 53, 121
    counting, 56
    Cro, 64, 109, 121
    epigenetics, 49

λ phage (cont.)
   induction by CI degradation, 52
   induction by RecA, 52
   induction by UV, 52, 55
   mutation to become virulent, 58
   N, 121
   OL, 71
   OL–OR looping, 71
   OR, 52
   OR affinities, 64
   PR, 66, 109
   PRM, 66, 109
   receptor, 50
   spontaneous induction, 56, 109
   spontaneous induction, barriers, 114
   stability of lysogen, 109
   switch, 52, 111
Langevin equation, 135, 296
lichen, 273
links in regulatory networks, 42
logic gates in gene regulation, 209
Lon, 29
loop entropy
   fractal globule, 76
   random coil, 74
lysis, 50
lysogen, 50
lysogenic state, 50
lysogeny, 50

macro-evolution
   co-existence matrix, 280
   steady state, 290
   transitions in geology, 294
metabolic
   flux balance, 219
   flux balance equation, 222
   flux = 1/recycling time, 24
   recycling time, 24
   stochiometry matrix, 221
metabolic flux, 219
metabolite
   flux, 162
metabolites, see small molecules, 162
Michaelis–Menten reaction, 212
mixed feedback, 47, 142
mobility, 80
modularity, 9
Monod growth law, 32, 35
morphological instability and tip-splitting dynamics,
     175
motif
   networks, 195
mRNA
   lifetimes in *E.coli*, 29, 31
   translation rates, 34
   translation speed, 30

N, an anti-terminator protein, 53
   role in timing CII and CIII, 121
network
   locality, 9
   modularity, 9
   scaling of regulation, 7
network motif
   adaptation, 216
network motifs, 162
   adaptation, 43
   collector, 165
   combined and mixed feedback, 164
   combining motifs, 166
   consumer, 165
   fashion, 165
   feed-forward, 195
   feedback with small molecules, 162
   socialist, 165
   umbrella motifs, 164
networks
   alignment, 177, 196
   amplification factor, 181
   Boolean, 177
   correlation profile, 194
   degree distribution, 179, 184, 207
   diameter, 181
   distance between nodes, 181
   ecosystem of phage and bacteria, 251
   enzyme kinetics, 212
   evolution by duplication and rewiring, 201
   evolution of molecular ones, 200
   evolution using a toolbox, 201
   heterogenous elements, 178
   history, 179
   limits on signaling, 179
   motif, feed-forward, 195
   one hub, one function considerations, 197
   preferential attachment, 234
   proper null models, 193
   protein–protein sequestration, 210
   randomization, 193
   reaction node, 206
   scale-free, 185
   separated timescales, 179
   signaling, 177, 179, 206, 211
   signaling cascades, 212, 215
   strong component, 207, 208
   threshold, 198
   toolbox model, 202
   transcription + protein:protein, 207
non-equilibrium, 10
non-specific interactions, 68
   buffering, 69
   co-operative, 70
nucleosome
   binding energy, 125

modifications, 126
recruitment, 126
recruitment games, 147
size, 125
states, 126
toolkit for epigenetics, 139
nucleosome systems
toolkit, 139

oct4, sox2, nanog, 125
off rates, 86
olfactoric differentiation, 142
on rates
diffusion-limited, 85
facilitated, 88
one molecule per *E. coli* equals 1.6 nM, 25
open complex, 26
operator, 37
acting at a distance, 71
operator bindings in phage lambda, 64
operator right (OR) in phage lambda, 53, 64
operon, 100, 201
minimization of relative noise, 102
polycistronic, 100
oscillations, 155

palindrome, 53
partition function
chemical binding, 60
method, 63
patchiness allows diversity, 276
pattern formation, 172
PCR, 12
persistence length, 74
phage
against phage, 250
and bacteria in oceans, 256
bacteria co-evolution, 248
bacteria dimensionless equations, 254
bacteria network, 251
burst sizes, 254
distribution, 245
capsid masses, 254
chosing lysogeny frequency, 270
co-transferred protein groups, 244
death rates, 254
F plasmid as a benign phage, 249
family tree, 243
filamentous, 243
genome sizes, 254
groups, 244
in oceans, 256
infection rates, 254
latency times, 254
numbers in world, 242
P2, 250
partially resistant hosts, 259

predator–prey, 254
resistant hosts, 257
susceptible and resistant hosts, 258
temperate, 242
temperate loss by going virulent, 271
temperate state as a hedging option, 268
temperate versus virulence, 270
temperate, hedging, 267
temperate, playing dice is good, 267
virulence number, 256
virulent, 244
phage defense systems
abortive infection, 247
CRISPR, 246
loss of receptors, 249, 259
restriction modification, 246
toxin–anti-toxin, 247
phage–bacteria systems may be eliminated in
transients, 255
pitchfork bifurcation, 130
PL promoter, 52, 54
plaque, 56, 57
plasmids, 249
as benign phages, 270
cell to cell variations, 95
plating, 56, 57
Poisson distribution, 96
polycistronic operon
co-occuring gene transfer, 100
noise minimization, 100, 121
polycomb, 125
polymeraze chain reaction, 12
polymers, 6
positive feedback
bistability, 52
cycles, 277
PR activity, 66, 69, 72
PR promoter, 53, 54
PR′ promoter, 54
PRE promoter, 53, 54
predator–prey as phage–bacteria, 254
preferential attachment
cumulative advantage, rich gets richer, 234
PRM activity, 66, 69, 72
PRM promoter, 53, 54
probability for complex operator occupancies, 63
production and decay of a protein, 104
prokaryotes
E. coli content, 23
E. coli growth rate, 32
co-existence with phages, 251, 256, 257, 259, 263
numbers on Earth, 49, 242
promoter, 37
interference, 90
interference equation, 92
open complex, 26

proteases, 21
　ClpX, 151
　HflB, 8, 51, 155
　Lon, 29
　proteasome in eukaryotes, 158
　regulatory dynamics, 42
protein, 20
　structure, 21
　protein–protein binding, 20
　stability, 20
　volume, 23
protein-protein sequestration, 210
protein–protein
　nearest and next-nearest binding, 211
　switch in phage TP901-1, 160
　UmuD:UmuD′ and UmuD:UmuC in SOS,
　　151
protein–protein binding
　gene regulation, 47
punctuated equilibrium, 280, 281, 287
push–pull, 8, 212

Q, an anti-terminator protein, 54

random walk
　biased, 221
　diffusion, 81
　time to locate a binding site, 84
read–write enzymes, 126
　barriers, 140
　distance dependence, 140
　limited supply, 140
　limited supply, limited spreading, 140
　local–global, 139
　silencers, 140
RecA
　λ phage induction, 52
　minus have small λ phage release, 111
　SOS, 151
RecA and RexB, 54
recruitment
　asymmetric, 137
　distance dependence, 140
　limits by barriers, 140
　limits by number of read–write complexes, 140
　local, 132
　local and global, 139
　of RNAP by CI in phage lambda, 65, 66
recruitment model, 128
Red Queen, 9, 281
Redfield ratio, 35
regulation
　cis, 124
　trans, 124
regulation by sequestration, 47, 211
regulatory dynamics
　activated reservoir, 43

activation, 42
　degradation, 42
　post-transcriptional, 42, 43
　repression, 42
　self-repression, 43
regulon, 201
repressilator, 157
　with breaks, 122
repressor, 37, 209
restriction modification, 246
　bacteria may compete using phages, 251
RexA and RexB, 247
ribosomes
　content, 28
　degradation and recycling, 29
　mass, 28
　numbers in *E. coli*, 33
RNA
　backbone, 19
　lifetime, 20
　structure, 19
RNAP, 24–26
　binding energies, 64
　collisions, 91
　numbers in *E. coli*, 25
robustness
　λ decision circuit, 121
　and taxonomic diversity, 285
rock–paper–scissors
　phage–bacteria–food, 253
　phage-susceptible, resistant hosts, 259
　tragedy of the commons, 236

scale-free
　gene expression, 30
　network degree distribution, 185
　self-organized criticality, 292, 293
　species extinctions, 287, 292
　taxonomic orders, 280
Schelling model, 230
segregation, 231
self-assembly, 3
　copying, 3
　phage release, 50
self-organization, 3
self-organized criticality, 289, 292
self-repression, 43
separated energy scales
　polymers, 6
separated timescales, 2
　evolution, 288, 293
　networks, 179
　social networks rewire slower than
　　communication, 233
separation of energy scales
　polymers, 17

Shea–Ackers formalism, 63
Shine–Delgarno, 29
signaling
    across networks, 54, 206, 207, 211, 212
    across tissue, 226
    protein–protein networks, 211
sir2, a read–write enzyme, 126, 147
small molecules
    Fe uptake in *E. coli*, 151
    galactose uptake in *E. coli*, 151
    turnover rate, 163
    umbrella motifs, 164
small RNA
    regulation, 44
    RyhB, 45
    RyhB in *E. coli*, 151
    Spot42, 45
    Spot42 in *E. coli*, 151
social network rewiring dynamics, 233
species extinctions
    and originations, 280
    correlated, 4
    random, 281
    scale-free, 287
sRNA, see small RNA, 44
stable fixed point, 104
steady-state
    macro-evolution, 290
stochastic behavior, 9
    bursty transcription initiation, 94
    determinstic equations, 174
    from single molecule fluctuations, 107
    plasmid copy number, 95
stochastic gene expression
    Gillespie algorithm, 106
    simulation with discrete timesteps, 104
stress response
    heat shock, 8
    negative feedback, 155
    oscillations, 155
success: sustainability or proliferation, 285
supercoiling
    distance-dependent contacts, 77, 140
    mediated promoter burstines, 99
    promoter sensitivity, 98
    transcription induced, 92
sustainability
    limits on evolvability, 9, 281
    or success on shorter timescale, 285

taxonomy that is scale free, 280
taxonomic diversity is not helping sustainability,
    285
temperate phages, 50
    characteristics, 254
    developmental decision, 51

genetic switch, 52
    leaving the bacteria, 52
    relation to virulent phages, 244
time delay
    cellular mechanisms, 157
    oscillations, 157
timescales
    existence times of genera, 281
    existence times of orders, 285
    geological, 280, 294
    phage infections and death, 254
    recruitment activity among nucleosomes,
        133
    replication, adaptation, punctuation, 288
    time to leave a binding site, 86
    time to locate a binding site in cell, 83
toolbox model, 202
toxin–anti-toxin, 247
tragedy of the commons, 235
    clonal selection, 239
    limits on virulence, 240
trans-regulation with diffusible factors, 124
transcription, 25
    bursty initiation, 94
    initiation, 26
    initiation, with interference, 90
    three-step initiation model, 27
    two-step initiation model, 90
transcription factor
    activator, 37
    co-operative operator binding, 41
    co-operativity, 38
    numbers in prokaryotes, 193
    operator binding, 37, 41
    repressor, 37
transcription regulation, 37
    combinatorial, 209
    networks, 190
transcription–translation, 24, 30
transcriptional activator, 41
transcriptional bursting
    RNAP recruits RNAP, 99
    slow transcription factors, 98
transcriptional repressor, 41
translation, 28
    speed, 30, 33
    speed differences between codons, 18, 30
    speed with growth rate, 34
    traffic, 30
tRNA, 20
turbid plaques, 57

umbrella motifs, 164

variance
    equal sum of variances for sub-processes,
        113

variance (cont.)
    Fano factor multiplied by mean, 96
Verhulst equation, 253
virulent phages
    burst size distribution, 245
    characteristics, 254
    co-existence with host, 256, 257
    examples, 243
    limits on virulence, 240
    predator–prey model, 253
    relation to temperate phages, 244

wave propagation, 170

yeast
    colony growth, 173

Z score
    networks, 194